*Walter Klöpffer and*
*Birgit Grahl*

**Life Cycle Assessment (LCA)**

## Related Titles

Reniers, G. L. L., Sörensen, K., Vrancken, K. (eds.)

**Management Principles of Sustainable Industrial Chemistry**

Theories, Concepts and Industrial Examples for Achieving Sustainable Chemical Products and Processes from a Non-Technological Viewpoint

2013
ISBN: 978-3-527-33099-7 (Also available in digital formats)

Hessel, V., Kralisch, D., Kockmann, N.

**Novel Process Windows**

Innovative Gates to Intensified and Sustainable Chemical Processes

2014
ISBN: 978-3-527-32858-1 (Also available in digital formats)

Houson, I. (ed.)

**Process Understanding**

For Scale-Up and Manufacture of Active Ingredients

2011
ISBN: 978-3-527-32584-9 (Also available in digital formats)

Imhof, P., van der Waal, J. C. (eds.)

**Catalytic Process Development for Renewable Materials**

2013
ISBN: 978-3-527-33169-7 (Also available in digital formats)

Jansen, R. A.

**Second Generation Biofuels and Biomass**

Essential Guide for Investors, Scientists and Decision Makers

2013
ISBN: 978-3-527-33290-8

Hites, R. A., Raff, J. D.

**Umweltchemie**

Eine Einführung mit Aufgaben und Lösungen

2014
ISBN: 978-3-527-33523-7

Klöpffer, W.

**Verhalten und Abbau von Umweltchemikalien**

Physikalisch-chemische Grundlagen
Zweite, vollständig überarbeitete Auflage

2012
ISBN: 978-3-527-32673-0

Hochheimer, N.

**Das kleine QM-Lexikon**

Begriffe des Qualitätsmanagements aus GLP, GCP, GMP und EN ISO 9000
Zweite, vollständig überarbeitete und erweiterte Auflage

2012
ISBN: 978-3-527-33076-8

*Walter Klöpffer and*
*Birgit Grahl*

# Life Cycle Assessment (LCA)

A Guide to Best Practice

Verlag GmbH & Co. KGaA

**The Authors**

*Prof. Dr. Walter Klöpffer*
LCA Consult & Review, Frankfurt, Germany
Am Dachsberg 56 E
60435 Frankfurt am Main
Germany

*Prof. Dr. Birgit Grahl*
Institut für Integrierte Umwelt
Forschung und Beratung
Schuhwiese 6
23858 Heidekamp
Germany

All books published by **Wiley-VCH** are carefully produced. Nevertheless, authors, editors, and publisher do not warrant the information contained in these books, including this book, to be free of errors. Readers are advised to keep in mind that statements, data, illustrations, procedural details or other items may inadvertently be inaccurate.

**Library of Congress Card No.:** applied for

**British Library Cataloguing-in-Publication Data**
A catalogue record for this book is available from the British Library.

**Bibliographic information published by the Deutsche Nationalbibliothek**
The Deutsche Nationalbibliothek lists this publication in the Deutsche Nationalbibliografie; detailed bibliographic data are available on the Internet at <http://dnb.d-nb.de>.

© 2014 Wiley-VCH Verlag GmbH & Co. KGaA, Boschstr. 12, 69469 Weinheim, Germany

All rights reserved (including those of translation into other languages). No part of this book may be reproduced in any form – by photoprinting, microfilm, or any other means – nor transmitted or translated into a machine language without written permission from the publishers. Registered names, trademarks, etc. used in this book, even when not specifically marked as such, are not to be considered unprotected by law.

**Print ISBN:** 978-3-527-32986-1
**ePDF ISBN:** 978-3-527-65565-6
**ePub ISBN:** 978-3-527-65564-9
**mobi ISBN:** 978-3-527-65563-2
**oBook ISBN:** 978-3-527-65562-5

**Cover-Design**  Formgeber, Mannheim, Germany
**Typesetting**  Laserwords Private Limited, Chennai, India
**Printing**  Strauss GmbH, Mörlenbach
**Binding**  Litges & Dopf GmbH, Heppenheim

Printed on acid-free paper

# Contents

**Preface** *XI*

**1 Introduction** *1*
1.1 What Is Life Cycle Assessment (LCA)? *1*
1.1.1 Definition and Limitations *1*
1.1.2 Life Cycle of a Product *2*
1.1.3 Functional Unit *3*
1.1.4 LCA as System Analysis *4*
1.1.5 LCA and Operational Input–Output Analysis (Gate-to-Gate) *5*
1.2 History *6*
1.2.1 Early LCAs *6*
1.2.2 Environmental Policy Background *7*
1.2.3 Energy Analysis *8*
1.2.4 The 1980s *8*
1.2.5 The Role of SETAC *9*
1.3 The Structure of LCA *10*
1.3.1 Structure According to SETAC *10*
1.3.2 Structure of LCA According to ISO *11*
1.3.3 Valuation – a Separate Phase? *12*
1.4 Standardisation of LCA *14*
1.4.1 Process of Formation *14*
1.4.2 Status Quo *16*
1.5 Literature and Information on LCA *17*
References *18*

**2 Goal and Scope Definition** *27*
2.1 Goal Definition *27*
2.2 Scope *28*
2.2.1 Product System *28*
2.2.2 Technical System Boundary *29*
2.2.2.1 Cut-Off Criteria *29*
2.2.2.2 Demarcation towards System Surrounding *32*
2.2.3 Geographical System Boundary *34*

| | | |
|---|---|---|
| 2.2.4 | Temporal System Boundary/Time Horizon | 35 |
| 2.2.5 | The Functional Unit | 37 |
| 2.2.5.1 | Definition of a Suitable Functional Unit and a Reference Flow | 37 |
| 2.2.5.2 | Impairment Factors on Comparison – Negligible Added Value | 40 |
| 2.2.5.3 | Procedure for Non-negligible Added Value | 41 |
| 2.2.6 | Data Availability and Depth of Study | 43 |
| 2.2.7 | Further Definitions | 44 |
| 2.2.7.1 | Type of Impact Assessment | 44 |
| 2.2.7.2 | Valuation (Weighting), Assumptions and Notions of Value | 45 |
| 2.2.7.3 | Critical Review | 46 |
| 2.2.8 | Further Definitions to the Scope | 47 |
| 2.3 | Illustration of the Component 'Definition of Goal and Scope' Using an Example of Practice | 47 |
| 2.3.1 | Goal Definition | 48 |
| 2.3.2 | Scope | 50 |
| 2.3.2.1 | Product Systems | 50 |
| 2.3.2.2 | Technical System Boundaries and Cut-Off Criteria | 53 |
| 2.3.2.3 | Demarcation to the System Surrounding | 53 |
| 2.3.2.4 | Geographical System Boundary | 54 |
| 2.3.2.5 | Temporal System Boundary | 55 |
| 2.3.2.6 | Functional Unit and Reference Flow | 55 |
| 2.3.2.7 | Data Availability and Depth of Study | 55 |
| 2.3.2.8 | Type of Life Cycle Impact Assessment | 56 |
| 2.3.2.9 | Methods of Interpretation | 57 |
| 2.3.2.10 | Critical Review | 57 |
| | References | 57 |
| | | |
| **3** | **Life Cycle Inventory Analysis** | **63** |
| 3.1 | Basics | 63 |
| 3.1.1 | Scientific Principles | 63 |
| 3.1.2 | Literature on Fundamentals of the Inventory Analysis | 64 |
| 3.1.3 | The Unit Process as the Smallest Cell of LCI | 65 |
| 3.1.3.1 | Integration into the System Flow Chart | 65 |
| 3.1.3.2 | Balancing | 67 |
| 3.1.4 | Flow Charts | 69 |
| 3.1.5 | Reference Values | 72 |
| 3.2 | Energy Analysis | 74 |
| 3.2.1 | Introduction | 74 |
| 3.2.2 | Cumulative Energy Demand (CED) | 77 |
| 3.2.2.1 | Definition | 77 |
| 3.2.2.2 | Partial Amounts | 77 |
| 3.2.2.3 | Balancing Boundaries | 79 |
| 3.2.3 | Energy Content of Inflammable Materials | 81 |
| 3.2.3.1 | Fossil Fuels | 81 |
| 3.2.3.2 | Quantification | 81 |

| | | |
|---|---|---|
| 3.2.3.3 | Infrastructure | 84 |
| 3.2.4 | Supply of Electricity | 85 |
| 3.2.5 | Transports | 88 |
| 3.3 | Allocation | 92 |
| 3.3.1 | Fundamentals of Allocation | 92 |
| 3.3.2 | Allocation by the Example of Co-production | 92 |
| 3.3.2.1 | Definition of Co-production | 92 |
| 3.3.2.2 | 'Fair' Allocation? | 93 |
| 3.3.2.3 | Proposed Solutions | 98 |
| 3.3.2.4 | Further Approaches to the Allocation of Co-products | 101 |
| 3.3.2.5 | System Expansion | 102 |
| 3.3.3 | Allocation and Recycling in Closed-Loops and Re-use | 105 |
| 3.3.4 | Allocation and Recycling for Open-Loop Recycling (COLR) | 107 |
| 3.3.4.1 | Definition of the Problem | 107 |
| 3.3.4.2 | Allocation per Equal Parts | 109 |
| 3.3.4.3 | Cut-off Rule | 111 |
| 3.3.4.4 | Overall Load to System B | 113 |
| 3.3.5 | Allocation within Waste-LCAs | 113 |
| 3.3.5.1 | Modelling of Waste Disposal of a Product | 114 |
| 3.3.5.2 | Comparison of Different Options of Waste Disposal | 116 |
| 3.3.6 | Summary on Allocation | 117 |
| 3.4 | Procurement, Origin and Quality of Data | 118 |
| 3.4.1 | Refining the System Flow Chart and Preparing Data Procurement | 118 |
| 3.4.2 | Procurement of Specific Data | 119 |
| 3.4.3 | Generic Data and Partial LCIs | 127 |
| 3.4.3.1 | Which Data are 'Generic'? | 127 |
| 3.4.3.2 | Reports, Publications, Web Sites | 129 |
| 3.4.3.3 | Purchasable Data Bases and Software Systems | 131 |
| 3.4.4 | Estimations | 132 |
| 3.4.5 | Data Quality and Documentation | 133 |
| 3.5 | Data Aggregation and Units | 134 |
| 3.6 | Presentation of Inventory Results | 136 |
| 3.7 | Illustration of the Inventory Phase by an Example | 137 |
| 3.7.1 | Differentiated Description of the Examined Product Systems | 138 |
| 3.7.1.1 | Materials in the Product System | 138 |
| 3.7.1.2 | Mass Flows of the Product after Use Phase | 140 |
| 3.7.1.3 | Handling of Sorting Residues and Mixed Plastics Fraction | 142 |
| 3.7.1.4 | Recovery of Transport Packaging | 143 |
| 3.7.2 | Analysis of Production, Recovery Technologies and Other Relevant Processes of the Production System | 143 |
| 3.7.2.1 | Production Procedures of the Materials | 143 |
| 3.7.2.2 | Production by Materials | 146 |
| 3.7.2.3 | Distribution | 148 |
| 3.7.2.4 | Collection and Sorting of Used Packaging | 148 |

| | | |
|---|---|---|
| 3.7.2.5 | Recovery Technologies (Recycling) | *149* |
| 3.7.2.6 | Recycling of Transport Packagings | *151* |
| 3.7.2.7 | Transportation by Truck | *152* |
| 3.7.2.8 | Electricity Supply | *152* |
| 3.7.3 | Elaboration of a Differentiated System Flow Chart with Reference Flows | *153* |
| 3.7.4 | Allocation | *153* |
| 3.7.4.1 | Definition of Allocation Rules on Process Level | *153* |
| 3.7.4.2 | Definition of Allocation Rules on System Level for Open-Loop Recycling | *157* |
| 3.7.5 | Modelling of the System | *157* |
| 3.7.6 | Calculation of the Life Cycle Inventory | *158* |
| 3.7.6.1 | Input | *159* |
| 3.7.6.2 | Output | *165* |
| | References | *170* |

| | | |
|---|---|---|
| **4** | **Life Cycle Impact Assessment** | *181* |
| 4.1 | Basic Principle of Life Cycle Impact Assessment | *181* |
| 4.2 | Method of Critical Volumes | *183* |
| 4.2.1 | Interpretation | *184* |
| 4.2.2 | Criticism | *185* |
| 4.3 | Structure of Impact Assessment according to ISO 14040 and 14044 | *187* |
| 4.3.1 | Mandatory and Optional Elements | *187* |
| 4.3.2 | Mandatory Elements | *187* |
| 4.3.2.1 | Selection of Impact Categories – Indicators and Characterisation Factors | *187* |
| 4.3.2.2 | Classification | *190* |
| 4.3.2.3 | Characterisation | *191* |
| 4.3.3 | Optional Elements of LCIA | *192* |
| 4.3.3.1 | Normalisation | *192* |
| 4.3.3.2 | Grouping | *197* |
| 4.3.3.3 | Weighting | *200* |
| 4.3.3.4 | Additional Analysis of Data Quality | *201* |
| 4.4 | Method of Impact Categories (Environmental Problem Fields) | *201* |
| 4.4.1 | Introduction | *201* |
| 4.4.2 | First ('Historical') Lists of the Environmental Problem Fields | *202* |
| 4.4.3 | Stressor-Effect Relationships and Indicators | *206* |
| 4.4.3.1 | Hierarchy of Impacts | *207* |
| 4.4.3.2 | Potential versus Actual Impacts | *209* |
| 4.5 | Impact Categories, Impact Indicators and Characterisation Factors | *212* |
| 4.5.1 | Input-Related Impact Categories | *212* |
| 4.5.1.1 | Overview | *212* |
| 4.5.1.2 | Consumption of Abiotic Resources | *214* |

| 4.5.1.3 | Cumulative Energy and Exergy Demand   *220* |
| 4.5.1.4 | Consumption of Biotic Resources   *222* |
| 4.5.1.5 | Use of (Fresh) Water   *224* |
| 4.5.1.6 | Land Use   *227* |
| 4.5.2 | Output-Based Impact Categories (Global and Regional Impacts)   *233* |
| 4.5.2.1 | Overview   *233* |
| 4.5.2.2 | Climate Change   *234* |
| 4.5.2.3 | Stratospheric Ozone Depletion   *240* |
| 4.5.2.4 | Formation of Photo Oxidants (Summer Smog)   *246* |
| 4.5.2.5 | Acidification   *254* |
| 4.5.2.6 | Eutrophication   *261* |
| 4.5.3 | Toxicity-Related Impact Categories   *268* |
| 4.5.3.1 | Introduction   *268* |
| 4.5.3.2 | Human Toxicity   *269* |
| 4.5.3.3 | Ecotoxicity   *279* |
| 4.5.3.4 | Concluding Remark on the Toxicity Categories   *285* |
| 4.5.4 | Nuisances by Chemical and Physical Emissions   *286* |
| 4.5.4.1 | Introduction   *286* |
| 4.5.4.2 | Smell   *286* |
| 4.5.4.3 | Noise   *287* |
| 4.5.5 | Accidents and Radioactivity   *289* |
| 4.5.5.1 | Casualties   *289* |
| 4.5.5.2 | Radioactivity   *290* |
| 4.6 | Illustration of the Phase Impact Assessment by Practical Example   *291* |
| 4.6.1 | Selection of Impact Categories – Indicators and Characterisation Factors   *293* |
| 4.6.1.1 | (Greenhouse) Global Warming Potential   *294* |
| 4.6.1.2 | Photo-Oxidant Formation (Photo Smog or Summer Smog Potential)   *295* |
| 4.6.1.3 | Eutrophication Potential   *296* |
| 4.6.1.4 | Acidification Potential   *297* |
| 4.6.1.5 | Resource Demand   *298* |
| 4.6.2 | Classification   *300* |
| 4.6.3 | Characterisation   *300* |
| 4.6.4 | Normalisation   *305* |
| 4.6.5 | Grouping   *310* |
| 4.6.6 | Weighting   *311* |
|  | References   *311* |

| **5** | **Life Cycle Interpretation, Reporting and Critical Review**   *329* |
| 5.1 | Development and Rank of the Interpretation Phase   *329* |
| 5.2 | The Phase Interpretation According to ISO   *331* |
| 5.2.1 | Interpretation in ISO 14040   *331* |
| 5.2.2 | Interpretation in ISO 14044   *331* |

| | | |
|---|---|---|
| 5.2.3 | Identification of Significant Issues | *332* |
| 5.2.4 | Evaluation | *333* |
| 5.3 | Techniques for Result Analysis | *334* |
| 5.3.1 | Scientific Background | *334* |
| 5.3.2 | Mathematical Methods | *335* |
| 5.3.3 | Non-numerical Methods | *338* |
| 5.4 | Reporting | *338* |
| 5.5 | Critical Review | *340* |
| 5.5.1 | Outlook | *342* |
| 5.6 | Illustration of the Component Interpretation Using an Example of Practice | *343* |
| 5.6.1 | Comparison Based on Impact Indicator Results | *343* |
| 5.6.2 | Comparison Based on Normalisation Results | *344* |
| 5.6.3 | Sectoral Analysis | *344* |
| 5.6.4 | Completeness, Consistency and Data Quality | *346* |
| 5.6.5 | Significance of Differences | *347* |
| 5.6.6 | Sensitivity Analyses | *348* |
| 5.6.7 | Restrictions | *350* |
| 5.6.8 | Conclusions and Recommendations | *351* |
| 5.6.9 | Critical Review | *351* |
| | References | *352* |
| **6** | **From LCA to Sustainability Assessment** | *357* |
| 6.1 | Sustainability | *357* |
| 6.2 | The Three Dimensions of Sustainability | *358* |
| 6.3 | State of the Art of Methods | *361* |
| 6.3.1 | Life Cycle Assessment – LCA | *361* |
| 6.3.2 | Life Cycle Costing – LCC | *364* |
| 6.3.3 | Product-Related Social Life Cycle Assessment – SLCA | *366* |
| 6.4 | One Life Cycle Assessment or Three? | *368* |
| 6.4.1 | Option 1 | *368* |
| 6.4.2 | Option 2 | *369* |
| 6.5 | Conclusions | *370* |
| | References | *371* |

**Appendix A Solution of Exercises**  *375*

**Appendix B Standard Report Sheet of Electricity Mix Germany (UBA 2000, Materials p. 179ff) Historic example, only for illustrative purposes**  *381*

**Acronyms/Abbreviations**  *385*

**Index**  *391*

## Preface

This book is the updated translation of a textbook and monograph written in German language by the same authors.[1] The first version emerged from lectures at the University of Mainz.

The topic of the book, life cycle assessment (LCA), developed from modest seeds in the 1970s and 1980s to become the only internationally standardised method of ecological product assessment. The development entered its decisive phase when the Society of Environmental Toxicology and Chemistry (SETAC) began to harmonise diverse older methods ('proto-LCAs'). This process culminated in 1993 in the publication of the *Guidelines for Life Cycle Assessment: A Code of Practice*, a result of the SETAC Workshop in Sesimbra, Portugal. In the same year started the standardisation by the International Standard Organization (ISO) involving 40 nations, resulting in the famous series of ISO LCA standards 14040ff (1997–2006). The authors of this book followed this development as members of the German mirror group 'Deutsches Institut für Normung-Normenausschuss Grundlagen des Umweltschutzes (DIN NAGUS)', discussing and commenting on the drafts developed by ISO/TC 207/SC 5 (TC, Technical Committee; SC, Sub Committee). In addition, German translations of the standards were checked and improved.

The topic 'valuation' caused heated discussions and turned out to be not consensual – surviving today as an optional element 'weighting' within the phase Life Cycle Impact Assessment (LCIA), and not as originally planned as an LCA phase of its own. Moreover, 'weighting' is strictly prohibited for comparative LCA studies intended to be made available to the public.

The revision of the LCA standards 2006 even enforced this, so that now the intention to use a comparative LCA publicly is sufficient for banning the 'weighting' of results and requiring strict regulations regarding publishing, documentation and critical review (panel method).

The authors have performed several critical reviews together and necessarily studied the standards in greater detail than possibly necessary for academic lectures alone. Most standards use cumbersome wording to some extent, which is why they are not ideally suited as teaching and learning material – a good reason to write this book that is expected to help beginners entering the field of LCA and

---

1) Klöpffer and Grahl (2009).

also offering advanced readers something new. The LCA standards are written in a spirit which shall prevent any misuse of the method, especially in marketing and advertisement. As a consequence, frequently we read what shall not be done and less details on how a real LCA is to be done correctly. To give an example, in the phase LCIA there is no list of impact categories, not even a default list, not to speak of indicators and characterisation factors. Therefore, LCIA is treated extensively in this book. Even so, no complete picture could be presented since several methods are still in development, cited in many references.

Equally important as reporting the mere facts seemed relating a deeper understanding of the LCA methodology including its limits. The same is true for the environmental problems forming the basis of the impact categories. The most important application of LCA is learning and understanding of environmental problems caused by product systems 'from cradle to grave', that is, from the raw materials to recycling and waste removal, respectively. This learning process cannot start without a good understanding of the processes and can be even worsened by thoughtlessly using software. The modern software offers great help in performing LCAs (hardly to dream of 10 years ago); it should not, however, replace the collection of original ('foreground') data, thorough system analysis, or selection and explanation of the impact categories.

There can be no doubt that LCA as an applied (simplified) system analysis offers much material for theoretical work, enriching the methodology. It is not, however, 'art pour l'art', but should rather achieve the learning effect mentioned above, the results of which should enter decision finding. Ecologically correct decisions during product development will lead to better products in the long range. The application of LCA is therefore of decisive importance. In order to demonstrate this point, the authors divided a 'real' LCA study into four parts, which were assigned to the four phases according to ISO 14040.

1. Goal and scope definition (Chapter 2)
2. Life cycle inventory analysis (Chapter 3)
3. Life cycle impact assessment (Chapter 4)
4. Interpretation (Chapter 5).

This 'real-life' LCA study in German has been provided by the Institut für Energie- und Umweltforschung (IFEU), Heidelberg, by courtesy of the commissioner Fachverband Getränkekarton (FKN), Wiesbaden. The translation of the recorded textual passages has been carried out by the authors of this book. We would like to point out explicitly that this specific LCA study was chosen as example for purely didactic reasons. A specific product system is always more descriptive compared to a theoretically constructed one. Specific conclusions included in the example LCA do not belong to the learning goal set by the authors of this book.

Textbooks on LCA are rare in any language, but even in English we remember only one, originating from Sweden.[2] We hope that this book will contribute

---

2) Baugmann and Tillmann (2004).

to academic lecturing as well as private studies and be of use in industry and governmental organisations.

We owe great thanks to Andreas Detzel (IFEU), who not only provided the example study but also carefully read and commented on the whole manuscript of the German version. Martina Krüger (IFEU) was a great help in adapting the example study to the didactic presentation needed in a textbook. Many of our friends in the LCA community contributed to the development of LCA and thus, finally, also to this book. To mention only few of them, Harald Neitzel (then at Umweltbundesamt (UBA), Berlin), the unforgettable chairman of DIN NAGUS in the 1990s; Isa Renner, main LCA practitioner at Battelle Frankfurt, later at C.A.U. Ltd. Dreieich; and Eva Schmincke, longstanding discussion partner, centrally involved in the development of environmental product declarations (EPDs) according to the ISO Type III declaration system. At the international level, the LCA-related activities of SETAC and the UNEP (United Nations Environmental Programme)/SETAC Life Cycle Initiative were of great help.

Almut B. Heinrich, the managing editor of the book series 'LCA Compendium – The whole world of Life Cycle Assessment' helped us with the translation of the German book doing final corrections in all chapters. She was managing editor of The International Journal of Life Cycle Assessment from 1996 to 2009 and contributed in that position to the proliferation of LCA world-wide.[3]

Last, but not least, we thank the editorial managers at Wiley-VCH for their patience and competence during the creation of this book.

Frankfurt am Main und Lübeck            Walter Klöpffer and Birgit Grahl
October 2013

## References

Baumann, H. and Tillman, A.-M. (2004) *The Hitch Hiker's Guide to LCA. An Orientation in Life Cycle Assessment Methodology and Application*, Studentlitteratur, Lund. ISBN: 91-44-02364-2.

Klöpffer, W. and Grahl, B. (2009) *Ökobilanz (LCA) – Ein Leitfaden für Ausbildung und Beruf*, Wiley-VCH Verlag GmbH, Weinheim. ISBN: 978-3-527-32043-1.

Klöpffer, W. and Curran, M.A. (2014) *LCA Compendium – The Whole World of Life Cycle Assessment*, Springer, Dordrecht.

---

[3] Klöpffer and Curran 2014.

# 1
# Introduction

To date life cycle assessment (LCA) is a method defined by the international standards ISO 14040 and 14044 to analyse environmental aspects and impacts of product systems. Therefore, the introduction of the methodology in Chapters 2–5 relates to these standards. As a prelude, the scope and development of the methodology are introduced here.

## 1.1
## What Is Life Cycle Assessment (LCA)?

### 1.1.1
### Definition and Limitations

In the introductory part of international standard ISO 14040[1] serving as a framework, LCA has been defined as follows:

> LCA studies the environmental aspects and potential impacts throughout a product's life (i.e. cradle-to-grave) from raw material acquisition through production, use and disposal. The general categories of environmental impacts needing consideration include resource use, human health, and ecological consequences.

A similar definition of LCA was adopted as early as 1993 by the Society of Environmental Toxicology and Chemistry (SETAC)[2] in the 'Code of Practice'.[3]

Similar definitions can also be found in the basic guidelines of[4] DIN-NAGUS as well as in the 'Nordic Guidelines'[5] commissioned by Scandinavian Ministers of the Environment. Those deliberate limitations of LCA to analysis and interpretation of *environmental impacts* have the consequence that the method is restricted to only quantify[6] the *ecological* aspect of sustainability (see Chapter 6). The exclusion

---

1) ISO (1997).
2) Foundation year 1979.
3) SETAC (1993a).
4) DIN-NAGUS (1994).
5) Lindfors *et al.* (1995).
6) Klöpffer (2003, 2008).

*Life Cycle Assessment (LCA): A Guide to Best Practice*, First Edition.
Walter Klöpffer and Birgit Grahl.
© 2014 Wiley-VCH Verlag GmbH & Co. KGaA. Published 2014 by Wiley-VCH Verlag GmbH & Co. KGaA.

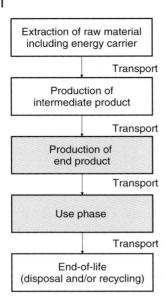

**Figure 1.1** Simplified life cycle of a tangible product.

of economical and social factors distinguishes LCA from product line analysis (PLA) (Produktlinienanalyse) and similar methods.[7] This separation was made to avoid a method overload, being well aware that a decision, for example, in the development of sustainable products, cannot and must not neglect these factors.[8]

### 1.1.2
### Life Cycle of a Product

The main idea of a *cradle-to-grave* analysis, that is, the life cycle of a product, is illustrated in a simplified manner in Figure 1.1. Usually, the starting point for building a product tree is the production of the end product and the use phase. Further diversification of the boxes in Figure 1.1 into singular processes, the so-called unit processes, as well as the inclusion of transports, diverse energy supply, co-products, and so on, turn this simplistic scheme, even with simple products, into very complex 'product trees' (diverse raw materials and energy supply, intermediate products, co-products, ancillary material, waste management including diverse disposal types and recycling).

Interconnected unit processes (life cycle or product tree) form a *product system*. The centre is a product, a process, a service or, in the widest sense, a human

---

7) Projektgruppe Ökologische Wirtschaft (1987) and O'Brien, Doig and Clift (1996).
8) Klöpffer (2008).

activity.[9] In an LCA, systems that serve a specific *function* and therefore have a specified performance are analysed.

*Therefore, the quantified performance (avail) of a product system is the intrinsic standard of comparison (reference unit). It is the sole correct basis for the definition of a 'functional unit'.*[10]

### 1.1.3
### Functional Unit

Besides the *cradle-to-grave* analysis (thinking in terms of systems, life cycles or production trees), the functional unit is the second basic term in an LCA and is therefore to be explained here.

The function of a beverage packaging, for example, is – besides shielding of the liquid – above all, transportability and storability. The functional unit is most frequently defined as the provision of 1000 l liquid in a way to fulfil the technical aspects of the performance. This function can, for instance, be mapped with different packaging specifications (the following examples are arbitrarily chosen):

- 5000 0.2 l[11] pouches
- 2000 0.5 l reusable bottles of glass
- 1000 1 l single-use beverage carton
- 500 2 l PET (polyethylene terephthalate) single-use bottles.

Thus, for a comparison of packaging systems, the life cycle of 5000 pouches, 2000 reusable glass bottles, 1000 cardboards and 500 2 l PET bottles, which are four product systems that roughly fulfil the same function, needs to be analysed and compared.

Slight variations in performance (convenience, e.g. weight, user friendliness, aesthetics, customer behaviour, suitability as advertising medium or other side effects of packaging systems) are not important in this simplistic example. It is, however, important to note that *systems* (not products) with matchable functions are compared.[12] This is the reason why tangible products (goods) can also be compared with *services*, as long as they have the same or a very similar function. Within an LCA, products are defined as goods and services. As with goods, services require energy, transport, and so on. Therefore, it is possible to define services as systems and compare them with tangible products on the basis of equivalent function by means of the functional unit.

---

9) SETAC (1993a).
10) Fleischer and Schmidt (1996); see ISO 14040 (2006a).
11) 1 l = 1 dm$^3$.
12) Boustead (1996).

### 1.1.4
### LCA as System Analysis

LCA is based on a simplified system analysis. The simplification consists of an extensive linearisation (see system boundaries and cut-off criteria in Section 2.2). Interconnections of parts of the life cycle of a product that always exist in reality lead to extremely complex relationships in the modelling, which are most difficult to handle. There are, nevertheless, possibilities to handle the formation of loops and other deviations from the linear structure, for example, by an iterative approach or matrix calculus.[13]

> **Example**
>
> LCA deals with the comparison of product systems, and not of products. This means the following:
>
> Within the product segment 'towel dispenser', for example, paper towels and cotton rolls are two possible variations. The cotton roll needs to be cleaned to fulfil its function. This means, the cleansing process (detergent, water and energy consumption) is part of the product system and must surely be considered. In addition, washing machines must be applied for cleaning.
> Has the production of washing machines to be considered as well?
> Their production requires, for example, steel. Steel is made from iron ore that needs to be transported, and so on. It is obvious that limitations need to be set, because every small product is linked to the entire industrial system. On the other hand, nothing essential shall be omitted.

System analysis and the meaningful selection and definition of system boundaries are therefore important and labour-intensive tasks within every LCA.

The main advantage of the life-cycle approach 'from cradle to grave' lies in its ability to easily detect the shifting of environmental burdens, the so-called *trade-offs*, which may, for example, occur owing to material substitutions. Therefore, it is of no use to seemingly solve an environmental problem if, later, in different life cycle stages or environmental media, the same or additional problems occur. The same applies when an unreasonable energy or resource consumption may be connected with the substitution. These kind of activities do not solve the problem at its base.

It is not arguable that in rare cases, especially those of health hazards (e.g. substitution of hazardous substances), such suboptimal decisions may be applicable.

---

13) Heijungs (1997) and Heijungs and Suh (2002).

## Example

As fossil resources diminish, substitution of the raw material base with renewable resources is an objective of science and development. For example, variants of loose-fill packaging chips made of polystyrene and potato starch[14] have been investigated through LCA. As the resources used and the production processes of both materials fundamentally differ, a thorough analysis of the product systems is necessary. For instance, on the one hand, the overall agricultural system including growth, maintenance and harvest needs to be considered during the production of renewable base products; on the other hand is the crude oil drilling or mining. Other life cycle stages of the loose-fill packaging systems differ fundamentally as well, depending on the raw material base. It cannot be decided at first sight whether substitution of the raw material base may have an ecological advantage for a product system.

### 1.1.5
### LCA and Operational Input–Output Analysis (Gate-to-Gate)

There is always a risk of problem shift when system boundaries that are too restrictive have been chosen. This is often the case when only operational input–output analyses have been conducted (frequently misused terms are *ecobalance of the enterprise*, *corporate-LCA* or *ecobalance* without additional explanation).

If, for instance, the system boundary is set equal to the fence around a factory (gate-to-gate), the fundamental concept of LCA is not satisfied: Neither the production of pre-products nor the disposal of end products is considered; the same is applicable with transports (e.g. *just in time*), outsourcing and parts of waste management activities (e.g. municipal waste water sewage plants).

## Example

### Pseudo improvement by outsourcing

A manufacturer of fine foods intended to not only advertise his products for taste and salubriousness but also for environmental aspects. For this purpose, data concerning energy and water consumption were gathered in an operational input–output analysis (gate-to-gate), which allowed the allocation of on-site environmental burdens to the production of different salads. It was striking that potato salad had an immense water supply. The reason was that potatoes, usually covered by earth, had to be washed. This waste water was then assigned to the potato salad. Some weeks later, the water supply per kilogramme salad had drastically diminished. This was not due to a technical innovation at the cleaning

---

14) BIfA/IFEU/Flo-Pak (2002).

> plant but due to outsourcing of the washing to another enterprise. For this reason, washing water was not a factor anymore in the operational input–output analysis within the system boundary of the investigated site.

Nevertheless, operational input–output analyses are useful for many applications, for example, as data bases in environmental management systems.[15]

A simple consideration shows that operational input–output analyses also provide data bases for the LCA of products: Every production process, for example, the production of 500 g of potato salad in a screw cap glass jar, takes place at a specific company, at a specific site. If data, for example, for energy and water consumption of the system '1000 screw cap glasses, each containing 500 g potato salad supplemented by cucumber, egg and yoghurt dressing' have to be procured, every company that is part of the production and transportation of the packed product as well as businesses involved in the waste management of the used packaging must have analysed their processes in such a way that the data can be allocated to the product under investigation. This is not simple: an agricultural corporation generally does not only produce milk and a dairy not only yoghurt; the manufacturer of glass jars provides glasses for diverse customers, and so on. If, however, all companies involved in manufacture, distribution and end-of-life management of the product (supply chain) had data from their specific operational input–output analysis in a product-related format, these results could be merged. Nevertheless, product-related data acquisition is not common practice in operational input–output analyses.

Coupling of such operational input–output analyses along the life cycle of products would provide the possibility of LCA chain management.[16] Companies that are part of a product system could explore and realise potentials for the optimisation in co-operation. There is the hope that, in this way, life cycle thinking and, in the end, also life cycle acting, may emerge (*Life Cycle Thinking* and *Life Cycle Management* – LCM).

## 1.2
## History

### 1.2.1
### Early LCAs

LCA is a relatively recent methodology, but not as recent as many believe. Approaches to life cycle thinking have already been reported in early literature. The Scottish economist and biologist Patrick Geddes has developed as early as in

---

15) Braunschweig and Müller-Wenk (1993), Beck (1993) and Schaltegger (1996).
16) Udo de Haes and De Snoo (1996, 1997).

the 1880s a procedure that can be considered as precursor for Life Cycle Inventory (LCI).[17] His interest focused on energy supply, especially on coal.

The first LCAs in the modern sense were conducted around 1970, termed *Resource and Environmental Profile Analysis (REPA)* at Midwest Research Institute in the United States.[18] As with nearly all early LCAs or 'proto-LCAs',[19] these were an analysis of resource consumption and emissions caused by product systems, the so-called inventories without impact assessment. To date, such studies are called *Life Cycle Inventory studies.*[20] The new methodology was first applied to compare beverage packaging. The same applies for the first LCA conducted in Germany[21] in 1972 under the leadership of B. Oberbacher at Battelle-Institute in Frankfurt, Main. The new method – originally proposed by Franklin and Hunt, USA – additionally captured costs (among others, those of disposal procedures). Interestingly, light polyethylene pouches, already in use at that time, obtained best results, similar to the results in more recent studies.[22]

Further, early LCAs were conducted by Ian Boustead in the United Kingdom[23] and Gustav Sundström in Sweden.[24] In addition, Swiss studies,[25] which can be considered as proto-LCAs, date back to the 1970s. They were conducted at the EMPA in St. Gallen; see memories of Paul Fink, former director of the EMPA.[26]

## 1.2.2
### Environmental Policy Background

Why did the development of LCA start in the early 1970s? At least two reasons can be determined:

1. Rising waste problems (therefore, studies on packaging)
2. Bottlenecks in energy supply and acknowledgement of limited resources.

While the former issue (i) was implemented into a just-emerging environmental policy by the authorities in most developed countries, public awareness of the latter (ii) was raised by the bestseller *The Limits to Growth* (the report to the Club of Rome).[27] Something must have been in the air because the book caused a sensation in 1972, the year of its publication. Did a change of paradigm occur? Was the throw-away mentality of post-war generation suddenly under scrutiny?

The theory in the 'Club of Rome' study was confirmed by reality through the first oil crisis in 1973/1974. Although the study was over-pessimistic with regard to the exhaustion of oil resources, it demonstrated the vulnerability of an industrial

---
17) Quoted by Suter and Walder (1995).
18) Hunt and Franklin (1996).
19) Klöpffer (1994, 1997, 2006).
20) ISO (1997).
21) Oberbacher, Nikodem and Klöpffer (1996).
22) Schmitz, Oels and Tiedemann (1995).
23) Boustead (1996) and Boustead and Hancock (1979).
24) Lundholm and Sundström (1985, 1986).
25) BUS (1984).
26) Fink (1997).
27) Meadows et al. (1972, 1973).

society which, to a great extent, relies on crude oil. To date, nothing has changed concerning this aspect, on the contrary.

System analysis, well known only to specialists, had its breakthrough as a commonly accepted method. The International Institute for Applied Systems Analysis (IIASA) at Laxenburg, Vienna, was founded. In Germany, car-free Sundays happened; an atmosphere of departure emerged, to date unimaginable, with a plethora of ideas on how to develop alternative energy sources as well as on how to use conventional forms of energy more efficiently. Some of them were realised, but most of them were not (yet).

### 1.2.3
### Energy Analysis

With this mainly energy-political background, it is not surprising that, from the theoretical side, *energy analysis or process chain analysis* was developed first, which today is an important integral part of the LCI[28] (see Chapter 3). In Germany, this development was mainly promoted by Professor Schäfer at the Technical University Munich[29] and in industry before.[30] The (primary) energy demand summarised through all stages of the life cycle is called *cumulative energy demand (CED)*.[31] It used to be an important part of the LCI in the time of the proto-LCAs and is still used in LCAs.

By way of political solutions to the oil crisis in the 1980s, interest in LCA with respect to its precursors declined but experienced an unexpected upswing at the end of the decade.

### 1.2.4
### The 1980s

Studies on LCA were sparse in the first half of the 1980s in the German language area. Exceptions are the study of BUS, later Federal Agency for Environment, Forestry and Agriculture, Bern,[32] which has already been named, a thesis by Marina Franke at TU Berlin[33] and the development of PLA by the Ökoinstitut.[34] PLA surpasses LCA as it is based on a needs assessment (NA) analysing the usefulness of a product and consumer behaviour. Here, the product-related environmental analysis is complemented by the investigation of social aspect (SA) and economical aspect (EA) of the product system:

$$PLA = NA + LCA + SA + EA$$

with LCA = inventory + environmental impact assessment.

---

28) Mauch and Schäfer (1996).
29) Mauch and Schäfer (1996) and Eyrer (1996).
30) Kindler and Nikles (1979, 1980).
31) VDI (1997).
32) BUS (1984).
33) Franke (1984).
34) Projektgruppe Ökologische Wirtschaft (1987).

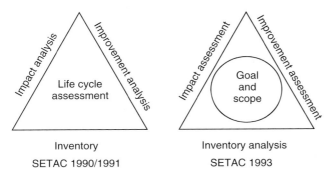

**Figure 1.2** The SETAC-triangle in LCA guidelines ('code of practice').[35]

PLA therefore comprises all three aspects of sustainability according to the Brundtland Commission[36] (see Chapter 6) and Agenda 21,[37] which was adopted at the UNO World Conference in Rio de Janeiro, 1992.

### 1.2.5
### The Role of SETAC

A strong upswing in the interest in LCA in Europe and North America – where the terms 'life cycle analysis' and 'life cycle assessment', originated – led to two international conferences that can be considered as the starting point for the newer development[38]:

In 1990, a workshop was organised by SETAC in Smugglers Notch, Vermont, on *A Technical Framework for Life Cycle Assessment*. One month later, a European workshop took place on the same topic in Leuven.[39]

In Smugglers Notch, the famous LCA triangle was conceptualised, and later persiflaged as 'holy triangle' (Figure 1.2). From 1990 to 1993 SETAC and SETAC Europe were leading agents in the development, harmonisation and early standardisation of LCA. Their reports[40] are part of the most important information sources concerning the development of the methodology. In the German-speaking part they were only equalled by the Swiss *Ecobalance of Packaging Materials 1990*,[41] updated in 1996 and 1998.[42] The UBA (Umweltbundesamt) (Berlin) study in 1992 also had a great influence.[43] A French adoption of history and methodology, *L'Ecobilan*, was published at about the same time.[44] The development of LCA in the United

---

35) SETAC (1993a).
36) World Commission on Environment and Development (1987).
37) UNO (1992).
38) Klöpffer (2006).
39) Leuven (1990).
40) SETAC (1991, 1993a,b, 1994), and SETAC Europe (1992).
41) BUWAL (1991).
42) BUWAL (1996, 1998).
43) UBA (1992).
44) Blouet and Rivoire (1995).

States[45)] and in Japan[46)] was presented in special issues of the *International Journal of Life Cycle Assessment*.

The special contributions from the Centre of Environment of University Leiden (Centrum voor Milieukunde Leiden, CML) under the leadership of Professor Helias Udo de Haes were appreciated in a study on sociology of scientific knowledge by Gabathuler[47)] and in a supplementary issue of the *International Journal of Life Cycle Assessment*.[48)] The greatest achievement of CML was, without any doubt, a stronger focus on the ecological aspects of LCA, compared to the earlier more technical ones. Nevertheless, a prior Swiss LCA had already featured a simple method of impact assessment.[49)] In practice, the CML method tended to overemphasise chemical releases in the impact assessment. At the same time – due to the absence of generally adhered indicators – it underestimated the impacts of the overuse of natural resources such as minerals, fossils, biota and land[50)] (see Chapter 4).

## 1.3
## The Structure of LCA

### 1.3.1
### Structure According to SETAC

A first attempt to structure LCA was by the SETAC triangle of 1990/1991 already quoted (Figure 1.2)

Inventory in the context of LCA (LCI) means material and energy analysis of the examined system from cradle to grave. The resulting inventory table contains a list of all material and energy inputs and outputs (see Figure 1.3 and Chapter 3).

These numbers of LCI need an ecological analysis or weighting. Inputs and outputs are sorted according to their impact on the environment. Thus, for instance, all releases into the air causing acid rain are aggregated (see Chapter 4). This procedure was formerly called *Impact Analysis* by SETAC, and later *Impact Assessment*.

The interpretation of the data procured in LCA has already been postulated in Smugglers Notch. It was called *Improvement Analysis*, later renamed *Improvement Assessment*. The introduction of this component was regarded as great progress because the interpretation of the data was conducted according to specific rules. The Environmental Agency Berlin (UBA)[51)] has included this task in 1992 in its recommendation to the conduct of LCAs as an option. The rules for interpretation were later modified during the standardisation process of ISO (see Section 1.3.2). To date this phase is named *interpretation*[52)] (see Figure 1.4).

---

45) Curran (1999).
46) Special issue Japan: Finkbeiner and Matsuno (2000).
47) Gabathuler (1998).
48) Huijbregts *et al.* (2006).
49) BUS (1984).
50) Klöpffer and Renner (2003).
51) German: Umwelbundesamt (UBA).
52) ISO (1997).

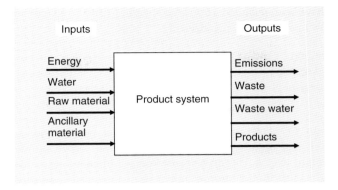

**Figure 1.3** Analysis of matter and energy of a product system.

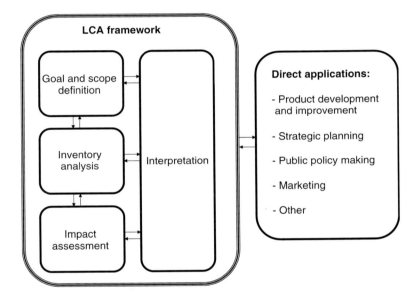

**Figure 1.4** LCA phases according to ISO 14040:1997/2006.

### 1.3.2
### Structure of LCA According to ISO

To date, the structure developed by SETAC has essentially been maintained by ISO[53] with the exception of *Improvement Assessment*, which was replaced by *Interpretation*. The optimisation of product systems was not adapted as standard content by ISO, but was listed besides other applications of the standard. The structure of the international standard is depicted in Figure 1.4.

---

53) ISO (1997, 2006a).

The phases of LCA have been renamed, compared to earlier structures, and the following terms are now internationally mandatory:

- Goal and Scope Definition
- Life Cycle Inventory Analysis
- Life Cycle Impact Assessment
- Interpretation.

The arrows in Figure 1.4 allow an iterative approach that is often necessary (see Chapter 2). Direct *applications of an LCA* lie out of scope of the standardised components of an LCA. This makes sense because, besides foreseeable applications during the standardisation process, others were developed in practice and have been summarised as 'other applications'. Examples can be found in Table 1.1.

### 1.3.3
### Valuation – a Separate Phase?

A special status is attached to the former component *valuation*,[54] which has not been assigned in the standardised structure. A valuation is always necessary when the results of a comparative LCA are not straightforward. A trade-off of system A against system B needs to be made when, for example, the former has lower energy consumption, but on the other hand has releases of substances leading to water eutrophication and to the formation of near-ground ozone: What is of greater importance? For these decisions, subjective and/or normative notions of value are necessary, common in daily life, for example, during purchase decisions.[55] For this reason, a valuation based on exact scientific methods cannot be made. Therefore, it was proposed by SETAC Europe at Leiden 1991[56] to introduce *valuation* as a component of its own. This proposition was seized by UBA Berlin[57] and by DIN-NAGUS[58] later on. However, because subjective notions of value cannot be standardised, a methodology was developed to support the process of conclusion. In the SETAC 'Code of Practice'[59] these methodological rules were subordinated to the phase 'Impact Assessment'. No changes were made by the standardisation process of ISO: Methodological rules are integrated into the phase 'Impact Assessment'[60] (see Section 4.3). The final survey of results that leads to a conclusion[61] is supposed to take place in the final phase of an LCA, 'Interpretation'[62] (see Chapter 5).

---

54) In German: Bewertung.
55) DIN-NAGUS (1994), Giegrich et al. (1995), Klöpffer and Volkwein (1995) and Neitzel (1996).
56) SETAC Europe (1992).
57) Schmitz, Oels and Tiedemann (1995).
58) (DIN NAGUS (1994) and Neitzel (1996).
59) SETAC (1993a).
60) ISO (2000a).
61) Grahl and Schmincke (1996).
62) ISO (2000b).

**Table 1.1** Examples of early LCA applications according to ISO 14040.

| Application | Query | Project |
|---|---|---|
| Environmental law and – policy | Packaging regulation | Beverage packaging[a] |
| | Waste oil regulation | Waste oil recovery[b] |
| | Genetically modified organisms (GMO) | GMO in agricultural LCA[c] |
| | Agriculture | Weed control in viticulture[d] |
| | PVC | PVC in Sweden[e] |
| | Public procurement | Cost-benefit analysis of environmental procurement[f] |
| | Integrated product policy | EuP directive[g] |
| Comparison of products | Surfactants | ECOSOL LCAs[h] |
| | Beverage packaging | Comparison of packagings[i] |
| | Food packaging | Comparison of packagings[j] |
| | Floor coverings | ERFMI survey[k] |
| | Insulating materials | Insulation of buildings[l] |
| Communication | Consumer consultation | ISO type III declaration[m] |
| | Chain management | PCR[n]: electricity, steam, water[o] |
| | Ecological building | EPD[p]: building products[q] |
| | Carbon footprint | PCR: product declaration[r] |
| | | Carbon-neutral enterprise[s] |
| Waste management | Concepts of disposal | Graphic papers[t] |
| | Recycling | Plastics[u] |
| Enterprise | Ecological valuation of business lines | Environmental achievement of an enterprise[v] |

[a] Schmitz et al. (1995) and UBA (2000b, 2002).
[b] UBA (2000a)
[c] Klöpffer et al. (1999).
[d] IFEU/SLFA (1998).
[e] Tukker Kleijn and van Oers (1996).
[f] Rüdenauer et al. (2007).
[g] Kemna et al. (2005).
[h] Stalmans et al. (1995) and Janzen (1995).
[i] IFEU (2002, 2004, 2007) and Detzel and Böß (2006).
[j] IFEU (2006) and Humbert et al. (2008).
[k] Günther and Langowski (1997, 1998).
[l] Schmidt et al. (2004).
[m] Schmincke and Grahl (2006).
[n] Product category rules.
[o] Vattenfall (2007).
[p] Environmental product declaration.
[q] Deutsches Institut für Bauen und Umwelt (2007).
[r] Svenska Miljöstyrningsrådet (2006) and BSI (2008).
[s] Gensch (2008).
[t] Tiedemann (2000).
[u] Heyde and Kremer (1999).
[v] Wright et al. (1997).
GMO, genetically modified organisms.

In Germany, the discussion on valuation has, during the final years of the 1990s, increased to such an extent that

- the former Minister of the Environment, Angela Merkel,[63] joined the discussion;
- the association of the German Industry (BDI) published a widely noticed policy brief[64] and
- UBA Berlin elaborated an ISO-conformal valuation methodology.[65]

## 1.4
## Standardisation of LCA

### 1.4.1
### Process of Formation

LCA standards ISO 14040 and 14044 belong to the ISO 14000 family concerning various aspects of environmental management (Figure 1.5).

The committee responsible for DIN in Germany is the NAGUS.[66] Similar committees existed in other countries. National propositions are brought together in the Technical Committee 207 (TC 207) of the ISO at international level. All nations that are members of TC 207 by their standardisation organisations participate and international standards are developed. Generally, this process takes several years.

LCA standardisation by national standardisation organisations[67] and, above all, by ISO has been conducted since the beginning of the 1990s with great effort.[68] This was difficult to achieve because individual phases of LCA – in particular, Impact Assessment and Interpretation – were still under technical/scientific development. On a national level, only two standardisation organisations have developed their own LCA standards before ISO 14040 was enacted: AFNOR (Association Française de Normalisation, France) and CSA (Canadian Standards Association, Canada). To date, a singular internationally accepted standardisation is aimed at promoting international communication, and this is why France and Canada have stepped into the ISO process.

The most important standardisation activity for LCA is therefore conducted by ISO. European Standardisations (Comité Européen de Normalisation, CEN) and their subsequent national organisations adapt ISO regulations and translate them into their individual languages (CEN 14040 standards are available in three

---

63) Merkel (1997).
64) BDI (1999).
65) Schmitz and Paulini (1999).
66) Normenausschuss Grundlagen des Umweltschutzes (Environmental Protection Standards Committee).
67) e.g. CSA (1992), DIN-NAGUS (1994) and AFNOR (1994).
68) ISO (1997, 1998, 2000a,b), Marsmann (1997), Saur (1997) and Klüppel (1997, 2002).
69) Normenausschuss Grundlagen des Umweltschutzes (NAGUS) in DIN Deutsches Institut für Normung e. V. (2013).

## 1.4 Standardisation of LCA | 15

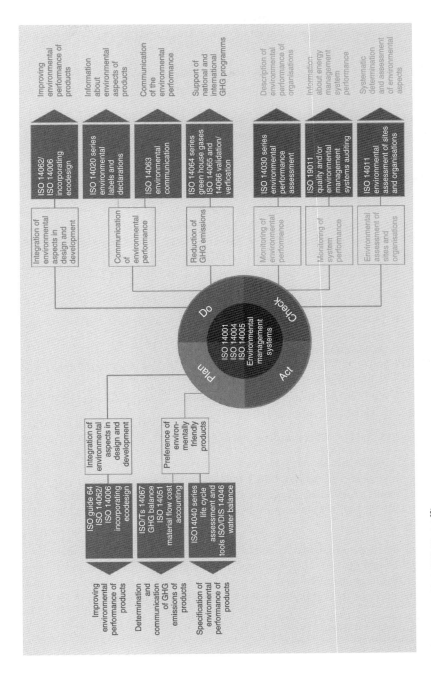

**Figure 1.5** ISO 14000 Model.[69]

official languages, English, French and German). DIN-NAGUS and similar national committees' activity is focused on preliminary work for the ISO work groups, the work-out and harmonisation of supplementary comments, the translation of ISO texts and supplementary standardisation for specifically national issues.

The first series of the international LCA standards closely followed the structure of Figure 1.4:

- ISO 14040 LCA – principles and framework; international standard 1997
- ISO 14041: LCA – goal and scope definition and inventory analysis; international standard 1998
- ISO 14042: LCA – life cycle impact assessment; international standard 2000;
- ISO 14043: LCA – interpretation; international standard 2000.

### 1.4.2
**Status Quo**

A revision of the international standards in 2001–2006 led to restructuring without any essential technical changes.[70] The basic standard continues to be called ISO 14040[71] with no mandatory directives. Directives have been summarised in a new standard ISO 14044[72] comprising all four LCA phases in Figure 1.4.

Two technical reports (TRs) and two technical specifications (TSs) have been added:

- ISO/TR 14047:2012 Illustrative example on how to apply ISO 14044 to impact assessment situations
- ISO/TS 14048:2002 Data documentation format
- ISO/TR 14049:2012 Illustrative examples on how to apply ISO 14044 to goal and scope definition and inventory analysis
- ISO/TS 14067:2013 Carbon footprint of products – Requirements and guidelines for quantification and communication.

Two new TS and one TR are under preparation:

- ISO TS 14071 Critical review processes and reviewer competencies – Additional requirements and guidelines to ISO 14044 (see Section 5.5)
- ISO TS 14072 Additional requirements and guidelines for organisations
- ISO/AWI TR 14073 Water footprint – Illustrative examples on how to apply ISO 14046.

TRs and TSs are non-mandatory documents but meant for help and support.

The methodology of LCA according to ISO 14040/44 is also the basis for a new standard under preparation aiming at the calculation of one specific impact (called *footprint* analogue to 'Carbon Footprint' in ISO/TS 14067 (see above)):

---

70) Finkbeiner *et al.* (2006).
71) ISO (2006a).
72) ISO (2006b).

- ISO 14046 Water footprint – Requirements and guidelines.

A detailed review of ISO standards in the context of ISO 14040 has been recently published by Finkbeiner (Finkbeiner, 2013).

As many as 24 national standardisation organisations participated in the first round of ISO standardisation talks and another 16 had the observer status. The final vote led to an overall acceptance of 95%. *LCA is therefore the only internationally accepted standardised method for analysing environmental aspects and potential impacts of product systems.*

Chapters 2–5 focus on the objective content of individual phases of LCA, their advantages and shortcomings.

## 1.5
### Literature and Information on LCA

Until the mid-1990s, almost only 'grey' LCA literature was available. Meanwhile, a series of books have been published.[73] Papers from national and international organisations provide essential information to LCA, mostly SETAC and SETAC Europe,[74] The Nordic Council,[75] US EPA (United States Environmental Protection Agency),[76] UBA Berlin[77] BUS/BUWAL Bern,[78] the European Environment Agency Copenhagen (EEA)[79] and the European Commission (EC).[80]

Recently, the Joint Research Centre 'Institute for Environment and Sustainability' of the EC (Ispra, Italy) published the ILCD Handbooks (*International Reference Life Cycle Data System*), which can be downloaded for free (*http://lct.jrc.ec.europa.eu/pdf-directory/ILCDHandbook.pdf*).

Since 1996, *The International Journal of Life Cycle Assessment* has been published by ecomed publishers Landsberg/Lech and Heidelberg and, since the beginning of 2008, by Springer, Heidelberg. Current information can be found at *http://www.springer.com/environment/journal/11367*.

The journal has rapidly developed into a leading publication organ of the promotion of LCA methodology.[81] *International Journal of Life Cycle Assessment* is also available electronically, and quite a number of articles as well as open access papers can be downloaded free of charge.

Further journals with regular contributions to LCA are the *Journal of Industrial Ecology* (MIT Press, part of Wiley-Blackwell since 2008), *Cleaner Production* (Elsevier)

---

73) Schmidt and Schorb (1995), Curran (1996), Eyrer (1996), Fullana and Puig (1997), Wenzel, Hauschild and Alting (1997), Hauschild and Wenzel (1998), Badino and Baldo (1998), Guinée *et al.* (2002), Baumann and Tillman (2004), Klöpffer and Grahl (2009) and Curran (2012).
74) Fava *et al.* (1991, 1993, 1994), SETAC (1993a), SETAC Europe (1992), Huppes and Schneider (1994), Udo de Haes (1996) and Udo de Haes *et al.* (2002).
75) Lindfors *et al.* (1994a,b, 1995).
76) EPA (1993) and SAIC (2006).
77) (UBA (1992, 1997), Klöpffer and Renner (1995) and Schmitz and Paulini (1999).
78) BUS (1984), BUWAL (1990, 1991, 1996, 1998).
79) Jensen *et al.* (1997).
80) EC (2010).
81) Heinrich (2013).

and *Integrated Environmental Assessment and Management,* IEAM (SETAC Press). Increasingly, *Environmental Science and Technology* (ACS) also publishes LCA-related papers. However, *International Journal of Life Cycle Assessment* is the only journal entirely devoted to LCA.

Other specialised journals also publish LCA literature. In 1995, for instance, the comprehensive ECOSOL – Surfactants-LCI of the European surfactants producers, conducted by Franklin Associates, published two issues of the journal *Tenside, Surfactants and Detergents*.[82]

A detailed treatment of all aspects of LCA and related methods such as Social Life Cycle Assessment (SLCA), Life Cycle Sustainability Assessment (LCSA) and LCM is being published in the LCA compendium (Springer 2014ff).[83] This work is conceived for 10 base volumes, and subvolumes as needed.

The significance of publication for propagation and discussion of methods, theories and results of research cannot be overestimated. Especially within new branches of science, peer reviews judge scientific validity on a day-to-day basis.[84] They serve as fine adjustments for the great principles of epistemology with special focus, according to Popper, on falsifiability,[85] which cannot be examined unambiguously for LCA. The scientific character of LCA is discussed critically in the following chapters dealing with the phases of LCA.

## References

Association Française de Normalisation (AFNOR) (1994) Analyse de cycle de vie. Norme NF X 30–300. 3/1994.

Badino, V. and Baldo, G.L. (1998) *LCA – Istruzioni per l'uso*, Progetto Leonardo, Esculapio Editore, Bologna.

Baumann, H. and Tillman, A.-M. (2004) *The Hitch Hiker's Guide to LCA. An Orientaion in LCA Methodology and Application*, Studentlitteratur, Lund. ISBN: 91-44-02364-2.

Bundesverband der Deutschen Industrie e. V. (BDI) (1999) *Die Durchführung von Ökobilanzen zur Information von Öffentlichkeit und Politik*, BDI-Drucksache Nr. 313, Verlag Industrie-Förderung, Köln. ISSN: 0407-8977.

Beck, M.(Hrsg.) (1993) *Ökobilanzierung im betrieblichen Management*, Vogel Buchverlag, Würzburg. ISSN: 3-8023-1479-4.

Berenbold, H. and Kosswig, K. (1995) A life-cycle inventory for the production of secondary alkane sulphonates (SAS) in Europe. *Tenside Surf. Det.*, **32**, 152–156.

Berna, J.L., Cavalli, L., and Renta, C. (1995) A lifecycle inventory for the production of linear akylbenzene sulphonates in Europe. *Tenside Surf. Det.*, **32**, 122–127.

BIfA/IFEU/Flo-Pak (2002) *Kunststoffe aus nachwachsenden Rohstoffen:Vergleichende Ökobilanz für Loose-fill-Packmittel aus Stärke bzw. aus Polystyrol*, Bayerisches Institut für Angewandte Umweltforschung und Technik, Augsburg.

Blouet, A. and Rivoire, E. (1995) *L'Écobilan: Les produits et leurs impacts sur l'environnement*, Dunod, Paris. ISBN: 2-10-002126-5.

---

82) Janzen (1995), Klöpffer, Grießhammer and Sundström (1995), Berna, Cavalli and Renta (1995), Stalmans *et al.* (1995), Hirsinger and Schick (1995a,b,c,d), Thomas (1995), Berenbold and Kosswig (1995), Postlethwaite (1995a,b), Schul, Hirsinger and Schick (1995) and Franke *et al.* (1995).
83) Klöpffer and Curran (2014) and Klöpffer (2014).
84) Klöpffer (2007).
85) Popper (1934).

Boustead, I. (1996) LCA – How it came about. The beginning in UK. *Int. J. Life Cycle Assess.*, **1** (3), 147–150.

Boustead, I. and Hancock, G.F. (1979) *Handbook of Industrial Energy Analysis*, Ellis Horwood Ltd, Chichester.

Braunschweig, A. and Müller-Wenk, R. (1993) *Ökobilanzen für Unternehmungen. Eine Wegleitung für die Praxis*, Verlag Haupt, Bern.

British Standards Institution (Ed.) (2008) Publicly Available Specification (PAS) 2050:2008. *Specification for the Assessment of the Life Cycle Greenhouse Gas Emissions of Goods and Services*, British Standards Institution.

Bundesamt für Umweltschutz (BUS) (Hrsg.) (1984) *Oekobilanzen von Packstoffen*. Schriftenreihe Umweltschutz, Nr. **24**, BUS, Bern.

BUWAL (1990) Ahbe, S., Braunschweig, A., and Müller-Wenk, R. *Methodik für Oekobilanzen auf der Basis ökologischer Optimierung*, Schriftenreihe Umwelt, Nr. **133**, Bundesamt für Umwelt, Wald und Landschaft (BUWAL), Bern (Hrsg.).

BUWAL (1991) Habersatter, K., and Widmer, F. *Oekobilanzen von Packstoffen. Stand 1990*, Schriftenreihe Umwelt, Nr. **132**, Bundesamt für Umwelt, Wald und Landschaft (BUWAL), Bern (Hrsg.).

BUWAL (1996, 1998) Habersatter, K., Fecker, I., Dall'Aqua, S., Fawer, M., Fallscher, F., Förster, R., Maillefer, C., Ménard, M., Reusser, L., Som, C., Stahel, U., and Zimmermann, P. *Ökoinventare für Verpackungen*, Schriftenreihe Umwelt, Nr. **250**/Bd. I und II. 2. erweiterte und aktualisierte Auflage, Bern 1998, 1. Auflage, ETH Zürich und EMPA, St. Gallen, BUWAL und SVI, Bern (Hrsg.).

Canadian Standards Association (CSA): (1992) *Environmental Life Cycle Assessment*. CAN/CSA-Z760, 5th Draft Edition, May 1992.

Curran, M.A. (ed) (1996) *Environmental Life-Cycle Assessment*, McGraw-Hill, New York. ISBN: 0-07-015063-X.

Curran, M.A. (1999) The status of LCA in the USA. *Int. J. Life Cycle Assess.*, **4** (3), 123–124.

Curran, M.A. (ed) (2012) *Life Cycle Assessment Handbook a Guide for Environmentally Sustainable Products*, John Wiley & Sons, Inc., Hoboken, NJ. ISBN: 978-1-1180-9972-8.

Detzel, A., and Böß, A. (2006) *Ökobilanzieller Vergleich von Getränkekartons und PETEinwegflaschen*, Endbericht, Institut für Energie- und Umweltforschung (IFEU), Heidelberg an den Fachverband Kartonverpackungen (FKN), Wiesbaden, August 2006.

Deutsches Institut für Bauen und Umwelt (2007) *TECU® – Kupferbänder und Kupferlegierungen*, KME Germany AG. Programmhalter Deutsches Institut für Bauen und Umwelt. Registrierungsnummer No: AUB-KME-30807-D. Ausstellungsdatum: 2007-11-01, verifiziert von: Dr. Eva Schmincke.

DIN-NAGUS (1994) DIN-NAGUS: Grundsätze produktbezogener Ökobilanzen (Stand Oktober 1993). *DIN-Mitteilungen*, **73** (3), 208–212.

EPA (1993) Vigon, B.W., Tolle, D.A., Cornaby, B.W., Latham, H.C., Harrison, C.L., Boguski, T.L., Hunt, R.G., and Sellers, J.D. *Life Cycle Assessment: Inventory Guidelines and Principles*, EPA/600/R-92/245, Office of Research and Development, Cincinnati, OH.

European Commission (2010) *ILCD Handbook*, Joint Research Centre, Institute for Environment and Sustainability, http://ies.jrc.ec.europa.eu (accessed 17 July 2007).

Eyrer, P.(Hrsg.) (1996) *Ganzheitliche Bilanzierung. Werkzeug zum Planen und Wirtschaften in Kreisläufen*, Springer, Berlin. ISBN: 3-540-59356-X.

Fink, P. (1997) LCA – How it came about. The roots of LCA in Switzerland: continuous learning by doing. *Int. J. Life Cycle Assess.*, **2** (3), 131–134.

Finkbeiner, M. (2013) The international standards as constitution of life cycle assessment: the ISO 14040 series and its offspring, in *Encyclopedia of Life Cycle Assessment*, vol. 1, Chapter 6 (ed. W. Klöpffer), Springer, Heidelberg.

Finkbeiner, M., Inaba, A., Tan, R.B.H., Christiansen, K., and Klüppel, H.-J. (2006) The new international standards for life cycle assessment: ISO 14040 and ISO

14044. *Int. J. Life Cycle Assess.*, **11** (2), 80–85.

Finkbeiner, M. and Matsuno, Y. (eds) (2000) LCA in Japan. *Int. J. Life Cycle Assess.*, **5** (5 Special Issue), 253–313.

Fleischer, G., and Schmidt, W.-P. (1996) Functional unit for systems using natural raw materials. *Int. J. Life Cycle Assess.*, **1** (1), 23–27.

Franke, M. (1984) *Umweltauswirkungen durch Getränkeverpackungen. Systematik zur Ermittlung der Umweltauswirkungen von komplexen Prozessen am Beispiel von. Einweg- und Mehrweg-Getränkebehältern*, EF-Verlag für Energie- und Umwelttechnik, Berlin.

Franke, M., Berna, J.L., Cavalli, L., Renta, C., Stalmans, M., and Thomas, H. (1995) A life-cycle inventory for the production of petrochemical intermediates in Europe. *Tenside Surf. Det.*, **32**, 384–396.

Fullana, P., and Puig, R. (1997) *Análisis del ciclo de vida*, Primera edición, Rubes Editorial, Barcelona. ISBN: 84-497-0070-1.

Gabathuler, H. (1998) The CML story. How environmental sciences entered the debate on LCA. *Int. J. Life Cycle Assess.*, **2** (4), 187–194.

Gensch, C.-O. (2008) *Klimaneutrale Weleda AG. Endbericht*, Öko-Institut Freiburg.

Giegrich, J., Mampel, U., Duscha, M., Zazcyk, R., Osorio-Peters, S., and Schmidt, T. (1995) *Bilanzbewertung in produktbezogenen Ökobilanzen. Evaluation von Bewertungsmethoden, Perspektiven*, UBA Texte 23/95, Endbericht des Instituts für Energie- und Umweltforschung Heidelberg GmbH (IFEU) an das Umweltbundesamt, Berlin, Heidelberg, März. ISSN 0722-186X.

Grahl, B., and Schmincke, E. (1996) Evaluation and decision-making processes in life cycle assessment. *Int. J. Life Cycle Assess.*, **1** (1), 32–35.

Guinée, J. B. (final editor); Gorée, M., Heijungs, R., Huppes, G., Kleijn, R., de Koning, A, van Oers, L., Wegener Sleeswijk, A., Suh, S., Udo de Haes, H. A., de Bruijn, H., van Duin, R., Huijbregts, M.A.J. (2002) *Handbook on Life Cycle Assessment – Operational Guide to the ISO Standards*. Kluwer Academic Publishers, Dordrecht. ISBN: 1-4020-0228-9.

Günther, A., and Langowski, H.-C. (1997) Life cycle assessment study on resilient floor coverings. *Int. J. Life Cycle Assess.*, **2** (2), 73–80.

Günther, A. and Langowski, H.-C. (eds) (1998) *Life Cycle Assessment Study on Resilient Floor Coverings. For ERFMI (European Resilient Flooring Manufacturers Institute)*, Fraunhofer IRB Verlag. ISBN: 3-8167-5210-1.

Hauschild, M., and Wenzel, H. (1988) *Environmental Assessment of Products*, Scientific Background, vol. 2, Chapman & Hall, London. ISBN: 0-412-80810-2.

Heijungs, R. (1997) *Economic drama and the environmental stage. formal derivation of algorithmic tools for environmental analysis and decision-support from a unified epistemological principle*. Proefschrift (Dissertation/PhD-Thesis). Leiden. ISBN: 90-9010784-3.

Heijungs, R., and Suh, S. (2002) *The Computational Structure of Life Cycle Assessment*, Kluwer Academic Publishers, Dordrecht. ISBN: 1-4020-0672-1.

Heinrich, A.B. (2013) Life cycle assessment as reflected by the international journal of life cycle assessment, in *Encyclopedia of Life Cycle Assess*, vol. 1, Chapter 7 (ed W. Klöpffer), Springer.

Heyde, M., and Kremer, M. (1999) *Recycling and Recovery of Plastics from Packaging in Domestic Waste. LCA-Type Analysis of Different Strategies*, vol. 5, LCA Documents, Ecoinforma Press, Bayreuth. ISBN: 3-928379-57-7.

Hirsinger, F., and Schick, K.-P. (1995a) A life-cycle inventory for the production of alcohol sulphates in Europe. *Tenside Surf. Det.*, **32**, 128–139.

Hirsinger, F., and Schick, K.-P. (1995b) A life-cycle inventory for the production of alkyl polyglucosides in Europe. *Tenside Surf. Det.*, **32**, 193–200.

Hirsinger, F., and Schick, K.-P. (1995c) A life-cycle inventory for the production of detergent grade alcohols. *Tenside Surf. Det.*, **32**, 398–410.

Hirsinger, F., and Schick, K.-P. (1995d) A life-cycle inventory for the production of oleochemical raw materials. *Tenside Surf. Det.*, **32**, 420–432.

Huijbregts, M.A.J., Guinée, J.B., Huppes, G., and Potting, J. (eds) (2006) Special issue honoring Helias A. Udo de Haes at the occasion of his retirement. *Int. J. Life Cycle Assess.*, **11** (1) Special Issue, 1–132.

Humbert, S., Rossi, V., Margni, M., Jolliet, O., and Loerincik, Y. (2008) Life cycle assessment of two baby food packaging alternatives: glass jars vs. plastic pots. *Int. J. Life Cycle Assess.*, **14** (2), 95–106.

Hunt, R., and Franklin, W.E. (1996) LCA – How it came about. Personal reflections on the origin and the development of LCA in the USA. *Int. J. Life Cycle Assess.*, **1** (1), 4–7.

Huppes, G. and Schneider, F. (eds) (1994) *Proceedings of the European Workshop on Allocation in LCA*, Leiden, February 1994, SETAC Europe, Brussels.

IFEU/SLFA (1998) *Ökobilanz Beikrautbekämpfung im Weinbau*. Im Auftrag des Ministeriums für Wirtschaft, Verkehr, Landwirtschaft und Weinbau Rheinland-Pfalz, Mainz, IFEU, Heidelberg, SLFA, Neustadt an der Weinstraße, Dezember 1998.

IFEU (2002) Ostermayer, A., and Schorb, A. *Ökobilanz Fruchtsaftgetränke Verbund-Standbodenbeutel 0,2 l, MW-Glasflasche, Karton Giebelpackung*. Im Auftrag der Deutschen SISI-Werke, Eppelheim, IFEU, Heidelberg, Juli 2002.

IFEU (2004) Detzel, A., Giegrich, J., Krüger, M., Möhler, S., and Ostermayer, A. *Ökobilanz PET-Einwegverpackungen und sekundäre Verwertungsprodukte*. Im Auftrag von PETCORE, Brüssel, IFEU, Heidelberg, August 2004.

IFEU (2006) Detzel, A., and Krüger, M. *LCA for Food Contact Packaging Made from PLA and Traditional Materials*. On behalf of Natureworks LLC, IFEU, Heidelberg, Juli 2006.

IFEU (2007) Krüger, M., and Detzel, A. *Aktuelle Ökobilanz zur 1,5-L-PET-Einwegflasche in Österreich unter Einbeziehung des Bottle-to- Bottle-Recycling*. Im Auftrag des Verbands der Getränkehersteller Österreichs, IFEU, Heidelberg, Oktober 2007.

ISO *(International Standard Organization)* (1997) ISO 14040. *Environmental Management – Life Cycle Assessment – Principles and Framework*, CEN European Committee for Standardization, Brussels.

ISO *(International Standard Organization)* (1998) ISO 14041. *Environmental Management – Life Cycle Assessment: Goal and Scope Definition and Inventory Analysis*, International Organization for Standardization, Genva.

ISO *(International Standard Organization)* (2000a) ISO 14042. *Environmental Management – Life Cycle Assessment: Life Cycle Impact Assessment*, International Organization for Standardization, Genva.

ISO *(International Standard Organization)* (2000b) ISO 14043. *Environmental Management – Life Cycle Assessment: Interpretation*, International Organization for Standardization, Genva.

ISO *(International Standard Organization)* (2006a) ISO 14040:2006, ISO/TC 207/SC 5. *Environmental Management – Life Cycle Assessment – Principles and Framework*, International Organization for Standardization, Genva.

ISO *(International Standard Organization)* (2006b) ISO 14044:2006, ISO/TC 207/SC 5. *Environmental Management – Life Cycle Assessment – Requirements and Guidelines*, International Organization for Standardization, Genva.

Janzen, D.C. (1995) Methodology of the European surfactant life-cycle inventory for detergent surfactants production. *Tenside Surf. Det.*, **32**, 110–121.

Jensen, A.A., Hoffman, L., Møller, B.T., Schmidt, A., Christiansen, K., Elkington, J., and van Dijk, F. (1997) *Life Cycle Assessment (LCA). A Guide to Approaches, Experiences and Information Sources*, Environmental Issues Series, vol. 6, European Environmental Agency.

Kemna, R., van Elburg, M., Li, W., and van Holsteijn, R. (2005) *Methodology Study Ecodesign of Energy-using Products – MEEUP Methodology Report for DG ENETR, Unit ENTR/G/3 in collaboration with DG TREN, Unit D1*. Delft.

Kindler, H., and Nikles, A. (1979) Energiebedarf bei der Herstellung und Verarbeitung von Kunststoffen. *Chem. Ing. Tech.*, **51**, 1125–1127.

Kindler, H., and Nikles, A. (1980) Energieaufwand zur Herstellung von Werkstoffen – Berechnungsgrundsätze und Energieäquivalenzwerte von Kunststoffen. *Kunststoffe*, **70**, 802–807.

Klöpffer, W. (1994) Environmental hazard assessment of chemicals and products. Part IV. Life cycle assessment. *Environ. Sci. Pollut. Res.*, **1** (5), 272–279.

Klöpffer, W. (1997) Life cycle assessment – From the beginning to the current state. *Environ. Sci. Pollut. Res.*, **4** (4), 223–228.

Klöpffer, W. (2003) Life-cycle based methods for sustainable product development. Editorial for the LCM Section in. *Int. J. Life Cycle Assess.*, **8** (3), 157–159.

Klöpffer, W. (2006) The Role of SETAC in the development of LCA. *Int. J. Life Cycle Assess.*, **11** (1 Special Issue), 116–122.

Klöpffer, W. (2007) Publishing scientific articles with special reference to LCA and related topics. *Int. J. Life Cycle Assess.*, **12** (2), 71–76.

Klöpffer, W. (2008) Life-cycle based sustainability assessment of products. *Int. J. Life Cycle Assess.*, **13** (2), 89–94.

Klöpffer, W. (ed) (2014) *LCA compendium*, vol. 1, Springer, Heidelberg.

Klöpffer, W., and Curran, M.A. (editors-in-chief.) (2014) *Background and Future Prospetcts in Life Cycle Assessment*, Springer, Heidelberg, p. 2014ff.

Klöpffer, W., and Grahl, B. (2009) *Ökobilanz (LCA) – Ein Leitfaden für Ausbildung und Beruf*, Wiley-VCH Verlag GmbH, Weinheim. ISBN: 978-3-527-32043-1.

Klöpffer, W., Grießhammer, R., and Sundström, G. (1995) Overview of the scientific peer review of the European life cycle inventory for surfactant production. *Tenside Surf. Det.*, **32**, 378–383.

Klöpffer, W., and Renner, I. (1995) *Methodik der Wirkungsbilanz im Rahmen von Produkt-Ökobilanzen unter Berücksichtigung nicht oder nur schwer quantifizierbarer Umwelt-Kategorien*, UBA-Texte 23/95, Bericht der C.A.U. GmbH, Dreieich, an das Umweltbundesamt (UBA), Berlin. ISSN: 0722-186X.

Klöpffer, W., and Renner, I. 2003. Life cycle impact categories – the problem of new categories and biological impacts – part I: systematic approach. SETAC Europe, 13th Annual Meeting, Hamburg, Germany, 27 April – 1 May 2003.

Klöpffer, W., Renner, I., Tappeser, B., Eckelkamp, C., and Dietrich, R. (1999) *Life Cycle Assessment gentechnisch veränderter Produkte als Basis für eine umfassende Beurteilung möglicher Umweltauswirkungen*, Monographien Bd. **111**, Federal Environment Agency Ltd, Wien. ISBN: 3-85457-475-4.

Klöpffer, W., and Volkwein, S. (1995) in *Enzyklopädie der Kreislaufwirtschaft, Management der Kreislaufwirtschaft* (Hrsg. K.J. Thomé-Kozmiensky), EF-Verlag für Energie- und Umwelttechnik, Berlin, S. 336–340.

Klüppel, H.-J. (1997) Goal and scope definition and life cycle inventory analysis. *Int. J. LCA*, **2** (1), 5–8.

Klüppel, H.-J. (2002) The ISO standardization process: Quo vadis? *Int. J. LCA*, **7**, 1.

Leuven (1990) Life cycle analysis for packaging environmental assessment. *Proceedings of the Specialised Workshop organized by Procter & Gamble, Leuven, Belgium, September 24/25*.

Lindfors, L.-G., Christiansen, K., Hoffmann, L., Virtanen, Y., Juntilla, V., Leskinen, A., Hansen, O.-J., Rønning, A., Ekvall, T., Finnveden, G., Weidema, B.P., Ersbøll, A.K., Bomann, B., and Ek, M. (1994a) LCA-NORDIC Technical Reports No. 10 and Special Reports No. 1–2. Tema Nord 1995:503. Nordic Council of Ministers, Copenhagen. ISBN: 92-9120-609-1.

Lindfors, L.-G., Christiansen, K., Hoffmann, L., Virtanen, Y., Juntilla, V., Leskinen, A., Hansen, O.-J., Rønning, A., Ekvall, T., and Finnveden, G. (1994b) LCA-NORDIC Technical Reports No. 1–9. Tema Nord 1995:502. Nordic Council of Ministers, Copenhagen.

Lindfors, L.-G., Christiansen, K., Hoffmann, L., Virtanen, Y., Juntilla, V., Hanssen, O.-J., Rønning, A., Ekvall, T., and Finnveden, G. (1995) *Nordic Guidelines on Life-Cycle Assessment*, Nord 1995:20, Nordic Council of Ministers, Copenhagen.

Lundholm, M.P., and Sundström, G. (1985) *Ressourcen und Umweltbeeinflussung. Tetrabrik Aseptic Kartonpackungen sowie Pfandflaschen und Einwegflaschen aus Glas*. Malmö 1985.

Lundholm, M.P., and Sundström, G. (1986) *Ressourcen- und Umweltbeeinflussung durch zwei Verpackungssysteme für*

Milch, Tetra Brik und Pfandflasche. Malmö 1986.

Marsmann, M. (1997) ISO 14040 – The first project. *Int. J. Life Cycle Assess.*, **2** (3), 122–123.

Mauch, W., and Schaefer, H. (1996) in *Ganzheitliche Bilanzierung. Werkzeug zum Planen und Wirtschaften in Kreisläufen* (Hrsg. Eyrer, P.), Springer, Berlin S. 152–180. ISBN: 3-540-59356-X.

Meadows, D.H., Meadows, D.L., Randers, J., and Behrens, W.W. III, (1972) *The Limits to Growth. A Report for the Club of Rome's Project on the Predicament of Mankind*, Universe Books, New York. ISBN: 0-87663-165-0.

Meadows, D.L., Meadows, D.H., Zahn, E., and Milling, P. (1973) *Die Grenzen des Wachstums. Bericht des Club of Rome zur Lage der Menschheit*, Ts. rororo Taschenbuch, neue Auflage im dtv Taschenbuchverlag, Hamburg, pp. 101–200. ISBN: 3-499-16825-1.

Merkel, A. (1997) Foreword: ISO 14040. *Int. J. Life Cycle Assess.*, **2** (3), 121.

Neitzel, H. (ed) (1996) Principles of product-related life cycle assessment. *Int. J. Life Cycle Assess.*, **1** (1), 49–54.

O'Brien, M., Doig, A., and Clift, R. (1996) Social and environmental life cycle assessment (SELCA) approach and methodological development. *Int. J. Life Cycle Assess.*, **1** (4), 231–237.

Oberbacher, B., Nikodem, H., and Klöpffer, W. (1996) LCA – How it came about. An early systems analysis of packaging for liquids which would be called an LCA today. *Int. J. Life Cycle Assess.*, **1** (2), 62–65.

Popper, K.R. (1934, 7. Auflage, Mohr, J.C.B. (Paul Siebeck), Tübingen, 1982, 1st English edn, *The Logic of Scientific Discovery*, Hutchison, London, 1959) *Logik der Forschung*, Julius Springer, Wien.

Postlethwaite, D. (1995a) A life-cycle inventory for the production of sulphur and caustic soda in Europe. *Tenside Surf. Det.*, **32**, 412–418.

Postlethwaite, D. (1995b) A life-cycle inventory for the production of soap in Europe. *Tenside Surf. Det.*, **32**, 157–170.

Projektgruppe Ökologische Wirtschaft (1987) *Produktlinienanalyse: Bedürfnisse, Produkte und ihre Folgen*, Kölner Volksblattverlag, Köln.

Rüdenauer, I., Dross, M., Eberle, U., Gensch, C., Graulich, K., Hünecke, K., Koch, Y., Möller, M., Quack, D., Seebach, D., Zimmer, W. et al. (2007) *Costs and Benefits of Green Public Procurement in Europe. Part 1: Comparison of the Life Cycle Costs of Green and Non Green Products*, Service Contract Number: DG ENV.G.2/SER/2006/0097r. Öko-Institut Freiburg.

Saur, K. (1997) Life cycle impact assessment (LCA-ISO activities). *Int. J. Life Cycle Assess.*, **2** (2), 66–70.

Schaltegger, S. (ed.) (1996) *Life Cycle Assessment (LCA) – Quo vadis?*, Birkhäuser Verlag, Basel und Boston, MA. ISBN: 3-7643-5341-4 (Basel), ISBN: 0-8176-5341-4 (Boston).

Schmidt, A., Jensen, A.A., Clausen, A., Kamstrup, O., and Postlethwaite, D. (2004) A comparative life cycle assessment of building insulation products made of stone wool, paper wool and flax. Part 1: background, goal and scope, life cycle inventory, impact assessment and interpretation. *Int. J. Life Cycle Assess.*, **9** (1), 53–66.

Schmidt, M., and Schorb, A. (1995) *Stoffstromanalysen in Ökobilanzen und Öko-Audits*, Springer-Verlag, Berlin. ISBN: 3-540-59336-5.

Schmincke, E., and Grahl, B. (2006) Umwelteigenschaften von Produkten. Die Rolle der Ökobilanz in ISO Typ III Umweltdeklarationen. *UWSF – Z Umweltchem Ökotox*, **18** (3), 185–192.

Schmitz, S., Oels, H.-J., and Tiedemann, A. (1995) *Ökobilanz für Getränkeverpackungen. Teil A: Methode zur Berechnung und Bewertung von Ökobilanzen für Verpackungen. Teil B: Vergleichende Untersuchung der durch Verpackungssysteme für Frischmilch und Bier hervorgerufenen Umweltbeeinflussungen*, UBA Texte 52/95, Umweltbundesamt, Berlin.

Schmitz, S., and Paulini, I. (1999) *Bewertung in Ökobilanzen. Methode des Umweltbundesamtes zur Normierung von Wirkungsindikatoren, Ordnung (Rangbildung) von Wirkungskategorien und zur Auswertung nach ISO 14042 und 14043*, Version '99. UBA Texte 92/99, Berlin.

Schul, W., Hirsinger, F., and Schick, K.-P. (1995) A life-cycle inventory for the production of detergent range alcohol ethoxylates in Europe. *Tenside Surf. Det.,* **32**, 171–192.

Scientific Applications International Corporation (SAIC) (2006) *Life Cycle Assessment: Principles and Practice,* U.S. EPA, Systems Analysis Branch, National Risk Management Research Laboratory, Cincinnati, OH.

Society of Environmental Toxicology and Chemistry (SETAC) (1991) Fava, J.A., Denison, R., Jones, B., Curran, M.A., Vigon, B., Selke, S. and Barnum, J. (eds) *SETAC Workshop Report: A Technical Framework for Life Cycle Assessments,* Smugglers Notch, Vermont, August 18–23 1990, SETAC, Washington, DC, January 1991.

Society of Environmental Toxicology and Chemistry – Europe (Ed.) (1992) *Life-Cycle Assessment,* Workshop Report, Leiden, 2–3 December 1991, SETAC Europe, Brussels.

Society of Environmental Toxicology and Chemistry (SETAC) (1993a) *Guidelines for Life- Cycle Assessment: 'A Code of Practice',* 1st edn. From the SETAC Workshop held at Sesimbra, Portugal, 31 March – 3 April 1993, SETAC Press, Brussels and Pensacola, FL, August 1993.

Society of Environmental Toxicology and Chemistry (SETAC) (1993b Fava, J., Consoli, F.J., Denison, R., Dickson, K., Mohin, T. and Vigon, B. (eds) *Conceptual Framework for Life-Cycle Impact Analysis.* Workshop Report, SETAC and SETAC Foundation for Environmental Education, Inc., Sandestin, FL, February 1–7, 1992..

Society of Environmental Toxicology and Chemistry (SETAC) (1994Fava, J., Jensen, A.A., Lindfors, L., Pomper, S., De Smet, B., Warren, J. and Vigon, B. (eds)) *Conceptual Framework for Life-Cycle Data Quality,* Workshop Report, SETAC and SETAC Foundation for Environmental Education, Inc., Wintergreen, VA, October 1992.

Stalmans, M., Berenbold, H., Berna, J.L., Cavalli, L., Dillarstone, A., Franke, M., Hirsinger, F., Janzen, D., Kosswig, K., Postlethwaite, D., Rappert, T., Renta, C., Scharer, D., Schick, K.-P., Schul, W., Thomas, H., and Van Sloten, R. (1995) European lifecycle inventory for detergent surfactants production. *Tenside Surf. Det.,* **32**, 84–109.

Suter, P., Frischknecht, R. (Projektleitung), Frischknecht, R. (Schlussredaktion), Bollens, U., Bosshart, S., Ciot, M., Ciseri, L., Doka, G., Frischknecht, R., Hirschier, R., Martin, A., Dones, R., and Gantner, U. (AutorInnen der Überarbeitung) (1996) *Ökoinventare von Energiesystemen,* ETH Zürich und Paul Scherrer Institut, Villingen, im Auftrag des Bundesamtes für Energiewirtschaft (BEW) und des Projekt- und Studienfonds der Elektrizitätswirtschaft (PSEL), 3. Auflage, Zürich.

Suter, P., Walder, E. (Projektleitung), Frischknecht, R., Hofstetter, P., Knoepfel, I., Dones, R., Zollinger, E. (Ausarbeitung), Attinger, N., Baumann, Th., Doka, G., Dones, R., Frischknecht, R., Gränicher, H.-P., Grasser, Ch., Hofstetter, P., Knoepfel, I., Ménard, M., Müller, H., Vollmer, M., Walder, E., and Zollinger, E., (AutorInnen) (1995) *Ökoinventare für Energiesysteme.* Villingen im Auftrag des Bundesamtes für Ener gie wirtschaft (BEW) und des Nationalen Energie-Forschungs-Fonds NEFF, 2. Auflage, ETH Zürich und Paul Scherrer Institut.

Svenska Miljöstyrningsrådet (2006) *Climate Declaration, Carbon Footprint for Natural Mineral Water.* Registration number: S-P-00123. Program: The EPD®system. Program operator: AB Svenska Miljöstyrningsrådet (MSR), Product Category Rules (PCR): Natural mineral water (PCR 2006:7), PCR review conducted by: MSR Technical Committee chaired by Sven-Olof Ryding (info@environdec.com), Third party verified: Extern Verifier: Certiquality *www.environdec.com* (accessed 10 September 2013).

Thomas, H. (1995) A life-cycle inventory for the production of alcohol ethoxy sulphates in Europe. *Tenside Surf. Det.,* **32**, 140–151.

Tiedemann, A. (Hrsg.) 2000) *Ökobilanzen für graphische Papiere,* UBA Texte 22/2000, Umweltbundesamt, Berlin.

Tukker, A., Kleijn, R., and van Oers, L. (1996) *A PVC Substance Flow Analysis for Sweden.* Report by TNO Centre for Technology and Policy Studies and Centre of Environmental Science (CML), TNO-Report STB/96/48-III. Leiden to Norsk Hydro, Apeldoorn, November 1996.

UBA (1992) *Arbeitsgruppe Ökobilanzen des Umweltbundesamts Berlin: Ökobilanzen für Produkte. Bedeutung – Sachstand – Perspektiven*, UBA Texte 38/92, Umweltbundesamt, Berlin.

UBA (2000a) Kolshorn, K.-U., and Fehrenbach, H. *Ökologische Bilanzierung von Altöl- Verwertungswegen*, UBA Texte 20/00. Abschlussbericht zum Forschungsvorhaben Nr. 297 92 382/01 des, Umweltbundesamtes, BerlinJanuar 2000. ISSN: 0722-186X.

UBA (2000b) Plinke, E., Schonert, M., Meckel, H., Detzel, A., Giegrich, J., Fehrenbach, H., Ostermayer, A., Schorb, A., Heinisch, J., Luxenhofer, K., and Schmitz, S. *Ökobilanz für Getränkeverpackungen II*, UBA Texte 37/00, Zwischenbericht (Phase 1) zum Forschungsvorhaben FKZ 296 92 504 des, Umweltbundesamtes, Berlin – Hauptteil, September 2000. ISSN: 0722-186X.

UBA (2002) Schonert, M., Metz, G., Detzel, A., Giegrich, J., Ostermayer, A., Schorb, A., and Schmitz, S. *Ökobilanz für Getränkeverpackungen II, Phase 2*, UBA Texte 51/02. Forschungsbericht 103 50 504 UBA-FB 000363 des, Umweltbundesamtes, BerlinOktober 2002. ISSN: 0722-186X.

Udo de Haes, H.A. (ed.) (1996) *Towards a Methodology for Life Cycle Impact Assessment*, SETAC Europe, Brussels. ISBN: 90-5607-005-3.

Udo de Haes, H.A., Finnveden, G., Goedkoop, M., Hauschild, M., Hertwich, E.G., Hofstetter, P., Jolliet, O., Klöpffer, W., Krewitt, W., Lindeijer, E., Müller-Wenk, R., Olsen, S.I., Pennington, D.W., Potting, J., and Steen, B. (2002) (eds.): *Life-Cycle Impact Assessment: Striving Towards Best Practice*. SETAC Press, Pensacola, FL. ISBN 1-880611-54-6

Udo de Haes, H.A., and de Snoo, G.R. (1996) Environmental certification. Companies and products: two vehicles for a life cycle approach? *Int. J. Life Cycle Assess.*, **1** (3), 168–170.

Udo de Haes, H.A., and de Snoo, G.R. (1997) The agro-production chain. Environmental management in the agricultural production- consumption chain. *Int. J. Life Cycle Assess.*, **2** (1), 33–38.

Umweltbundesamt Berlin (1997) *Materialien zu Ökobilanzen und Lebensweganalysen: Aktivitäten und Initiativen des Umweltbundesamtes; Bestandsaufnahme Stand März 1997*, UBA Texte 26/97, Umweltbundesamt, Berlin. ISSN: 0722-186X.

UNO (1992) *Agenda 21 in deutscher Übersetzung. Konferenz der Vereinten Nationen für Umwelt und Entwicklung im Juni 1992 in Rio de Janeiro – Dokumente*, www.sustainabledevelopment.un.org/content/documents/Agenda21.pdf (accessed 17 December 2013).

Vattenfall (2007) *Product Category Rules: PCR for Electricity, Steam, and Hot and Cold Water Generation and Distribution*. Registration no: 2007:08, Publication date: November 21 2007, PCR documents: pdf-file; download from www.environdec.com. Prepared by: Vattenfall AB, British Energy, EdF – Electricite de France, Five Winds International, Swedpower and Rolf Frischknecht – esu-services Switzerland, Enel Italy. PCR moderator: Caroline Setterwall, Vattenfall AB, Sweden.

VDI-Richtlinie VDI 4600 (1997) *Kumulierter Energieaufwand (Cumulative Energy Demand). Begriffe, Definitionen, Berechnungsmethoden. deutsch und englisch*. Verein Deutscher Ingenieure, VDI-Gesellschaft Energietechnik Richtlinienausschuss Kumulierter Energieaufwand, Düsseldorf.

Wenzel, H., Hauschild, M., and Alting, L. (1997) *Environmental Assessment of Products Vol. 1: Methodology, Tools and Case Studies in Product Development*, Chapman & Hall, London. ISBN: 0-412-80800-5.

World Commission on Environment and Development (1987) *Our Common Future (The Brundtland Report)*, Oxford University Press Deutsche Übersetzung: *Der Brundtland-Bericht der Weltkommission für Umwelt und Entwicklung*. Eggenkamp, Greven, 1987.

Wright, M., Allen, D., Clift, R., and Sas, H. (1997) Measuring corporate environmental performance. The ICI environmental burden system. *J. Indust. Ecol.*, **1**, 117–127.

# 2
# Goal and Scope Definition

## 2.1
## Goal Definition

The 'Definition of goal and scope' must be present in any standard life cycle assessment (LCA) study as the first component.[1] Here, the fundamental concepts of the study are specified within the framework of the standard. While an iterative approach within the standard is explicitly intended (see double arrows in Figure 1.4), any change of goal and scope must be documented during the conduct of an LCA.

The International Standard 14044[2] reads:

> *The goal and scope of an LCA shall be clearly defined and shall be consistent with the intended application. Due to the iterative nature of LCA, the scope may have to be refined during the study.*

The goal definition is a declaration made by the organisation (such as companies, industry or trade associations, environmental offices, NGOs, etc.) commissioning an LCA, by providing an explanation to the following[3]:

- Range of application: *What is the objective of the study?*
- Interest of realisation: *Why is an LCA study conducted?*
- Target group(s): *For whom will an LCA study be conducted?*
- Publication or other accessibility for the public: *Are comparative assertions intended in the study?*[4]

The depth and accuracy of the study have to be considered during the goal definition. The fundamental standard ISO 14040 explicitly points out that the goal definition and therefore also the application of an LCA represent the commissioner's free will decision and as such shall not be challenged by the critical review (see Section 2.2.7.3 and Chapter 5).[5] Thus, a multiplicity of possible applications (for

---

1) ISO (2006a).
2) ISO (2006b, Section 4.2.1).
3) SETAC (1993), DIN NAGUS (1994), Neitzel (1996) and ISO (2000b,c).
4) Comparative assertions in the sense of ISO standards mean that product A under environmental aspects is alike or better than product B; products in the sense of LCA standards are any goods *and* services.
5) This of course will not apply for ethically non-acceptable goals!

*Life Cycle Assessment (LCA): A Guide to Best Practice*, First Edition.
Walter Klöpffer and Birgit Grahl.
© 2014 Wiley-VCH Verlag GmbH & Co. KGaA. Published 2014 by Wiley-VCH Verlag GmbH & Co. KGaA.

examples, see Section 1.3.2, Table 1.1) are feasible – among others, those preparing environmental policy measures. Since the international standards are quite flexible with regard to the details of the conduct of an LCA (this is valid in particular for the phase of life cycle impact assessment, LCIA; see Chapter 4), first an adaptation of the general methodology to the problem in question must be specified. This is achieved by defining the scope of the study.

## 2.2
## Scope

### 2.2.1
### Product System

First, the examined product system or, in case of comparative LCAs product systems must be clearly described. This includes, above all, the functions of the systems as basis for the definition of the functional unit (fU) (see Section 2.2.5). The description should be brief, but as precise as possible in this early phase.

A product system is best described in a system flow chart. Figure 2.1 shows a simplified system flow chart of a poly(vinyl chloride) (PVC) window.

In a system flow chart, unit processes and their interrelations are usually represented by boxes. The entire, often very complex, pattern reminds of a tree and is therefore often called *product tree*. Since an essentially linear system definition is aimed at, branches occur only at the boxes (by several inputs with pre-chains

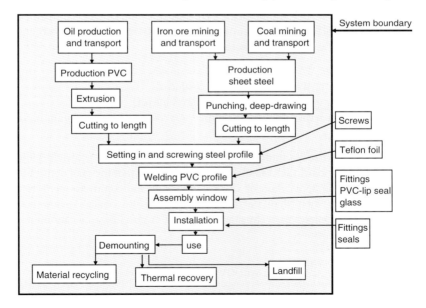

**Figure 2.1** Simplified flow chart of the product system PVC window.

or by several outputs in waste treatment), but no network. An exception is the treatment of recycling, which is discussed in Section 3.3. Within a complete LCA, the presentation ends at the disposal or at a point where co-products, by-products or waste for reutilisation exceed the system boundary (thus leaving the product system).

A special problem arises during the omission of parts of the life cycle. This can, by all means, be justified, if, for example, a provisional estimation showed that the overall system contribution is only very small (criteria: mass, energy, environmental relevance) (see Section 2.2.2.1). However, it must always be examined whether, within comparative studies, and thus in the majority of cases, no asymmetry of systems results from omission. Here, particular attention should be paid to the LCIA, because, compared to mass, very small emissions can nevertheless show large effects. Within comparative LCAs, large parts of the life cycle may be omitted in principle if they match accurately in all systems compared (*black box method*).

In Figure 2.1, for example, the construction elements on the right of the system boundary (screws, Teflon foil, fittings, etc.) are not considered. This is adequate, if, for instance, different windows (PVC, wood or aluminium windows) are to be compared with each other and if these construction elements are used similarly in all variants regarded. An estimation of relevance of the omitted sections should nevertheless be made so that comparison of systems is not based on completely insignificant differences. If, for example, two systems only differ in waste treatment (*End-of-Life* Stage) and if these can be neglected in both, the 'black box'[6] approach is inadmissible: both systems are – within an error limit – identical in their environmental behaviour analysed in the specific LCA.

A precise description and quantification of material and energy flow is conducted in the stage 'life cycle inventory analysis' (LCI) (see Chapter 3). Should details in the context of LCI analysis indicate an inadequate description of the product system, the description of the scope must be iteratively modified.

## 2.2.2
## Technical System Boundary

### 2.2.2.1 Cut-Off Criteria
The specification of system boundaries is one of the most important steps in an LCA. When two studies on a similar topic (e.g. single-use vs re-usable packaging) contradict themselves, which may incidentally be the case, usually one or several of the following reasons are responsible:

1. different methodology,
2. different data quality,
3. different system boundaries.

---

[6] 'Black box' means a life cycle stage or unit process that may be omitted within comparative LCAs because it is identical within all life cycles to be compared. It should nevertheless be applied sparingly because some most important environmental aspects may be blinded out.

To the first point, great progress has already been made by the International Standardisation (see Section 1.4). To the second, a uniform data format for data bases and data communication has been initiated.[7] To the third criterion, no general presetting can be provided because system boundaries depend on the specific problem in question. If, for example, a product is manufactured only in Italy using native raw materials and pre-products and distributed solely in Italy, the European Union as geographical system boundary makes little sense. Nevertheless, in the component LCIA, transnational emissions and their respective potential impacts have to be considered (see Chapter 4). In this context, as in LCA everywhere, *transparency* is very important (see Section 5.4).

The necessity for *cut-off criteria*, regulating the exclusion of insignificant inputs into the product system, results from the following consideration:

> Product systems are embedded into the large systems 'technosphere' and 'environment'.[8] It is a fundamental realisation of system analysis that all subsystems are linked, even though more or less intensely. To be able to study a subsystem for itself, numerous less important links must be broken. For this, rules are necessary. An important rule states that, for example, the infrastructure (roads, rails, etc.) is usually neglected (there are important exceptions,[9] however). Something similar is true for capital goods (e.g. the production of machines to manufacture the products), provided these are not the ones to be compared in a study.

ISO 14044[10] states three cut-off criteria applied for the entire product system as well as for individual unit processes:

1. mass
2. energy
3. environmental relevance.

Often, a proportion of 1% (mass, energy, etc.) of the overall system is chosen as the cut-off criterion. If a first analysis has, for example, shown that for the manufacture of a product 12 different materials are needed, their percentage ratio is determined in a first step. In the fictitious example of Figure 2.2, component ratios of 5, 6, 9 and 12 are below 1%. The cut-off criterion 'mass < 1%' alone entails that these components are not balanced over their entire life cycle. However, a first estimation of the energy consumption shows that component 9 has a mass ratio of only 0.2%, although for its production, 2.7% of the total energy is needed. Therefore, component 9 would be examined through its entire life cycle.

In addition, the rule is often applied that the portion to be cut off shall not exceed 5% per unit process (one box in the product tree). In Figure 2.3, a unit process with

---

7) ISO (2002) and EC (2010).
8) Both together result in the world in which we live; the technosphere, according to this functional definition, is 'everything under human control', and the environment is 'all that is not technosphere'. Frische et al. (1982) and Klöpffer (1989, 2001).
9) Frischknecht et al. (2004, 2005).
10) ISO (2006b, Section 4.2.3.3.3).

## 2.2 Scope

| | | Mass fraction (%) | Energy (%) |
|---|---|---|---|
| Raw material Pre-product | 1 | 73.8 | 12.0 |
| | 2 | | 54.7 |
| | 3 | | 23.3 |
| Ancillary material | 4 | 1.2 | 0.9 |
| | 5 | 0.1 | 0.1 |
| | 6 | 0.1 | < 0.1 |
| | 7 | 1.7 | 0.6 |
| | 8 | 1.4 | 0.7 |
| | 9 | 0.2 | 2.7 |
| | 10 | 19.8 | 4.5 |
| | 11 | 1.7 | 0.4 |
| | 12 | < 0.1 | < 0.1 |
| | 13 | | |
| Sum | | 100.0 | 99.9 |

Cut-off rules prevent arbitrariness in the choice of system boundaries. Example: Analysing material input.

**Figure 2.2** Application of the cut-off criteria 'mass' and 'energy'.

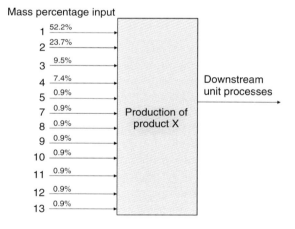

**Figure 2.3** Application of cut-off criteria: the 5% rule.

13 inputs is presented. The first analysis shows that the mass ratios of the inputs 5–13 are below 1% each. However, the cumulative mass ratio adds up to 7.2%, which would not be traced back to the raw materials in case cut-off criterion of 1% is applied. Therefore, the sole application of the 1% rule would result in large asymmetries when in a second variant, for example, just 1.5% would be the overall cut-off result.

In systems with high energy or mass throughput and simultaneous long life time of the product, the cut-off of less important branches of the product tree,

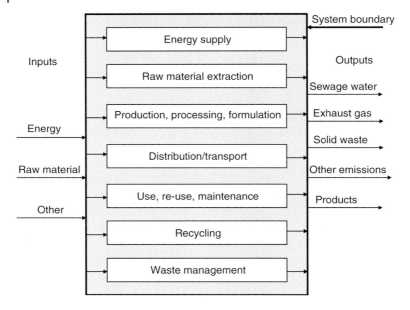

**Figure 2.4** System boundary of the inventory modified according to Society of Environmental Toxicology and Chemistry (SETAC) (1991).

infrastructure and so on, contributing less than 1% related to the entire life cycle is usually without problems.[11] In any case, an error estimation is required.

The cut-off criterion 'environmental relevance' is to prevent, for example, the omission of highly toxic emissions (say polychlorinated dibenzodioxins) in the investigated product systems due to too small masses.

The primarily highly interlaced systems of the complete system analysis become one-dimensional approximations by cutting off links. Interlaced subsystems (loops) are either calculated iteratively or by other suitable mathematical tools.[12] Branches without feedback represent no deviation of the linear sequence; they may, however, represent allocation problems (see Section 3.3).

### 2.2.2.2 Demarcation towards System Surrounding

The system surrounding[13] is composed by the ecosphere ('environment'; see Section 2.2.2.1, 'all that is not technosphere') *plus* the large remainder of the technosphere not included in the analysis. In Figure 2.4, this boundary is called[14] *system boundary*. The system under examination receives input from the system surrounding and delivers output to it.

---

11) Hunt, Sellers and Franklin (1992).
12) Heijungs and Frischknecht (1998) and Heijungs and Suh (2002).
13) The system surrounding is often called 'system environment', which can be misleading.
14) Society of Environmental Toxicology and Chemistry (SETAC) (1991).

Inputs specified in Figure 2.4 originate 'from the earth' (+ air + water) or are directly or indirectly made available by solar power. The following inputs from the environment are to be considered:

- All processes necessary for the extraction of raw materials (mining industry, oil production, forestry, etc.) belong to the investigated system ('exploitation of raw materials').
- In addition, inputs which, due to the cut-off criteria, are not traced over their entire life cycle must be considered ('miscellaneous'). These inputs may be pre-products, ancillary material or lubricant produced in the remainder of technosphere not included into the system under study.
- The entry 'energy' on the input side should actually be named 'energy raw materials', because the energy is produced from fossil, nuclear and regenerating raw materials. Exceptions are solar energy, potential energy of water (hydro power) and kinetic energy of wind (wind power). The energy supply, for example, in power stations, is within the system boundary.

Outputs, such as usable products and releases into the environment, are delivered to the system's surrounding. The investigated product in the centre of the system remains within the system boundary.

- 'Usable products' are the product under study, *co-products and secondary raw materials* (see below), which remain in the technosphere.
- Material emissions are delivered into the ecosphere by waste water and exhaust air. The plants for waste water treatment and exhaust air purification are within the system boundary.
- The allocation of solid *wastes* (landfill) has in former times occasionally been rated as 'releases into soil', which means they would leave the system. Today, controlled landfills are regarded as part of the technosphere and thus lie within the system boundaries. Only degassing and contamination of the groundwater due to leaky landfills are regarded as outputs into the environment. For waste incineration, analogous considerations apply. In the early days of LCA ('proto-LCAs'),[15] the sum of solid wastes has been an important aggregated parameter of the inventory.[16]
- 'Other emissions' can be radiation, biological releases, noise and similar non-chemical emissions.

The handling of co-products and secondary raw materials requires special attention during the definition of the system boundary.

**2.2.2.2.1 Co-products** During a chemical synthesis (or any other production process), besides the desired output within the examined product system, further useful products, materials or substances may be generated and covered by the generic term *co-products*.[17] In particular, co-products are frequently formed in the

---

15) Klöpffer (2006).
16) BUWAL (1991).
17) Riebel (1955).

chemical industry, but agriculture and its downstream industries are known for their co-product problem as well. Thus, for example, with the production of grain, straw is produced as a co-product that is transferred as 'usable product' to the system surrounding. In this case, environmental loads of the processes must be allocated, according to defined rules, both to the examined product and the co-product (see also Section 3.3, 'Allocation', and particularly Section 3.3.2.5, 'System Expansion'). Co-products can play a role in different unit processes of a product tree.

**2.2.2.2.2 Secondary Raw Material** Non-directly usable by-products are usually called *residual material*. Depending upon the recycling potential, distinction is drawn between 'secondary raw materials' (after cleaning or other processing) and 'wastes'. The 'Closed Substance Cycle and Waste Management Act'[18] in Germany has resulted in different designations for the same issue: wastes for reutilisation and wastes for disposal. Secondary raw materials that are gained from waste for reutilisation leave the system and are used as input in other product systems.

Recycling of materials, which lead to new products, where the materials thus become parts of other systems, is called *open loop recycling*.[19] The respective secondary raw materials (e.g. scrap, waste paper, waste glass, plastic wastes, etc.) leave the product system under study, of which where they are the residual material, as respective wastes for reutilisation.

Within the system boundaries remain materials in those recycling processes that lead back to the same product (the one under investigation), that is, it remains in the product system (*closed loop recycling*).[20] Moreover, in the case of product re-use, these materials remain in the investigated system (usually after cleaning). Examples of closed loop recycling are the re-feed of plastic shreds, punching, cutting-off, and so on, into the extruder. A good example of *re-use* is the refilling of returnable bottles.

Rules to be applied for allocation (Allocation, see Section 3.3.4) shall already be specified within the goal and scope definition; if not avoidance of allocation, for example, by system expansion, becomes compulsory in a specific case[21] (see Section 3.3.2.5).

The system boundary requires further explanations, the most important concern being the geographical and temporal system boundary.

## 2.2.3
### Geographical System Boundary

The geographical system boundary results from the economic context and from the product definition:

- Is the concerned special product manufactured in factory A, at site B, and so on, or a group of very similar products manufactured in multiple factories all over

---

18) German: 'Kreislaufwirtschaftsgesetz'.
19) Klöpffer (1996), Hunt, Sellers and Franklin (1992) and Boustead (1992).
20) SETAC Europe (1992), Curran (1996), Klöpffer (1996) and Hunt, Sellers and Franklin (1992).
21) ISO (1998, 2006b).

Europe (North America, Japan, etc. world-wide)? Similar considerations are valid for agricultural products, services, and so on.
- Even if a relatively close framework is selected, for example, production and sales in only one country, the geographical system boundary always has extensions beyond the selected range, because certain raw materials may be missing in the concerned country and thus have to be imported. Therefore, pollution of the environment also occurs in the countries of origin and in transportation from these. For export products, it must be noted that transportation, use and disposal predominantly take place in other countries. The international distribution of tasks in the context of progressive globalisation of the world economy (supplier) must also be considered within the geographical system boundary.
- In LCIA (see Chapter 4), global effects are considered for some impact categories (e.g. climate change/greenhouse effect, stratospheric ozone depletion), while for others regional or local effects (e.g. eutrophication potential) are considered. Local boundaries can, however, be clearly assigned only in rare cases, for instance, if a special product is manufactured in one factory only. In this case, at least one point in the life cycle can unambiguously be assigned geographically. Something similar is valid in agriculture if the farming region can be determined.

Altogether, the definition of the geographical system boundary is straightforward; it is a question of data availability. Commodities (e.g. metals, mass plastics, chemicals of very large production volume) often do not reveal their origin; in these cases, a regional allocation of impacts is difficult, if not impossible (see Chapter 4).

### 2.2.4
**Temporal System Boundary/Time Horizon**

The temporal system boundary is more difficult to define than the geographical boundary.

The minimum specification to the system boundary 'time' is a year of reference or another time period for data acquisition. For long-lived products, a determined or estimated lifetime or time of use provides a boundary shifted into the future of the inventory: disposal or re-use will only occur in the future. Accordingly, the modelling of these life cycle phases is difficult and uncertain.

For long, those problems with time have not been sufficiently considered in LCA research.[22] This did not play a role as long as predominantly short-lived products were examined such as packaging. Problems related to time became evident when LCAs of building materials, buildings and other long-lived products were carried out:

- How may LCA experts know which (perhaps not even yet invented) methods of waste disposal will predominate in 50 a, how recycling will be organised, and so on?

---
22) Hofstetter (1996) and Held and Klöpffer (2000).

- On the side of impact assessment, the so-called Eigenzeiten[23] and 'rhythms' of ecosystems, which are affected by product systems,[24] are to be considered. Today, it is not yet clear how time can be better integrated into LCA without making the method too complicated.

Some impact categories require the specification of a time horizon for the selection of suitable characterisation factors (see Chapter 4): With the greenhouse effect, for example, usually a time horizon of 100 a is assumed. Longer time horizons are very uncertain because of our total ignorance of the far future, but are, however, often demanded for reasons of justice towards coming generations.[25] There is a close link to the question whether negative effects on the environment – in analogy to financial computations – may be discounted.[26]

Statements to the time horizon must be provided at the beginning of the study; if necessary, modifications can be made during the progress of the study.

---

**Exercise: System Analysis**

Provide a first system flow chart for the product system 'Strawberry Yogurt in a polypropylene (PP)-cup with aluminium cover – 150 g'. In the initial phase of an LCA, it is always useful to have the product in reality and to weigh individual components, if necessary.

- Set up a list of the materials contained in the product.
- Make a first assessment of materials that may not be examined by their entire life cycle using the cut-off criterion 'mass'.
- Draw the flow chart and indicate the technical system boundary in your sketch.
- Make sure that in your flow chart the boxes (unit processes) indicate processes and not substances or materials.
- Name in further sketches all inputs and outputs qualitatively for each process you considered. Since no detailed research has yet taken place, the degree of detail in this step is dependent on your background knowledge of the processes you defined.
- Consider usual recycling pathways (open loop and closed loop) in your flow chart.
- Explain the meaning of 'co-products' using the example of the unit process 'milk production'.
- Define the geographical and temporal system boundaries.

---

23) 'Eigenzeit' is difficult to translate; it means a time specific or intrinsic for the system under consideration.
24) Held and Geißler (1993, 1995) and Held and Klöpffer (2000).
25) World Commission on Environment and Development (1987).
26) Hellweg, Hofstetter and Hungerbühler (2003).

## 2.2.5
## The Functional Unit

### 2.2.5.1 Definition of a Suitable Functional Unit and a Reference Flow

Although data acquisition initially does not need a fU – conversion of other reference units to the fU can be made later – it is urgently recommended to specify a fU already at the beginning of an LCA and, if necessary, to make adjustments later. If serious problems already arise within the definition of the fU, this indicates an unsatisfactory understanding of the system, serious data leaks or that LCA may not be a suitable method for solving the individual problem.

The quantitative determination of fU and reference flow is, to a certain degree, arbitrary. In the example of Section 1.1.3 (beverage packaging), it is unimportant whether the fU 'defined supply quantity beverage for customer consumption' is defined for 1000, 100 or 1 l[27] packed beverage.[28] Results would only vary in their numerical values by the appropriate factors (1000 : 1 respectively 100 : 1). This is without consequences if different packaging systems using the same fU are compared with one another. A variant that has been recommended in Dutch guidelines[29] for detailed LCA is the use of annual quantities or similar realistic data as a basis for determining the fU. In the example of beverage packaging, this would mean the annual quantity (e.g. in million litres) of a certain kind of beverage to be eligible for that type of packaging. According to the point of interest, a special product of the commissioner or the sum of similar products would be concerned. In the first case, the manufacturer or bottler would be interested, whereas in the second case, probably a trade association or a national environmental protection authority would be interested.

In comparative studies of products of high life time, such as, for example, floor coverings, a length of use time must be included into the description of function. It is important that early in the study, functions and performance of the product systems are correctly defined. The functions of a floor covering consist of providing the ground of an interior with specific characteristics (protection of supporting surface, accessibility, etc.). Thus the fU can be defined as follows: $1\,m^2$ floor is covered for a period of 30 a, for a defined stress.

$$fU = \text{area of the floor covering (e.g.} 1\,m^2) \text{ for one period (e.g. 30 a)}$$

In the next step, the defined fU must be applied to the product variants to be examined, and thus a reference flow of data acquisition be defined. The following fictitious example clarifies this approach by the example of two plastics ('plastic A' and 'plastic B'). As floor coverings contain beside the base material, for example, fillers, softeners, and so on, and as production as well as disposal of the floor covering can be neglected here, this arithmetical example for illustrating

---

27) One litre (SI: $1\,l \equiv 1\,dm^3$) equals 1.75 pints (UK) or 2.13 pints (US); in international LCA studies *only* SI units should be used (ISO 1000).
28) A practical unit should be chosen; micro pint would not be wrong but absurd.
29) Guinée et al. (2002).

the methodical approach must not be interpreted as a 'comparison by ecological criteria'!

On the basis of the defined fU, the next step is to determine how much floor covering is needed for the fulfillment of this function: A thick floor covering (if not foam material is concerned) will, by its higher weight per unit area, usually cause higher resource consumption and emissions than a thinner one. The weight of 3 mm plastic A (density $\rho = 1.2\,\mathrm{g\,cm^{-3}}$) amounts to

$$3\,\mathrm{mm} \times 1.2\,\mathrm{g\,cm^{-3}} = 3.6\,\mathrm{kg\,m^{-2}}$$

For illustration purposes, in the following cumulative energy demand (CED) will be applied for both variants. The CED is the sum of the total energy demand over the entire life cycle regarded (see Section 3.2.2). The CED for the production of plastic A as a European average may amount to 60 MJ kg$^{-1}$ (under neglect of fillers, floor covering production and disposal). Considering the energy demand for the production of plastic A, the CED amounts to

$$\mathrm{CED} \approx 3.6\,\mathrm{kg\,m^{-2}} \times 60\,\mathrm{MJ\,kg^{-1}} = 216\,\mathrm{MJ\,m^{-2}}$$

A thinner covering of plastic B with $d = 2\,\mathrm{mm}$ and $\rho = 0.9\,\mathrm{g\,cm^{-3}}$ weighs only

$$2\,\mathrm{mm} \times 0.9\,\mathrm{g\,cm^{-3}} = 1.8\,\mathrm{kg\,m^{-2}}$$

so that, despite a higher CED for the production of plastic B of approximately 90 MJ kg$^{-1}$ (same neglects as with plastic A), the CED amounts to

$$\mathrm{CED} \approx 1.8\,\mathrm{kg\,m^{-2}} \times 90\,\mathrm{MJ\,kg^{-1}} = 162\,\mathrm{MJ\,m^{-2}}$$

As is often true, the lighter product would thus do much better, at least concerning the sum parameter CED, than its heavier counterpart. This is, however, only valid as long as an equal use time is presumed. If, for example, the thicker quality has a longer use time, which is very probable, the result will differ. (Life times here are hypothetical and do not correspond to any real product performance):

Assumption:

Life time of plastic A (thick): 30 a
Life time of plastic B (thin): 15 a.

It can be directly deduced from the above example that while two thin coverings are needed (2 mm plastic B) for use over 30 a, only one of the thick covering (3 mm plastic A) is needed. Because the fU must relate to area *and* time (here: 30 a), the following reference flows result for the two variants, which must form the basis of data acquisition:

Plastic A : $3.6\,\mathrm{kg\,m^{-2}} \times 1 \Rightarrow 3.6\,\mathrm{kg\,fU^{-1}}$
Plastic B : $1.8\,\mathrm{kg\,m^{-2}} \times 2 \Rightarrow 3.6\,\mathrm{kg\,fU^{-1}}$

The reference flow is equal to the mass of the product that corresponds to the fU and is thus the basis for further work in LCA (see Chapter 3).

Therefore, the CEDs of the two variants amount to

CED plastic A (3 mm) = $3.6 \, \text{kg fU}^{-1} \times 60 \, \text{MJ kg}^{-1} = 216 \, \text{MJ fU}^{-1}$

CED plastic B (2 mm) = $3.6 \, \text{kg fU}^{-1} \times 90 \, \text{MJ kg}^{-1} = 324 \, \text{MJ fU}^{-1}$

This example is fictitious and, together with the above accomplished rough estimates, only serves a better understanding of the fU and the reference flow. From this simple example, however, it can already be perceived that, in view of the use time as *assumed*, the thinner product fares substantially worse regarding the single criterion CED. This relationship would not change if either any different area or a different life time is used in the fU (e.g. $13.4 \, \text{m}^2/26.7 \, \text{a}$).

If, however, a very short use time, below 15 a, is presumed in the fU in the example above, another reference flow would result as a basis of the calculation. Very short use times are, however, not reasonable assumptions related to building products such as floor coverings, apart from rapidly changing products, for example, in halls for trade fairs. The practical problem lies in the determination of realistic use times of long-lived products, including the buildings themselves, which set an upper limit for many building products. If use times as a function of thickness cannot be determined, there can be a way out by restricting the comparison to 'light commodity A versus light commodity B, C, and so on' and to 'heavy commodity A versus heavy commodity B, C, and so on'. It is thus assumed that life times within the groups are similarly long.

A discussion of the correct use of fUs can be found in Technical Report ISO TR 14049.[30] In addition, a set of product examples is provided, for example, lamps, paints and hand-drying systems, and the limits of comparability of technical systems are pointed out.

In view of the central importance of the fU in LCA, its correct determination has the highest priority. It should be unambiguously feasible in all those cases, where product systems to be compared have the same or a similar performance and fulfil a similar function, respectively. Border cases are treated in the next two sections.

---

**Exercise: Functional Unit and Reference Flow**

Define a meaningful fU and describe the way to specify an appropriate reference flow for the following products or services:

- Ball-point pen
- Window
- Disposal of polyethylene (PE) foil
- Spreading of a daily message.

Guideline:

- Which performance do these specified products or services have?
- Which variants can be considered to generate the performance?

---

30) ISO (2000a).

> **Example**
>
> **Often several variants can be assumed for the definition of the functional unit**
>
> The definition of the fU has substantial influence on the result of the study. A comparison of different diapers serves as a descriptive example from consumer products. There are multiple comparative LCAs on disposable and fabric diapers; the former represents a relevant domestic waste fraction.
>
> Results of LCAs occasionally differ relevantly, depending on the commissioner.
>
> Sometimes disposable diapers, then again the fabric diapers, are environmentally more compatible. There are many possible reasons for a mismatch of the results in different studies. One reason could be the choice of different fUs.
>
> What functions do diapers have? At first, the answer seems to be simple. The child is to be kept dry. This thought may result in comparing one cellulose diaper with one fabric one. However, the fabric diaper is used several times such as 200 times. Washing of the diaper (individual or diaper service) must be considered for each circulation. The production of the diaper is, however, only considered by 1/200.
>
> The comparison may look different as soon as the function of a diaper is defined to hold the child dry for 1 day. It can be argued that per day more fabric than cellulose diapers are needed, because the child in fabric diapers feels uncomfortable more quickly. Under this consideration it can be reasonable to compare, for example, 1.2 fabric diapers with one cellulose diaper. The result will certainly be different from the 1:1 comparison.
>
> Further variants can be thought of to define the function of diapers, for example, it is the function of diapers to keep a child dry for life. Under this aspect, it is thinkable that, during the entire swaddling time of the child, more cellulose diapers than fabric diapers are needed because in the first case the child needs to be swaddled until it gets older. Fabric diapers are less comfortable, and the child starts going on the potty earlier.
>
> This example of quite a simple product shows that the definition of the fU is by no means trivial.

### 2.2.5.2 Impairment Factors on Comparison – Negligible Added Value

On the rarest occasions, two products have exactly the same performance, even if they have, technically speaking, the same function. The reason often lies within aesthetic side benefit, in varying fulfillment of comfort, owner's pride, and so on. Such subtle differences, as important as they may be in marketing, are difficult to capture in LCA, if these cannot be related to measurable parameters, such as, for example, use time, weight or fuel consumption of cars. At any rate, however, they can and should be verbally described and be verbally considered in the Interpretation phase (Chapter 5).

It gets more difficult when the technical functions differ. So, in the example of floor coverings, the highly varying variants, for example, parquet versus tile, can be applied alternatively in certain areas (vestibule) but in others such as in the bath (humidity) or in the living room (well-being, at least in northern latitudes without floor heating), their application is nearly exclusive.

Different cleaning requirements can be integrated into the fU and be quantified in the use phase of LCI analysis. If, however, experience shows that the coverings require practically the same care, the use phase, within comparative studies, can be treated as black box and be excluded from comparison (see Section 2.2.1). However, omission of a life cycle stage is not recommended in principle because an optimisation analysis that may follow is based on incomplete information: perhaps the most important part of the life cycle was omitted. Who can say, without applying LCA, whether the care of a floor covering over 30 a requires more or less energy or raw materials, and so on, than the floor covering production, the installation or the disposal? If only bad data are available, at least an estimation of the omitted life cycle section should be attempted. If within the goal definition, product optimisation has the highest priority, a life cycle stage must not be omitted, under any circumstances.

Genuine added value is not easily traceable in the simple examples of beverage packaging and floor coverings. Here, the qualitative description in the comparative discussion of the results is usually sufficient. Should an energy recovery occur within a variant with heating value, it can be considered as bonus in the Inventory Analysis; not every small difference of performance needs to be considered by system expansion (see Section 3.3).

### 2.2.5.3 Procedure for Non-negligible Added Value

In some system comparisons, one of the regarded systems show a substantial added value that has to be taken into consideration and accordingly to be assessed.[31]

As has already been mentioned, 'Products' in ISO language are both goods and services. An important application field of LCA is the comparison of different options for waste management services (LCAs in waste management).

Within a comparison of different waste disposal methods (fU = disposal of a certain mass, e.g. 1 ton of domestic waste), thermal disposal with energy generation supplies a quantity of electricity and/or steam and/or hot water that corresponds to the calorific value of the waste (× efficiency of energy conversion). With disposal by landfilling, however, only in the most favourable cases a part of the dump gas can be collected and used energetically (see Figure 2.5). In Germany, this applies to old facilities only, because landfilling of domestic waste is just permitted after pre-treatment (e.g. thermal).

To adequately compare both systems, the fU must be extended. It then reads as follows (for example):

fU = disposal of a given mass of domestic waste plus supply of energy

(reference flow : $1 t + x$ MJ energy)

This fU is fulfilled by system A (thermal disposal with energy recovery) in Figure 2.5, whereby a characteristic number not specified here may characterise

---

31) Fleischer and Schmidt (1996).

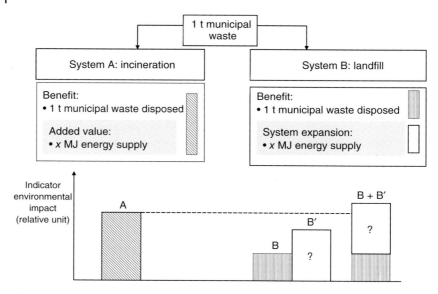

**Figure 2.5** System expansion to obtain equal performance.

the environmental impact by system A. System B (disposal without energy recovery) may have a smaller environmental impact. To fulfil the extended fU, system B must be *expanded* by the environmental impacts resulting from the supply of $x$ MJ energy. The total impact (B + B′) may, but does not, have to be higher than that by system A. This procedure is called *system expansion* for the achievement of an approximately equal benefit. The method of fU expansion for the achievement of an equal benefit is also called *basket of benefit method* (see also Section 3.3).

In simpler cases, the inequality between two systems can also be balanced by the subtraction of a bonus in the system with additional benefit (system A in Figure 2.5). The system expansion can be avoided by a credit entry to system A that corresponds to the effort necessary for supplying $x$ MJ of energy. The decision of which method shall be applied, system expansion or crediting, has to be targeted towards better results in view of the smallest possible additional data requirement. A disadvantage with use of system expansion may be very large, unmanageable systems with an accordingly high data requirement. Besides, this approach can make additional assumptions necessary, which may lead to doubts concerning its usefulness (see also Section 3.3).

The system expansion was strongly recommended by the standard ISO 14044,[32] in order to avoid allocation at nearly all costs. The Dutch guidelines[33] recommend system expansion more reservedly than the international standard.

---

32) ISO (1998, 2006b).
33) Guinée et al. (2002) and Klöpffer (2002).

> **Exercise: Provide an Equal Benefit of System Variants**
>
> The following two packaging systems for food are compared:
>
> A) Polyethylene (PE)-containers with aluminium cover
> B) Glass with screw-cap.
>
> For simplification, only the main materials PE and glass are considered.
> Make the following assumptions:
> The non-negligible added value in the case of the PE container consists of the fact that PE in the Duales System Deutschland (DSD)[34] system is collected and can be used as energy source in the cement plant. Here, for example, light fuel oil can be replaced. In the case of the glass container, 90% of the collected waste glass is used for other glass products as secondary raw material.
> Outline, in analogy to Figure 2.5, the overall systems to be examined, under consideration of all benefits.

## 2.2.6
## Data Availability and Depth of Study

Data availability and quality[35] are fundamental issues of the LCI Analysis and are discussed in Chapter 3. In the goal definition it has to be decided which data will presumably be available for the study, who collects or computes them and how, if necessary, information on competitive products is to be procured. Further, it has to be specified for which processes primary data are to be procured, for which processes recourse to already existing data is possible and which data have to be approximated by estimated values (see Section 3.4). Data availability is the most important criterion to decide on the level of detail to which a study can be conducted actually. Among the options to obtain meaningful statements with smaller data effort, mainly two are discussed[36]:

1. Screening LCA
2. Simplified or streamlined LCA.

While formerly both terms were used rather synonymously, today screening LCAs for the determination of the so-called *hot spots* are distinguished from simplified LCAs (e.g. internal assessments). Both variants, which in practice are often not distinguished, predominantly operate with easily available or estimated data, also under omission of some life cycle stages, if necessary. This discussion has been intensely conducted, particularly in the States,[37] but was also the topic of a working group of the Society of Environmental Toxicology and Chemistry

---

34) *German* – 'Duales System Deutschland', dual system.
35) Fava et al. (1994).
36) Christiansen (1997).
37) Curran (1996), Curran and Young (1996), Weitz et al. (1996), Canter et al. (2002), Mueller, Lampérth and Kimura (2004), Hochschorner and Finnveden (2003, 2006) and Rebitzer (2005).

(SETAC) Europe.[38] A study of Franklin Associates on behalf of the United States Environmental Protection Agency (US-EPA)[39] showed that the omission of whole life cycle stages is not recommendable; it is better to proceed through those phases or unit processes, for which only few or no data are available, with estimated data and to examine the result with sensitivity analyses (see Chapter 5). Only then should a decision be taken upon their omission.

In goal definition, the desired level of detail should be specified. It cannot be viewed independently of the demanded capacity of LCA results. Thus, for internal orienting studies less detail will be required than for LCAs that have internal decision-making processes or external comparative assertions as a goal. For decision making in the phase of design (ecodesign), where there is not much time available, a simplified LCA and further tools such as life cycle costing (LCC)[40] are indispensable. A step-wise, computer-assisted combination of such tools was developed by the euroMat methodology.[41] Such methods are destined to serve the development of sustainable products (not only 'environmentally friendly' ones; see Chapter 6).

### 2.2.7
### Further Definitions

In this section, occasionally terms are used which are explained later on in Chapters 3–5. Appropriate section references are inserted. In an actual LCA, however, these definitions are already to be incorporated into the component 'Definition of Goal and Scope' and, for these reasons, will be specified as follows.

#### 2.2.7.1 Type of Impact Assessment

In LCIA, the data procured during the LCI analysis are assigned to impact categories (e.g. global warming/climate change, acidification). This process is called *classification* (see Chapter 4). It has to be specified earlier within the scope definition which impact categories, indicators and characterisation factors (for terminology, see Chapter 4) shall be used in the study. The selection should be justified, as it may have influence on the results. Additional aspects to be processed, for example, risk assessment in special situations, are already to be specified in this phase. The type of impact assessment influences the data procurement, as a simple example shows: the impact category 'acidification' cannot be quantified without data on emission of acids (HCl, HF, etc.) and acid-forming gases ($SO_2$, $NO_x$, $NH_3$).

In a number of studies only the phases

- Definition of Goal and Scope,
- LCI Analysis and
- Interpretation

are considered, and the LCIA is omitted.

---

38) Christiansen (1997).
39) Hunt et al. (1998).
40) Hunkeler, Lichtenvort and Rebitzer (2008) and SETAC (2011).
41) Fleischer et al. (1999).

An analysis shortened in such a manner by omission of the LCIA (then called *LCI study*), must not be designated as LCA according to the standard ISO 14040.[42] The same is true when only *one* impact category is used, for example, 'climate change', quantified by the global warming potential (GWP). Such studies have recently been named 'carbon footprint (CF)' studies[43] (see Section 4.5.2.2).

Such an analysis can, however, prove to be of greatest importance as data base for complete LCAs. Thus, for example, Plastics Europe (former Association of Plastics Manufacturers in Europe, APME) compiled LCI studies for its technically most important plastics. These studies started from the extraction of raw materials and ended with the production of the polymer (*cradle-to-factory gate*). The data records contain average values of, for example, different refineries or production locations, because with purchase of a polymer, the exact origin of the crude oil molecules from which it was produced cannot be retraced. Such data sets are called *generic data sets*. The author (practitioner) of a specific inventory (e.g. a plastic floor covering of company X) can insert generic data into the LCA if there are no contradictions to the selected system boundaries of the study.

#### 2.2.7.2 Valuation (Weighting), Assumptions and Notions of Value

LCA valuation in German standardisation discussions was intended to become a separate component of an LCA[44]; it had been termed *valuation* by SETAC Europe.[45] The international discussion in SETAC and later during the ISO standardisation process has led to a refusal of a formal component (phase) *valuation*. The problem nevertheless remains, if system A is *not* superior, by all impact categories, to B (or vice verse) or alike within the margins of error. What is to be done? According to which method should a system be valuated,[46] if at all? Who valuates in the context of which decision process?[47] In Chapter 5, 'Interpretation', the standard variants are discussed.

According to ISO 14044,[48] the weighting of different impacts and their centralisation to one 'environmental indicator' are illegal for LCAs that are intended to provide comparative assertions to be disclosed to the public. A comparative assertion implies that product A, under environmental aspects, is better or worse than Product B or that both products are equivalent regarding the environment. This very restrictive decision was made to exclude subjective notions of value from LCA as far as possible, The reason is that, for example, it would have to be decided in which way 'climate change' and 'acidification' are to be balanced, and for this it would have to be decided whether both effects are equally important or whether one is more important than the other. The regulation in ISO 14044 (Section 4.4.5) literally reads:

---

42) ISO (1997, 2006a).
43) Finkbeiner (2009).
44) UBA (1992).
45) SETAC Europe (1992).
46) Giegrich *et al.* (1995), Klöpffer and Volkwein (1995), Volkwein, Gihr and Klöpffer (1996), Klöpffer (1998), BUWAL (1990) and Volkwein, Gihr and Klöpffer (1996).
47) IWÖ (1996) and Grahl and Schmincke (1996).
48) ISO (2006b).

> *Weighting ... shall not be used in LCA studies intended to be used in comparative assertions intended to be disclosed to the public.*

The German Federal Environment Agency (Umweltbundesamt, UBA) developed a valuation method on the basis of the then valid standards for LCIA, ISO 14042, and Interpretation, ISO 14043.[49] This valuation method is mandatory for LCAs commissioned on behalf of the environment agency. Details are discussed in Chapter 4. It is essential for the phase 'Definition of Goal and Scope' to decide on weighting type (if any) at the beginning of the study. This presupposes a definition of the later use of the LCA study; it can, however, be changed in written form in the sense of an iterative approach.

Frequently, the commissioner requests a weighting for internal communication and decision making. If this is the case, the basic notions of value used have to be described. An option to describe them exists by defining typical human characters and behaviour patterns to be consulted for a (necessarily very schematic) weighting.[50]

### 2.2.7.3 Critical Review

Comparative LCAs that are intended for publication in accordance with its goal definition must, according to ISO 14040/44, be reviewed critically by an independent panel. The critical reviewers have to answer the following questions:

- Are the methods used to carry out the LCA consistent with the international standards ISO 14040/44?
- Are the methods used scientifically and technically valid?
- Is the data used appropriate and reasonable in relation to the goal of the study?
- Does the interpretation reflect the goal of the study and the limitations identified?
- Is the study report transparent and consistent?

There are two options for the conduct of a critical review[51]:

1. accompanying (interactive) critical review,
2. critical review 'a posteriori'.

Option (1) has advantages regarding consent between commissioner, author (practitioner) and review team and is therefore recommendable. The accompanying critical review, proposed first by SETAC (1993) as 'Interactive peer review', usually starts after conclusion of the phase 'Definition of Goal and Scope' on the basis of an interim report of this phase.

The decision for publication of an LCA can also be made at an advanced state of the study when an accompanying review is no longer possible (modification of goal definition during the work-out of the project). In this case, it is advisable to accomplish the critical review in the phase of the writing of the study report. For this, a draft of the final report with the phases 1–3 (Definition of Goal and Scope,

---

49) UBA (1999) and ISO (2000b,2000c).
50) Hofstetter (1998).
51) SETAC (1993), Klöpffer (2000, 2005, 2012) and ISO (2013).

Inventory Analysis and Life Cycle Impact Assessment; see Figure 1.4) should be present; the critical reviewers will then be in a position to accomplish interactively at least the important phase of interpretation.

Alternatively, the critical review has to be accomplished really 'a posteriori' (afterwards) on the basis of the final study report.

According to the revised standard ISO 14044, a critical review team must consist of a minimum of three individuals. The commissioner appoints the chair and the appointment of co-referees usually takes place in consent with the commissioner and the author (practitioner) of the LCA.

For internal LCAs a critical review is optional and can be conducted either by internal or external independent experts. When internal experts are invited, great care should be taken to provide independence of those and of all parties interested in the results. This could, for instance, be a member of quality control or similar specialist teams or headquarters such as product stewardship.

A detailed discussion of the critical review process can be found in Chapter 5.

### 2.2.8
### Further Definitions to the Scope

Besides these issues that were discussed relatively in detail above, the international standard 14044[52] additionally requires the following specifications for the task of 'Definition of Goal and Scope':

- allocation procedures,
- methods for interpretation,
- restrictions,
- type and structure of the report intended.

With the exception of allocation, which has already been addressed in Section 2.2.2.2 (treatment in depth in Section 3.3), these issues are of particular interest for the phase 'Interpretation'. They are discussed in Chapter 5.

## 2.3
## Illustration of the Component 'Definition of Goal and Scope' Using an Example of Practice

In practice, LCAs are very extensive studies. For an illustration of a gradual introduction to the methodology of LCA, tasks as discussed are presented after every leading paragraph on the basis of an LCA study published as 'Comparative LCA of Beverage Carton and PET single-use bottles'. The study was conducted by the Institute for Energy and Environmental Research (IFEU Heidelberg) commissioned by the Association of Carton Packaging for liquid foods (Fachverband Getränkekarton

---

52) ISO (2006b).

(FKN)[53] Wiesbaden).[54] We gratefully acknowledge the possibility to illustrate the theoretical demonstrations in this book by this practical example.

In this section, the annotations in the above-mentioned study to work steps specified in Section 2.2 are presented. The selected example will also illustrate the work steps 'Life Cycle Inventory Analysis' (Section 3.7), 'Life Cycle Impact Assessment' (Section 4.6) and 'Interpretation' (Section 5.6). Regardless of its environmental-political background, it exclusively serves *didactic purposes*. For the same reason, shortenings and simplifications have been made by the authors of this book, which were approved by the practitioners of the LCA study. As a number of details which need to be specified in the work step 'Definition of Goal and Scope' are discussed later in this book, appropriate references to these later sections are inserted. Quoted texts from the study,[55] marked by a grey bar, already indicate what should be implemented into which level of detail.

### 2.3.1
### Goal Definition

As described in Section 2.1, the goal definition must answer questions, which, for every LCA conducted according to ISO 14040/44, ensure the necessary transparency of the framework. An answer to the first two questions

- What is the objective of the study?
- Why is an LCA study conducted?

is provided in the goal definition of the example study as follows:

> Since the mid of the 1990s, PET single-use bottles as packaging system are increasingly significant on the German beverage market. Only recently PET single-use bottles have been applied in this country for sensitive, $CO_2$-free filled nutrition such as fruit juice and fruit nectar, ice tea or milk mixture beverages (MMB).
>
> Contrary to PET single-use bottles, beverage carton have been classified as 'ecologically favourable' in the amended version of the German packaging ordinance of January 2005 and are thus excluded from the mandatory deposit system.[56] For single-use packaging in the beverage segments of fruit juices and fruit nectars as well as milk and MMB, where a deposit regulation is

---

53) 'Fachverband Getränkekarton'.
54) IFEU (2006).
55) Translated from the original report written in German language.
56) In Germany a deposit system, regulated by the packaging ordinance (VerpackVO), is valid for beverage packaging, which regulates an ecological classification of the packaging of quantitatively most important beverages. This classification was politically specified on the basis of LCAs.

generally not mandatory, the above-mentioned ecological classification of cartons has no direct steering effect on trade and consumers.[57]

Particularly, this effect is regarded as a challenge by the Association of Carton Packaging for liquid foods (FKN) in analysing the LCA profiles of PET single-use bottles and beverage carton. *Both beverage packaging systems are thereby to be compared with one another in all competing deposit-free market segments of these two packaging systems in Germany in the year 2005. Besides, in the comparison all market-relevant sizes of packaging are to be considered.*

For the first time, with this study an ISO-conformal LCA for a comparison of PET single-use bottles and beverage carton is conducted.

The study is elaborated closely according to the methodology of studies conducted by the Environmental Agency (UBA) Berlin for an ecological comparison of beverage packaging.[58]

Answers to the two additional questions

- For whom will an LCA study be conducted?
- Are comparative assertions intended in the study?

are provided in the study as follows:

- The study addresses primarily the commissioner and the represented members of the association. The system comparisons conducted here are to provide information on the life cycle perspective of the beverage carton in relation to PET single-use systems and thus support the ecologically oriented adjustment of market strategies and packaging developments.
- Further, the commissioner team and the member firms involved are to assess the relevance of their area of responsibility for the overall beverage carton system and to derive starting points for an optimisation.
- Beyond that, derivable facts from the study can represent important information for decision makers in beverage industry and trade.
- Finally, the insights are meant to promote an object-driven dialogue, based on transparent and current data, on the ecological valuation of the examined beverage packaging. Target groups are consumer and environmental organisations, but in particular political decision makers.

The reason for conducting the study, addressees and goals are explicitly pointed out. The results of the LCA must be measured according to these goals and the

---

[57] Evidence for the fact that VerpackVO aims at a steering effect can be seen on the basis of §1 Waste Management Objectives, VerpackVO. The goal definition reads 'the portion of beverage filled in returnable beverage packing as well as in ecologically favourable single-use beverage packaging shall be strengthened by this regulation with the objective to achieve a portion of at least 80%'.

[58] UBA et al. (2000) and UBA (2002).

achievement has to be reflected. The goal definition clearly states that comparative assertions on the ecological valuation of the examined packaging systems are made and that these are to be disclosed to the specialist public.

The organisational structure of the study is described in the following section; it illustrates that extensive preliminary work and preliminary talks are necessary, before an LCA can be tackled. Since the manufacturing industry for carton packaging was the commissioner of this study, good data availability could be presumed in this case.

> The study was commissioned by the Association of Carton Packaging for liquid foods (FKN, Wiesbaden).[59] In the project, the FKN is represented by Dr. Wallmann. Member firms of FKN are Tetra Pak GmbH & CO, (Hochheim), SIG Combibloc (Linnich) and Elopak (Speyer). In the project group, the enterprises mentioned are represented by Mrs. Babendererde (Tetra Pak), Dr. Bohmel (SIG Combibloc) and Mrs. Deege (Elopak).
> The project was conducted by the IFEU Heidelberg (IFEU GmbH). In IFEU, project leaders are Mr. Detzel and Mr. Boß.
> A technical advisory group was assigned, composed of Mrs. von Bremerstein (DSD), Professor Strobl (FH Wiesbaden), Mr Geiger (Campina company) and Mr Lentz (Emig company), beside the persons already mentioned.

The selection of the technical advisory group demonstrates that those representatives of the economy can be integrated whose interests are not unequivocally aligned to one of the compared systems, here, companies operating in waste management and bottling enterprises.

### 2.3.2
### Scope

ISO 14040/44 requires the clear exemplification of all methodical rules relevant in an LCA in the phase 'scope definition'. Therefore, much of what is defined here according to the iterative approach of the method will only be specified in detail in later work steps (inventory analysis, impact assessment, valuation). Only substantial statements of the example are summarised in this chapter; the rules for allocation in particular will be explained later, in 'Life Cycle Inventory Analysis' (Section 3.3).

#### 2.3.2.1 **Product Systems**
Since transparency is a central requirement for all LCA studies, the product system, on the basis of the goal definition, is clearly described verbally and by a system flow

---

59) Now (2011) Berlin.

## 2.3 Illustration of the Component 'Definition of Goal and Scope' Using an Example of Practice

chart (Figure 2.6). Equality of benefit of the regarded variants is stressed, which is a fundamental condition for an LCA comparison of product systems:

> The examined packaging systems are supposed, according to the goal definition, to cover beverage segments relevant for both packaging variants. As a further aspect of product performance, the resealability of the packaging as well as the product visibility can be named among others. The primary function, however, is the protection of the filled good; therefore, an equivalent functionality of the cartons examined and PET bottles can be presumed. Performance equality as a basic condition for system comparison by LCA is thus provided.

According to the goal definition, different groups of filled goods with different filling capacities are examined. The selected variant for illustration purposes is the storage for fruit juices and fruit nectar in a 1 l container.

1. *Fruit juices and fruit nectars*
   a. Storage
      Beverage carton: 1000, 1500 ml
      PET single-use bottle: 1000, 1500 ml
   b. Instant consumption
      Beverage carton: 200, 500 ml
      PET single-use bottle: 330, 500 ml
2. *Ice tea*
   a. Storage
      Beverage carton: 1500 ml
      PET single-use bottle: 1500 ml
3. *Fresh milk beverages*[60]
   a. Storage (fresh milk)
      Beverage carton: 1000 ml
      PET single-use bottle: 1000 ml
   b. Instant consumption milk mixture beverages (MMB)
      PET single-use bottle: 500 ml
      Beverage carton: 500 ml.

Figure 2.6 shows the qualitative system flow chart and the system boundary for the two examined variants, beverage carton and PET single-use bottle:

---

[60] In this study, dairy products that are sold over the cold chain are designated as fresh milk beverages. This includes fresh milk (pasteurized milk and ESL (extended shelf life) milk, but *not* preserved milk) as well as different MMB. Because of the appropriate market relevance, instant consumption (500 ml packaging) refers to MMB, whereas for storage purchase fresh milk prevails. Packaging for milk mixture beverages and fresh milk are approximately comparable, whereby fresh milk represents the more sensitive filling material.

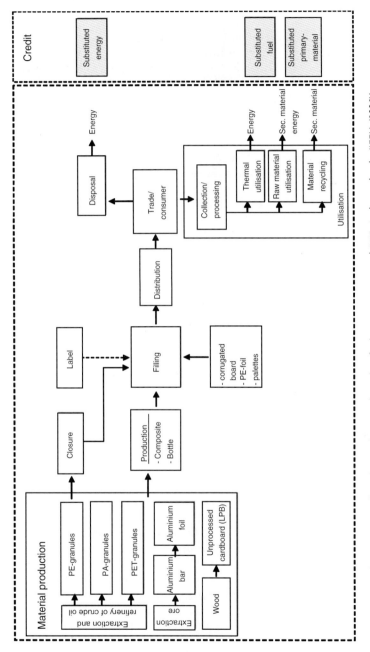

**Figure 2.6** Qualitative system flow chart and system boundary for beverage carton and PET single-use bottle (IFEU (2006)).

Unit processes that are not considered are named explicitly.

> Not to be considered:
>
> - Production and disposal of the infrastructure (machines, aggregates, means of transportation) and their maintenance
> - Production and sterilisation of the respective filled good as well as its cooling
> - Environmental aspects that result from activities of the consumer (transport to the shop, refrigeration processes)
> - Environmental effects by accidents.

#### 2.3.2.2 Technical System Boundaries and Cut-Off Criteria

As cut-off criteria the '1% and the 5% rule' are specified for the process level, which is common practice.

> The goal is to consider input materials in product systems if they cover, within the respective sub-process of the life cycle, more than 1% *of the mass* of the desired output in the process. At the same time, however, the sum of the neglected materials within a process should amount to no more than 5% *of the output*.

Here also an explanation is given that production and sterilisation of the filled good are not considered, but that the comparison refers to the packaging alone.

> The system boundaries only include environmental impacts due to the packaging material. The expenditures and emissions caused by the production of the filled good (plus its upstream-processes) are, in analogy to UBA studies,[61] not included in the LCA. This also applies to the transportation of the packed filled good in the course of distribution. Since, however, the distribution as a process step lies within the system boundary of the LCA, an allocation of the emissions between beverage packaging and filled good is necessary.

#### 2.3.2.3 Demarcation to the System Surrounding

The rules, according to which the demarcation to the system surrounding was treated in the example, are described. In the study, the handling of allocation is already defined within the description of the scope definition. According to ISO 14040/44, this is intended. Since, however, the discussion of these work steps takes place in Chapter 3 'Life Cycle Inventory Analysis', appropriate detailed definitions are explained there. The handling of system expansion and credits within the

---

61) Plinke *et al.* (2000) and UBA (2002).

modelling of waste treatment has already been addressed in Section 2.2.5.3, and the necessity of allocations in Section 2.2.2.2.

> As illustrated in Figure 2.6, the system boundary also covers collection and processing of used packaging. Credits are provided for the formation of secondary materials and usable energy from thermal waste treatment.
> The modelling of the regarded product systems requires, at various places, the application of so-called allocation rules. Two systematic levels are to be differentiated: An allocation can be required on the level of the individual processes within the examined product system or between the level of the examined product system and other product systems.
> In the case of process-related allocations multi-input and multi-output processes are distinguished. The question of the system-related allocation arises if a product system, apart from the actual performance defined by the fU, provides added values. This is the case, when the examined product system provides energy and material flows available for other product systems or when it processes wastes.

#### 2.3.2.4 Geographical System Boundary

The geographical system boundary for different unit processes is described as precisely and as differentiated as possible. By this, the geographical scope of the packaging comparison is characterised. Data acquisition must be related to the geographical system boundary characterised accordingly. If data as specified is not available, this would have to be noted expressly.

> The geographical scope of this study is the packaging production and the packaging disposal in Germany.
> Some of the raw materials applied in the packaging systems under examination are produced and traded on a Europe-wide market and from there also purchased by the German industry. This is valid in particular for the composite raw materials aluminium and polyethylene as well as for PET granulate. For these materials, European average data are used.
> Liquid packaging board (LPB), used for the examined beverage carton, originates from north European countries. Production in the countries of origin and transportation of the packaging materials to Germany are considered in the modelling.
> Concerning production of carton composite material and PET bottles as well as filling and distribution, the process data are modelled in such a way as if the appropriate processes had been established in Germany exclusively. Beverage import and export, which to a certain degree, exists in reality, is not considered.

## 2.3.2.5 Temporal System Boundary

Here, it is specified for which period the data holds. This is particularly important as the production period in synopsis with the geographical system boundary provides a background to the technical state of the art to be expected.

As an example, it is inadmissible to compare, non-commented data of 1975 from the production in an Asian country with data of 2005 from European production. To prevent this, strong emphasis is put on a precise specification of the temporal system boundary.

> For the comparison of packaging systems those packages are to be used which were available on the German market in 2005. The weights and the material composition of the examined packaging applied are expected to appropriately reflect that.
>
> For process data a reference time between 2002 and 2005 is valid. This means it is aspired that the validity of the used data applies to the period mentioned or is as near as possible.

## 2.3.2.6 Functional Unit and Reference Flow

This defines exactly what is to be considered for a determination of the reference flow of the entire packing system on the basis of the defined fU.

> In analogy to the UBA LCAs, the packaging necessary for the supply of 1000 l filling material to the point-of-sale[62] is defined as fU. The reference flow of the product system consists of the actual beverage packaging, thus, the composite carton or PET bottle, the labels and caps as well as transport packaging (cardboard tray, shrink-wrap and pallets) necessary for filling and distribution of 1000 l filled goods.

## 2.3.2.7 Data Availability and Depth of Study

Requirements for the data are discussed in relation to the impact categories to be included in LCIA. Thus, the consideration of the impact category 'acidification' implies the quantification of emissions that contribute to acidification (see Section 2.2.6).

As a further important aspect of data availability it is indicated that statistic valuation is usually not possible for data from operational data acquisition, and thus the accuracy of the data records can only insufficiently be defined.

> As a result of using the UBA methodology, there are requirements concerning the data categories to be considered. In principle, all input and output flows of product systems that provide a relevant contribution to impact categories according to the UBA LCA must be considered.

---

62) Plinke et al. (2000, S.5) and UBA (2002, S.6).

An evaluation of the accuracy of data records is difficult, as process data are usually not available with variants, error margins or standard deviations. The evaluation is thereby essentially based on qualitative expert knowledge. Therefore, for a descriptive evaluation of the data material, available information such as, the average of a used technology, the year of reference, and so on, has to be consulted. Thus, above all, information is obtained on how representative the data are. Exceptions are packaging specifications of the beverage cartons. Here, a mapping of qualified band widths is possible.

### 2.3.2.8 Type of Life Cycle Impact Assessment

At this stage impact categories to be considered are already defined. From this results the data set to be procured in the inventory analysis (see Table 2.1).

**Table 2.1** Correlation of inventory parameters procured in the project and impact categories (for explanation of the impact categories, see Chapter 4).

| Impact category | Substantial LCI parameter[a] | Unit of category indicator |
| --- | --- | --- |
| Resource demand | Crude oil, raw gas, brown coal, hard coal | kg of crude oil equivalents |
| Land use | Spatial categories II–V | $m^2$ |
| Greenhouse effect[b] | $CO_2$ fossil, $CH_4$, $CH_4$ regenerative, $N_2O$, $C_2F_6$, $CF_4$, $CCl_4$ | kg $CO_2$-equivalent |
| Eutrophication (terrestrial) | $NO_x$, $NH_3$ | kg $PO_4^{3-}$-equivalent |
| Acidification | $NO_x$, $SO_2$, $H_2S$, HCl, HF, $NH_3$, TRS | kg $SO_2$-equivalent |
| Photo smog ~ ozone formation (near surface) | NMVOC, VOC, benzene, $CH_4$, acetylene, ethanol, formaldehyde, hexane, toluene, xylene, aldehydes unspecified. | kg of ethene equivalents |
| Eutrophication (aquatic) | P-total, CSB, N-total, $NH_4^+$, $NO_3^-$, $NO_2^-$, N unspecified | kg $PO_4^{3-}$-equivalent |

[a] The complete lists of inventory parameters considered in the example can be found in Sections 3.7 and 4.6 for illustration of the phases Inventory Analysis and Impact Assessment.
[b] The impact category 'climate change' is designated as 'greenhouse effect' in the example.
TRS, total reduced sulphur; NMVOC, non-methane volatile organic substances; VOC, volatile organic compounds; CSB, chemischer Sauerstoff-Bedarf.

---

63) Since in the examined systems ozone-destroying substances are not released in relevant quantities, the impact category 'stratospheric ozone depletion' is omitted.

The impact assessment in the study is conducted on the basis of the following impact categories[63]:

A) Resource-related categories
- demand of fossil resources,
- land use (forest),

B) Emission-related categories
- greenhouse effect,
- terrestrial eutrophication,
- acidification,
- summer smog (as Photo-oxidant creation potential, POCP),
- aquatic eutrophication.

### 2.3.2.9 Methods of Interpretation

According to the definition of rules referring to the demarcation of the examined system to the system surrounding, rules have also be defined concerning the weighting of environmental loads in the phase 'Interpretation'. Since these rules are only discussed in Chapters 4 and 5, the definitions in the sample study are detailed there.

### 2.3.2.10 Critical Review

Since the critical review is discussed in Chapter 5, annotations in the sample study on its implementation are given in Section 5.6.

## References

BUWAL (1990) Ahbe, S., Braunschweig, A., and Müller-Wenk, R. *Methodik für Oekobilanzen auf der Basis ökologischer Optimierung*. Schriftenreihe Umwelt, Nr. 133, Bundesamt für Umwelt, Wald und Landschaft (BUWAL), Bern (Hrsg.).

Boustead, I. (1992) *Eco-Balance Methodology for Commodity Thermoplastics*. Report to The European Centre for Plastics in the Environment (PWMI)*, Brussels, APME/PWMI, December 1992 (*later APME, now Plastics Europe).

Boustead, I. (1994) Eco-Profiles of the European Polymer Industry. Report 6: Polyvinyl Chloride. Report for APME's Technical and Environmental Centre, Brussels, April 1994.

Canter, K.G., Kennedy, D.J., Montgomery, D.C., Keats, J.B., and Carlyle, W.M. (2002) Screening stochastic life cycle assessment inventory models. *Int. J. Life Cycle Assess.*, 7, 18–26.

Christiansen, K. (1997) Simplifying LCA: Just a Cut? Final Report of the SETAC Europe LCA Screening and Streamlining Working Group, SETAC, Brussels.

Curran, M.A. (ed.) (1996) *Environmental Life-Cycle Assessment*. McGraw-Hill, New York. ISBN: 0-07-015063-X.

Curran, M.A. and Young, S. (1996) Report from the EPA conference on streamlining LCA. *Int. J. Life Cycle Assess.*, 1, 57–60.

DIN-NAGUS (1994) Grundsätze produktbezogener Ökobilanzen (Stand Oktober 1993). *DIN-Mitteilungen*, 73 (3), 208–212.

European Commission, Joint Research Centre, Institute for Environment and Sustainability (2010) *International Reference Life Cycle Data System (ILCD) Handbook – General Guide for Life Cycle Assessment – Detailed Guidance*. 1st, EUR 24708 EN edn, Publications Office of the European Union, Luxembourg, March 2010.

Society of Environmental Toxicology and Chemistry (SETAC) (1991) Fava, J.A., Denison, R., Jones, B., Curran, M.A., Vigon, B., Selke, S., and Barnum, J. eds. *SETAC Workshop Report: A Technical Framework for Life Cycle Assessments*. August 18–23 1990, Smugglers Notch, Vermont, SETAC, Washington, DC.

Fava, J., Jensen, A.A., Lindfors, L., Pomper, S., De Smet, B., Warren, J. and Vigon, B. eds. (1994) Conceptual Framework for Life-Cycle Data Quality. Workshop Report. SETAC and SETAC Foundation for Environmental Education, SETAC, Wintergreen, VA.

Finkbeiner, M. (2009) Carbon footprinting – opportunities and threats. *Int. J. Life Cycle Assess.*, **14** (2), 91–94.

Fleischer, G., Becker, J., Braunmiller, U., Klocke, F., Klöpffer, W., and Michaeli, W. (Hrsg.) (1999) *Eco-Design. Effiziente Entwicklung nachhaltiger Produkte mit euroMat*. Springer-Verlag, Berlin.

Fleischer, G., and Schmidt, W.-P. (1996) Functional unit for systems using natural raw materials. *Int. J. Life Cycle Assess.*, **1** (1), 23–27.

Frische, R., Klöpffer, W., Esser, G., and Schönborn, W. (1982) Criteria for assessing the environmental behavior of chemicals: selection and preliminary quantification. *Ecotox. Environ. Safety*, **6**, 283–293.

Frischknecht, R., Jungbluth, N., Althaus, H.-J., Doka, G., Dones, R., Heck, T., Hellweg, S., Hischier, R., Nemecek, T., Rebitzer, G., and Spielmann, M. (2004) Overview and Methodology. Ecoinvent Report No. 1, Swiss Centre for Life Cycle Inventories, Dübendorf.

Frischknecht, R., Jungbluth, N., Althaus, H.-J., Doka, G., Dones, R., Heck, T., Hellweg, S., Hischier, R., Nemecek, T., Rebitzer, G., and Spielmann, M. (2005) The ecoinvent database: overview and methodological framework. *Int. J. Life Cycle Assess.*, **10** (1), 3–9.

Giegrich, J., Mampel, U., Duscha, M., Zazcyk, R., Osorio-Peters, S., and Schmidt, T. (1995) *Bilanzbewertung in produktbezogenen Ökobilanzen. Evaluation von Bewertungsmethoden, Perspektiven*. UBA Texte 23/95. Endbericht des Instituts für Energie- und Umweltforschung Heidelberg GmbH (IEFU) an das Umweltbundesamt, Berlin, Heidelberg, März 1995. ISSN: 0722-186X.

Grahl, B., and Schmincke, E. (1996) Evaluation and decision-making processes in life cycle assessment. *Int. J. Life Cycle Assess.*, **1**, 32–35.

Guinée, J.B. (final editor), Gorée, M., Heijungs, R., Huppes, G., Kleijn, R., de Koning, A., van Oers, L., Wegener Sleeswijk, A., Suh, S., Udo de Haes, H.A., de Bruijn, H., van Duin, R., and Huijbregts, M.A.J. (2002) *Handbook on Life Cycle Assessment – Operational Guide to the ISO Standards*. Kluwer Academic Publishers, Dordrecht. ISBN: 1-4020-0228-9.

BUWAL (1991) Habersatter, K., and Widmer, F. *Oekobilanzen von Packstoffen. Stand 1990*. Schriftenreihe Umwelt, Nr. **132**, Bundesamt für Umwelt, Wald und Landschaft (BUWAL), Bern (Hrsg.).

Heijungs, R., and Frischknecht, R. (1998) On the nature of the allocation problem. A special view on the nature of the allocation problem. *Int. J. Life Cycle Assess.*, **3** (6), 321–332.

Heijungs, R., and Suh, S. (2002) *The Computational Structure of Life Cycle Assessment*. Kluwer Academic Publishers, Dordrecht ISBN: 1-4020-0672-1.

Held, M. and Geißler, K.A. (Hrsg.) (1993) *Ökologie der Zeit*. Edition Universitas, S. Hirzel, Stuttgart. ISBN: 3-8047-1264-9.

Held, M., and Geißler, K.A. (1995) *Von Rhythmen und Eigenzeiten. Perspektiven einer Ökologie der Zeit*. Edition Universitas, S. Hirzel, Stuttgart. ISBN: 3-8047-1414-5.

Held, M., and Klöpffer, W. (2000) Life cycle assessment without time? Time matters in life cycle assessment. *Gaia*, **9**, 101–108.

Hellweg, S., Hofstetter, T.B., and Hungerbühler, K. (2003) Discounting and the environment. Should current impacts be weighted differently than impacts harming future generations? *Int. J. Life Cycle Assess.*, **8**, 8–18.

Hochschorner, E., and Finnveden, G. (2003) Evaluation of two simplified life cycle assessment methods. *Int. J. Life Cycle Assess.*, **8**, 119–128.

Hochschorner, E., and Finnveden, G. (2006) Life cycle approach in the procurement process: the case of defence material. *Int. J. Life Cycle Assess.*, **11**, 200–208.

Hofstetter, P. (1996) Time in life cycle assessment, in *Developments in LCA Valuation*. IWÖ-Diskussionbeitrag Nr. 32, Final Report of the Project Nr. 5001-35066 from the Swiss National Science Foundation, Swiss Priority Programme Environment. ISBN: 3-906502-31-7. St. Gallen, pp. 97–121

Hofstetter, P. (1998) *Perspectives in Live Cycle Assessment. A Structured Approach to Combine Models of the Technosphere, Ecosphere and Value Sphere.* Kluwer Academic Publishers, Boston, FL. ISBN: 0-7923-8377-X.

Hunkeler, D., Lichtenvort, K., and Rebitzer, G. (eds) (2008) *Environmental Life Cycle Costing*. CRC Press, Boca Raton, FL and SETAC.

Hunt, R.G., Boguski, T.K., Weitz, K., and Sharma, A. (1998) Case studies examining streamlining techniques. *Int. J. Life Cycle Assess.*, **3**, 36–42.

Hunt, R.G., Sellers, J.D., and Franklin, W.E. (1992) Resource and environmental profile analysis a life cycle environmental assessment for products and procedures. *Environ. Impact Assess. Rev.*, **12**, 245–269.

IFEU (2006) Detzel, A., and Böß, A *Ökobilanzieller Vergleich von Getränkekartons und PET-Einwegflaschen.* Endbericht, Institut für Energie und Umweltforschung (IFEU) Heidelberg an den Fachverband Kartonverpackungen (FKN), Wiesbaden.

ISO (International Organization for Standardization) (1997) ISO 14040. *Environmental Management – Life Cycle Assessment – Principles and Framework*, International Organization for Standardization (ISO).

ISO (International Organization for Standardization) (1998) ISO 14041. *Environmental Management – Life Cycle Assessment: Goal and Scope Definition and Inventory Analysis*, International Organization for Standardization (ISO).

ISO (International Organization for Standardization) (2000a) Technical Report ISO TR 14049. *Life Cycle Assessment – Examples of the Application of Goal and Scope Definition and Inventory Analysis*, International Organization for Standardization (ISO).

ISO (International Organization for Standardization) (2000b) ISO 14042. *Environmental Management – Life Cycle Assessment: Life Cycle Impact Assessment*, International Organization for Standardization (ISO).

ISO (International Organization for Standardization) (2000c) ISO 14043. *Environmental Management – Life Cycle Assessment: Interpretation*, International Organization for Standardization (ISO).

ISO (International Organization for Standardization) (2002) Technical Specification ISO/TS 14048, ISO/TC 207/SC 5. *Environmental Management – Life Cycle Assessment. Data Documentation Format*, International Organization for Standardization (ISO).

ISO (International Organization for Standardization) (2006a) ISO 14040:2006-2010, ISO TC 207/SC 5. *Environmental Management – Life Cycle Assessment – Principles and Framework*, International Organization for Standardization (ISO).

ISO (International Organization for Standardization) (2006b) ISO14044:2006-2010, ISO TC 207/SC. *Environmental Management – Life Cycle Assessment – Requirements and Guidelines*, International Organization for Standardization (ISO).

ISO (International Organization for Standardization) (2013) ISO PDTS 14071, ISO TC207/SC5/WG9. *Environmental Management – Life Cycle Assessment – Critical Review Processes and Reviewer Competencies – Additional Requirements and Guidelines to ISO 14044:2006*, International Organization for Standardization (ISO), Geneva.

IWÖ (1996) *Developments in LCA Valuation*. IWÖ-Diskussionbeitrag Nr. 32, Final Report of the Project Nr. 5001–35066 from the Swiss National Science Foundation, Swiss Priority Programme Environment, St. Gallen. ISBN: 3-906502-31-7.

Klöpffer, W. (1989) Persistenz und Abbaubarkeit in der Beurteilung des Umweltverhaltens anthropogener Chemikalien. *UWSF-Z. Umweltchem. Ökotox.*, **1**, 43–51.

Klöpffer, W. (1996) Allocation rules for open loop recycling in life cycle assessment – A review. *Int. J. Life Cycle Assess.*, **1**, 27–31.

Klöpffer, W. (1998) Subjective is not arbitrary. Editorial in Number 2. *Int. J. Life Cycle Assess*, **3**, 61.

Klöpffer, W. (2000) in *Ökobilanzen & Produktverantwortung. Dokumentation* Stiftung Arbeit und Umwelt (Hrsg.), Buchwerkstätten Hannover GmbH, pp. 37–42. ISBN: 3-89384-041-9.

Klöpffer, W. (2001) Kriterien für eine ökologisch nachhaltige Stoff- und Gentechnikpolitik. *UWSF- Z. Umweltchem. Ökotox.*, **13**, 159–164.

Klöpffer, W. (2002) The second Dutch LCA guide, published as book (Guinée et al. 2002). Book review. *Int. J. Life Cycle Assess*, **7**, 311–313.

Klöpffer, W. (2005) The critical review process according to ISO 14040-43: An analysis of the standards and experiences gained in their application. *Int. J. Life Cycle Assess*, **10**, 98–102.

Klöpffer, W. (2006) The role of SETAC in the development of LCA. *Int. J. Life Cycle Assess*, **11** (Suppl. 1), 116–122.

Klöpffer, W. (2012) The critical review of life cycle assessment studies according to ISO 14040 and 14044: origin, purpose and practical performance. *Int. J. Life Cycle Assess*, **17** (9), 1087–1093.

Klöpffer, W., and Volkwein, S. (1995) in *Enzyklopädie der Kreislaufwirtschaft, Management der Kreislaufwirtschaft* (Hrsg. K.J. Thomé-Kozmiensky), EF-Verlag für Energie- und Umwelttechnik, Berlin, pp. 336–340.

Mueller, K.G., Lampérth, M.U., and Kimura, F. (2004) Parameterised inventories for life cycle assessment. *Int. J. Life Cycle Assess*, **9**, 227–235.

Neitzel, H. (ed.) (1996) Principles of product-related life cycle assessment. *Int. J. Life Cycle Assess*, **1**, 49–54.

UBA (2000) Plinke, E., Schonert, M., Meckel, H., Detzel, A., Giegrich, J., Fehrenbach, H., Ostermayer, A., Schorb, A., Heinisch, J., Luxenhofer, K., and Schmitz, S. *Ökobilanz für Getränkeverpackungen II.* UBA Texte 37/00. Zwischenbericht (Phase 1) zum Forschungsvorhaben FKZ 296 92 504 des Umweltbundesamtes, Berlin – Hauptteil. ISSN: 0722-186X.

Rebitzer, G. (2005) Enhancing the application efficiency of life cycle assessment for industrial uses. Thèse No. 3307, École Polytechnique Féderale de Lausanne.

Riebel, P. (1955) *Die Kuppelproduktion. Betriebs- und Marktprobleme.* Westdeutscher Verlag, Köln.

UBA (1999) Schmitz, S., and Paulini, I. *Bewertung in Ökobilanzen. Methode des Umweltbundesamtes zur Normierung von Wirkungsindikatoren, Ordnung (Rangbildung) von Wirkungskategorien und zur Auswertung nach ISO 14042 und 14043.* Version '99. UBA Texte 92/99, Berlin.

UBA (2002) Schonert, M., Metz, G., Detzel, A., Giegrich, J., Ostermayer, A., Schorb, A., and Schmitz, S. *Ökobilanz für Getränkeverpackungen II, Phase 2.* UBA Texte 51/02, Forschungsbericht 103 50 504 UBA-FB 000363 des, Umweltbundesamtes, Berlin. ISSN: 0722-186X.

Society of Environmental Toxicology and Chemistry – Europe (ed.) (1992) *Life-Cycle Assessment* Workshop Report, 2–3 December 1991, Leiden SETAC Europe, Brussels.

Society of Environmental Toxicology and Chemistry (SETAC) (1993) *Guidelines for Life- Cycle Assessment: A Code of Practice.* 1st edn. From the SETAC Workshop held at Sesimbra, Portugal, 31 March – 3 April 1993, SETAC, Brussels and Pensacola, FL.

SETAC (2011) Swarr, T., Hunkeler, D., Klöpffer, W., Pesonen, H.-L., Ciroth, A., Brent, A., and Pagan, B. *Environmental Life Cycle Costing: A Code of Practice.* SETAC Press, Pensacola, FL.

UBA (1992) *Arbeitsgruppe Ökobilanzen des Umweltbundesamts Berlin Ökobilanzen für Produkte. Bedeutung – Sachstand – Perspektiven.* UBA Texte 38/92, Umweltbundesamtes, Berlin.

Volkwein, S., Gihr, R., and Klöpffer, W. (1996) The valuation step within LCA. Part II: A formalized method of prioritization by expert panels. *Int. J. Life Cycle Assess.*, **1**, 182–192.

Volkwein, S., and Klöpffer, W. (1996) The valuation step within LCA. Part I: general principles. *Int. J. Life Cycle Assess.*, **1**, 36–39.

Weitz, K.A., Todd, J.A., Curran, M.A., and Malkin, M.J. (1996) Streamlining life cycle assessment. Considerations and a report on the state of practice. *Int. J. Life Cycle Assess.*, **1**, 79–85.

World Commission on Environment and Development (1987) *Our Common Future (The Brundtland Report)* Oxford University Press.

# 3
# Life Cycle Inventory Analysis

## 3.1
### Basics

### 3.1.1
#### Scientific Principles

The revised ISO standard 14040:2006[1] defines 'life cycle inventory LCI' analysis – as a

> *phase of life cycle assessment involving the compilation and quantification of inputs and outputs for a product throughout its entire life cycle.*

LCI is a material and an energy analysis based on a simplified (linear) systems analysis, whereby loops can only be solved approximately by iteration. Calculation procedures based on matrix inversion can also assess loops.[2] However, calculation procedures most frequently used so far are based on spread-sheet analysis to be found in software programmes of the type 'Microsoft Excel'.

A figurative presentation of the product system is the 'product tree' consisting of process units. The product tree should have been developed, at least roughly, in the first phase of life cycle assessment (LCA), the 'Goal and scope Definition' (Chapter 2), and has now to be refined. Software packages[3] for conducting LCAs help to elaborate the system flow charts.

LCI in its scientific part – by and large – is based on the following laws of nature:

1. Conservation of mass.
2. Conservation of energy (first principle of thermodynamics)
   The following applies for the conversion of thermal energy into other forms of energy – part of almost all LCAs – as well as to chemical thermodynamics.

---

1) ISO (2006a Section 3.3).
2) Heijungs and Suh (2002, 2006).
3) For example, Gabi (PE-International), *www.gabi-software.com*; SimaPro (Pré Consultants) *www.pre.nl/software.htm*; TEAM (Ecobilan), *www.ecobalance.com/fr team.php*; Umberto (ifu), *www.umberto.de*. Pionier-Software programs see Vigon (1996), Rice, Clift and Burns (1997) and Siegenthaler, Linder and Pagliari (1997).

*Life Cycle Assessment (LCA): A Guide to Best Practice*, First Edition.
Walter Klöpffer and Birgit Grahl.
© 2014 Wiley-VCH Verlag GmbH & Co. KGaA. Published 2014 by Wiley-VCH Verlag GmbH & Co. KGaA.

3. Increase of entropy (second principle of thermodynamics)
Relevant for the explicit examination of chemical reactions (very frequent in LCI analyses, among others for the production and transformation of chemicals and, for example, the determination of $CO_2$ loads by incineration of fossil fuels)
4. Principles of stoichiometry (basis for all chemical reactions)
Transitions of mass into energy and vice versa are only relevant in nuclear reactions, thus providing an exception to the first and second principle.
5. $E = mc^2$ (equivalence of mass ($m$) and energy ($E$) according to Einstein).[4]

*These principles (1–5) belong to the scientifically best proved laws and thus provide a solid framework for processes analysed within LCIs.*[5]

These principles can be used as estimations for what quantity of a product can maximally be formed, how much energy can maximally be released or is necessary as a *minimum* amount for a chemical reaction to occur, how much usable ('free') energy can be produced from combustion heat, and so on. Technically attainable yields, efficiencies, and so on, are usually lower than those theoretically predicted, never higher. In the praxis of LCI this means that in the absence of specific, that is, measured data, respective estimations[6] can also be made by the use of technical handbooks and manuals or by technical information from the Internet.[7] This is often suited for the estimation of main flows (mass, energy) but fails, however, with trace emissions, which often stem from uncontrolled side reactions. The database in the centre of the inventory with the most important mass and energy flows is usually much more extensive than the data converted into impact categories (classification) in life cycle impact assessment (LCIA, see Section 3.7 and Chapter 4).

The laws of conservation of mass and energy can be used for a strict balance[8] (input = output), which, however, most LCAs do not use; for example, oxygen as a seemingly inexhaustible resource on the input side is mostly not assessed and waste heat on the output side is practically not quantitatively recorded.

## 3.1.2
### Literature on Fundamentals of the Inventory Analysis

Fundamentals of a method are often better described in older texts than in newer ones, where much is already assumed to be known. Classical descriptions of LCA have been given by, for example, William Franklin, Robert Hunt and co-workers,[9]

---

4) The use of Einstein's equation in LCA as a basis for an estimation of energy equivalence has also been discussed (Heijungs and Frischknecht, 1998).
5) Hunt, Sellers and Franklin (1992) and Hau, Yi and Bakshi (2007).
6) Boustead and Hancock (1979).
7) Greatest care is to be taken to ensure legitimacy of data.
8) Ecobalance (in German "Ökobilanz" is still the correct translation of LCA) was used prior to the term life cycle assessment. It is derived from the Italian expression bilancio; a reference to economical balance is straightforward.
9) Hunt *et al.* (1992), Janzen (1995) and Boguski *et al.* (1996).

James Fava and co-workers,[10] and Ian Boustead.[11] In a more recent text, Fleischer and Hake[12] discuss LCIs in detail. The international standard dealing with LCI has been, from 1998 to 2006, ISO 14041[13]; since October 2006 LCI has become part of ISO 14044.[14] Regional guidelines and standards have been elaborated in Scandinavia, in the USA, France,[15] and Canada.[16] Scandinavian guidelines are documented in the 'Nordic Guidelines on Life Cycle Assessment'[17] and have been further elaborated in the Danish EDIP programme (Environmental Design of Industrial Products).[18] US-EPA has commissioned guidelines for the conduct of LCAs by the Battelle Memorial Institute and by Franklin Associates.[19] A newer publication is available from the European Joint Research Centre (JRC).[20]

In German-speaking countries, the Swiss publications of BUWAL[21] have for long almost been standard, in particular for Inventory Analysis and its relevant data. The original, German version of the monograph and textbook 'Ökobilanz (LCA)'[22] contains some official variants of LCA, which are typical for Germany.

### 3.1.3
### The Unit Process as the Smallest Cell of LCI

#### 3.1.3.1 Integration into the System Flow Chart

A system flow chart as a diagram of the examined product system consists of small boxes where the processes involved are specified and their mutual dependencies are indicated by one- or two-sided arrows (Figure 3.1).

As long as a linear approximation is adequate, branching will occur at some boxes (see Section 3.1.4) but there will be no formation of networks. The small boxes (1, 2, 3, 4 ... $n$, $m$), which can, for instance, designate production or processing steps of a product, are called *unit processes*.

According to ISO 14040, these are the smallest elements considered in LCI for which input and output data are quantified (Figure 3.2). With large data resolution the unit process can correspond to a printing process, a transportation procedure, a metal deformation, a filling, a cleansing, a single chemical reaction, and so on; if less data are available (or for small data resolution), these can refer to a plant or to a side chain, for example, 'production of electricity' (see Section 3.4.3).

---

10) SETAC (1991).
11) Boustead and Hancock (1979) and Boustead (1992, 1995b).
12) Fleischer and Hake (2002).
13) International Standard Organization (ISO) (1998a).
14) International Standard Organization (ISO) (1998a) and ISO (2006b).
15) Association Française de Normalisation (AFNOR) (1994).
16) Canadian Standards Association (CSA), 1992.
17) Lindfors *et al.* (1994a,b, 1995).
18) Wenzel, Hauschild and Alting (1997) and Hauschild and Wenzel (1998).
19) EPA, 1993; EPA, 2006; see also EPA's LCA Web site: *www.lcacenter.org/InLCA*.
20) *http://lct.jrc.ec.europa.eu/pdf-directory/ILCD-Handbook-General-guide-for-LCA-DETAIL-online-12March2010.pdf*, pp. 70–87.
21) BUWAL (1991) and BUWAL (1996, 1998).
22) Klöpffer and Grahl (2009).

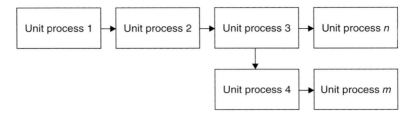

**Figure 3.1** Linear section of a system flow chart.

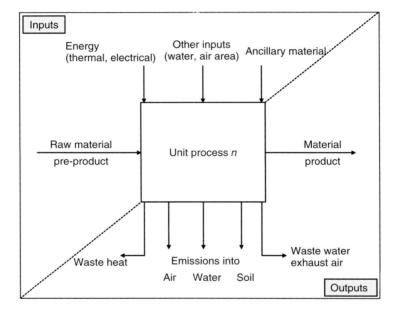

**Figure 3.2** Schematic illustration of a unit process (without co-products).

As unit processes are also used for data organisation, they are referred to as *data collection template*.[23] For reasons of transparency and data quality, the unit processes should map accurately defined and specific processes. Any time, and if necessary, these unit processes can be aggregated into larger units and be successively averaged; however, not vice versa.

Another problem arises if unit processes are too large because necessary attributions will cause problems, for example, the distribution of the total electricity consumption of a plant and the attribution to a single product analysed. It is, however, easier to obtain data for a complete manufacturing plant (especially for emissions) than for a single process. Site-specific data may be available, for

---

23) Boguski et al. (1996).

instance, via an operational input–output analysis of a factory in the context of an environmental management system (see Section 1.1.5). In general, however, a factory produces *several* products; inputs and outputs need to be attributed to these products according to defined rules. For data acquisition in a factory in line with the processes leading to a specific product, assignment is unnecessary because the data are already available. This procedure is recommended by ISO but requires much more data collection work.

Data acquisition is one of the most complex phases of LCA (see the Pellston Workshop on Global Guidance Principles for LCA Databases[24]), especially when site-specific upstream and downstream data are required.

### 3.1.3.2 Balancing

Theoretically, a complete energy and mass assessment (input and output) should be conducted for every unit process. In praxis this often fails due to the inadequacy of the data: Usually the waste heat is not measured, the waste water output is set equal to the fresh water input, the greenhouse gas $CO_2$ formed during combustion is usually not measured but calculated assuming an approximate stoichiometry, for example, for long-chain aliphatic hydrocarbons. In the simplest case, the combustion of methane (main part of natural gas), Equation 3.1a is valid:

$$CH_4 + 2O_2 \rightarrow CO_2 + 2H_2O \qquad (3.1a)$$

For petrol, for example, Equation 3.1b is valid on the simplified assumption that it contains pure Octane.[25]

$$2C_8H_{18} + 25O_2 \rightarrow 16CO_2 + 18H_2O \qquad (3.1b)$$

According to this equation, the combustion of 1 l petrol (average density $740\,g\,l^{-1}$) results in the release of 2.28 kg $CO_2$.

The principle conservation of mass cannot be applied in such cases as its validity is a prerequisite of the equation. If the empirical basis of the chemical equation is known, calculations as quoted are very precise. This, for instance, applies for the formation of sulphur dioxide from the sulphur content of fuels, as it can very securely be presumed that, via combustion, one molecule of $SO_2$ is formed from every single sulphur atom (Equation 3.1c).

$$S + O_2 \rightarrow SO_2 \qquad (3.1c)$$

---

24) Sonnemann and Vigon (2011).
25) Falbe and Reglitz (1995, p. 351).

## Exercise: Sample case for a calculation of $CO_2$-emissions

An energy concern supplies natural gas to its customers (original data). The following figures are known (even though the unit kilowatt hour should only be used for electricity, it is also applied in technical contexts, as in this case, to indicate low and high heat values):

| Natural gas component | | Average fraction | Unit |
|---|---|---|---|
| Methane | $CH_4$ | 87.535 | mol% |
| Ethane | $C_2H_6$ | 5.545 | mol% |
| Propane | $C_3H_8$ | 2.000 | mol% |
| i-Butane | $C_4H_{10}$ | 0.248 | mol% |
| n-Butane | $C_4H_{10}$ | 0.351 | mol% |
| i-Pentane | $C_5H_{12}$ | 0.056 | mol% |
| n-Pentane | $C_5H_{12}$ | 0.004 | mol% |
| Nitrogen | $N_2$ | 3.260 | mol% |
| Carbon dioxide | $CO_2$ | 0.960 | mol% |
| Other data | Average value | | Unit |
| High heat value | 11.580 | | kWh m$^{-3}$ |
| Low heat value | 10.457 | | kWh m$^{-3}$ |
| Density | 0.821 | | kg m$^{-3}$ |

Calculate the $CO_2$ emissions in g MJ$^{-1}$ that are released due to the incineration of the natural gas. Use the low heat value. Energy expenditure for extraction and transport of the natural gas to the customer (upstream) is not considered here.

If detailed data procurement is possible, it should be made. As for data procured in the factory, the primary data (sometimes called *foreground data*[26]) can be combined with an operational input–output analysis or be taken from it as the same data are required at the process level. An operational input–output analysis,[27] however, does not require an allocation of inventory parameters to particular products.

Besides, it should be considered that many unit processes do not refer to industrial products as such, but to agri- or silvicultural processes or to those of disposal or to those of use/consumption of a product. The latter depend on consumers' attitudes and behaviours in daily life, which is a field that has rarely been investigated quantitatively.

---

26) According to our knowledge, a distinction between foreground and background data was first made in a SETAC Europe Working Group on Life Cycle Inventory Analysis with Roland Clift as chair (unpublished, about 1997).
27) Hulpke and Marsmann (1994), Schaltegger (1996), Schmidt and Schorb (1995), Finkbeiner, Wiedemann and Saur (1998) and Rebitzer (2005).

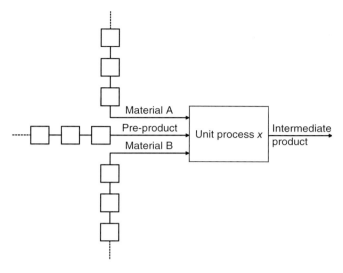

**Figure 3.3** Branching due to several main inputs (multi-input process).

### 3.1.4
### Flow Charts

Each box in a flow chart represents a unit process that requires full attention from the LCA experts. Less important unit processes have already been cut-off in the first phase (see Section 2.2.2.1). However, an iterative approach is preferable whereby, in an LCI overview, unit processes to be neglected and side chains are first determined with the use of estimated data. At the beginning of data acquisition at the latest, a decision must be made concerning side chains to be cut-off and those to be considered using estimated values.

The distinction between main and side chains cannot easily be made in complex product systems.[28] Starting with the use phase the main chain follows the production of the product upstream: production of product, production of intermediate products, and finally reaches the extraction of raw material (the *cradle*). The disposal chain runs in the opposite direction ('downstream') until the final destruction, for example, by incineration (the *grave*).[29]

A flow chart as a 'bead thread' according to Figure 3.1 is too simplistic. Real flow charts always produce branches. Two fundamental process types, multi-input and multi-output processes, can be distinguished.

1. Several materials, pre-products and intermediate products, and so on, enter the main chain by a unit process. This is called a *multi-input process*. In Figure 3.3 a pre-product and two materials, A and B, enter the unit process X. A and B are

---

28) Fleischer and Hake (2002), Lichtenvort (2004) and Kougoulis (2007).
29) The methaphor *cradle to grave* has strongly contributed to a fundamental understanding of LCA, see Chapter 1.

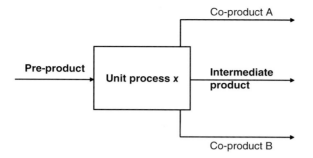

**Figure 3.4** Branching due to several main outputs (multi-output process).

essential, not negligible ancillary materials and cannot be cut off. For clarity, in this and the following illustrations all further inputs and outputs such as energy, ancillary materials and emissions are omitted.

2. A unit process yields several usable products of which only one is further processed within the product system (multi-output process). Besides an intermediate product, which is necessary for the product assessment under study, in Figure 3.4, two further products A and B are released for use in other production chains. These are called *co-products* (see Section 2.2.2.2), because the formation processes of the intermediate product as well as the products A and B are necessarily coupled.

In systems analysis each unit process must be examined with respect too its co-products. The data are needed either to allocate material and energy demand as well as emissions to the intermediate products and co-products or to be able to make an adequate system expansion (see Sections 2.2.2.2 and 3.3). Co-products of the product system under examination are *not* integrated into the system flow chart; they *leave* the system and can be presented outside its boundary (Figure 3.5, case A). This is different in the case of system expansion; here co-products remain within the system boundary, which can lead to very large systems (Figure 3.5 case B), especially if such a system expansion has to be performed more than once in a given product system.

Another possibility of branching in a product tree occurs if several processes are considered as an output (Figure 3.6). This is true for the life cycle phase 'disposal' if there are several ways of disposal or recycling. Closed-loop recycling (CLR) occurs if waste from the production is re-inserted into the production; open-loop recycling (OLR) occurs if waste is used in other production processes. As in the case of co-products, a decision has to be made concerning the position of the system boundary. The quantitative handling of recycling processes is discussed in Sections 3.3.3 and 3.3.5.

A real, although highly simplified, flow chart is depicted in Figure 3.7. It describes the production of linear alkylbenzene sulfonate (LAS) (sodium-*n*-dodecyl benzene

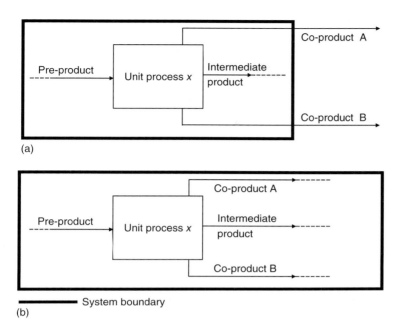

**Figure 3.5** Allocation or system expansion with multi-output processes. (a) Case A: allocation necessary and (b) Case B: system expansion.

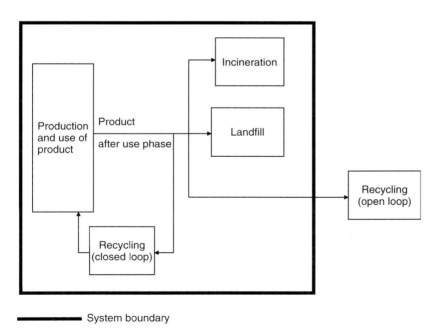

**Figure 3.6** Branching through several process options for an output.

72 | 3 Life Cycle Inventory Analysis

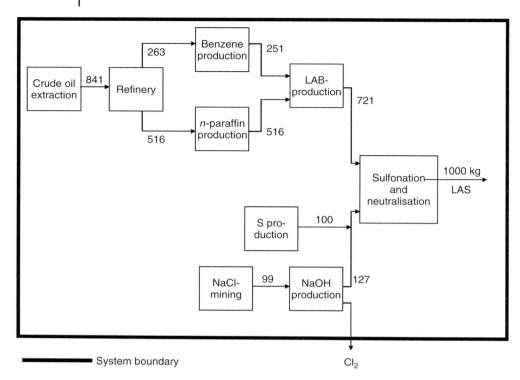

**Figure 3.7** Flow chart of LAS production. LAS – linear alkyl benzene (predominantly n-dodecylbenzene) – is transferred by sulfonation into LAS. Numbers without unit refer to kilogram of the substance specified in the left unit process. Chlorine ($Cl_2$) is the co-product of sodium hydroxide (NaOH) in the electrochemical production from common salt (NaCl) and leaves the system (allocation necessary); a further co-product is hydrogen (not indicated).

sulfonate). In this synthesis four components (n-paraffin, benzene, sulphur trioxide via sulphur and sodium hydroxide) are gradually produced and set into reaction.[30]

3.1.5
**Reference Values**

With the exception of Figure 3.7, the definition of unit processes and their integration into flow charts have so far not implied quantification. However, in business practice basic information on unit processes of production systems are often procured related to operating time (per hour or per annum) or are related to various other reference values depending on the cause of measurement or

---
30) Janzen, 1995; Berna, Cavalli and Renta, 1995.

procurement of the data. In LCI data have to be related to the part of the output relevant to the production of the assessed product, and therefore the original data used in the inventory have to be converted accordingly.

The most frequent unit with goods (contrary to services) is a certain mass of the final product, for example, 1 (metric) ton = 1000 kg = 1 Mg, as in the LAS example. The numbers in Figure 3.7 indicate European averages similar to those of APME plastics[31]; and current updates. ECOSOL data are completely and openly published[32] as opposed to APME data, which are only accessible as short-cuts with restricted transparency.[33] The study, conducted by Franklin Associates (USA) and commissioned by ECOSOL on behalf of the European industries, is best described as inventory analysis with the *cradle-to-factory gate* as the technical system boundary. With these data, surfactants, the surface-active ingredients of detergents, can make an entry as a complete unit process (from raw material extraction to surfactant) to LCIs of divers detergents. Depending upon goal definition, these must then also assess further components of the detergents, packaging, distribution, washing and the ultimate destination of the chemicals (waste water purification plant, degradation, etc.). The surfactant data of the ECOSOL study are typical generic data which are highly valuable as background data to LCA practitioners. They are presently (2013) updated within a project coordinated by CEFIC, the European association of the chemical industry.

The mass data in Figure 3.7 give an overview on the quantitative flow of material that is necessary to produce 1 Mg LAS: for an environmental assessment of loads related to the production of 1 Mg LAS, 127 kg of NaOH, 100 kg S (by $SO_3$) and 721 kg LAB must be included, which means that these production lines must be traced back, step by step, to the raw materials. Thus, for example, 251 kg benzene and 516 kg *n*-paraffin are needed for the production of 721 kg LAS . As processes usually do not have a 100% yield and the figures already consider allocations (see Sections 2.2.2.2 and 3.3), figures cannot be simply added. The quantitative data in Figure 3.7 are discussed more precisely in Section 3.3 (allocation).

The aggregation of data, which are standardized on a certain mass of the final product, is done by simple multiplication and addition which can be accomplished with spread-sheet programmes of the type Microsoft Excel.

In doing this, the data for partial aggregations for the individual unit processes *must not be lost*, because the processes causing the load cannot be deduced from the aggregated values. The analysis of final results on the basis of unit processes or groups of unit processes (= sectors, for example, all transportation or all waste disposal units) is called *sectoral analysis*. It can be accomplished during the inventory analysis or following the impact assessment (see Chapter 5).

In comparative LCA studies, the data calculated per mass unit can easily be converted into the functional unit (fU) or reference flow.

LCI s can be seen as a special case of material flow analysis (MFA). MFA with other system boundaries and reference values (usually not 'from cradle to grave'

---

31) Boustead, 1993a,b,c, 1994a,b, 1995a,b, 1996a,b, 1997a,b,c; Boustead and Fawer, 1994.
32) *www.plasticseurope.org*.
33) Complete quote in Section 1.5.

and not product-related), can be applied for regional issues, for example, in waste management, and as a tool in industrial ecology.[34] Contrary to LCA, no impact assessment is usually conducted in MFA (that would correspond to an inventory or LCI). There are, however, exceptions to this rule (MFA is not standardised), for example, the study 'poly(vinyl chloride) (PVC) in Sweden', where an MFA including system boundary = state border of Sweden was supplemented by an impact assessment according to CML.[35]

## 3.2
## Energy Analysis

### 3.2.1
### Introduction

Energy analysis based on process chain analysis is, together with the material flow analysis, one of the centrepieces of the inventory analysis. For this, three reasons are indicated by Boustead and Hancock[36]:

1. Environmental problems are frequently coupled with energy supply and 'energy consumption'.[37]
2. The availability of resources (above all fossil resources like oil, natural gas and, to a smaller degree, coal) is limited. This aspect has been dramatically described in the report to the Club of Rome 'Limits to Growth'.[38]
3. Energy prices rising on a long-term basis (energy as commodity) leads to dependence on politically uncertain regions.

Despite the fact that today the task of LCA is substantially broader defined, energy analysis has remained one of the central instruments of LCA. The most important forms of energy which, according to the first principle of thermodynamics, can be transformed into one another, are listed in Table 3.1.

If energy is regarded as commodity, there is a primary interest in the *final energy* which is bought by customers (industry, private consumers, agriculture, etc.). In LCA this definition of energy is only relevant as input: how much energy is necessary for a specific unit process for the production of a defined amount of output? However, for environmental assessments the *primary energy* expenditure is of interest. Production and transport of energy carriers, efficiency of plants for

---

34) Baccini and Brunner, 1991; Ayres and Ayres, 1996; Baccini and Bader, 1996; Brunner and Rechberger, 2004.
35) Tukker, Kleijn and van Oers, 1996.
36) Boustead and Hancock, 1979.
37) According to the first principle of thermodynamics, only energy conversions occur. This applies for the physical notion of energy. Energy in different forms is, however, economically a commodity which is traded to be used, and therefore the expression 'energy consumption' is justifiable.
38) Meadows et al., 1973.

**Table 3.1** The most important forms of energy.[39]

| Energy form | Example | MJ (example) |
| --- | --- | --- |
| Kinetic energy | Mass of 1 kg moving with 60 km h$^{-1}$ | $10^{-4}$ (100 J) |
| Potential energy | Mass of 1 kg 500 m above point of reference | $5 \times 10^{-3}$ (5 kJ) |
| Heat | 1 kg water at the boiling point, related to 20 °C | 0.34 |
| Electric current | 1 A, 1 h at 230 V | 0.83 |
| Light | Sunlight reaching 1 m$^2$ on a sunny day at noon | 3.4 |
| Chemical energy | 1 kg oil burned to $CO_2$ and $H_2O$ | 45 |
| Nuclear energy | 1 kg uranium 235 (nuclear fission) | $80 \times 10^6$ (80 TJ) |

[a] Note on units:
The SI unit of energy is the Joule (J).
Mechanical definition (J) (N = Newton, SI unit of force):
$1 J = 1 N m = 1 kg m s^{-2} m = 1 kg m^2 s^{-2}$.
Combination (J) with power and electrical units:
(W = Watt, SI unit of power; V = volt, SI unit of voltage; A = ampere, SI unit of electric current):
$1 J = 1 W s; 1 W = 1 V A$.
In LCAs the megajoule (MJ) is the most common energy unit.
The watt hour (Wh) and the kilowatt hour (kWh) do not belong to 'the Système International', but according to ISO 1000ISO (1981), they are permitted as supplementary energy units within special applications. In LCA, (kWh) is often used for *electrical energy*.
A kilowatt hour converts to exactly:
$1 kWh \equiv 3.6 MJ$.
The conversion factor of 3.61, frequently used in US-American publications, is wrong.

energy conversion (e.g. power plant, heating), and in the case of electricity the grid losses are considered here. The same applies to feedstock energy, for example, raw oil, which is a constituent of polyethylene or other synthetic plastics after chemical transformation. The primary energy can serve as a meaningful measure that, especially in the case of fossil fuels, directly correlates to the resources demand.

In simple cases, a minimum energy demand can be extracted from material data if not provided by specific process data. This is demonstrated for aluminium melt by Boustead and Hancock[40].

To transfer 1 kg aluminium metal of 290 K (about room temperature) into melt, the metal first has to be heated up to the melting temperature (932 K). For this, an amount of heat Q ($c_p$ = specific heat capacity of aluminium: 913 J kg$^{-1}$ K$^{-1}$) is needed.

$$Q = c_p \times \Delta T = 913 (J\,kg^{-1}K^{-1}) \times (932-290)(K) = 586\,146\,J\,kg^{-1} \approx 0.586\,MJ\,kg^{-1}$$

For melting the metal at the melting point, 0.397 (MJ kg$^{-1}$) × m (kg) is needed, or 0.397 MJ for 1 kg.

For 1 kg aluminium the total energy for this process thus amounts to at least

$$0.586\,MJ + 0.397\,MJ = 0.983\,MJ \approx 1\,MJ$$

---

39) Boustead and Hancock, 1979.
40) Boustead and Hancock, 1979.

It is easy to realize that the *real* (final) energy demand will be higher due to losses. If the process energy was spent in the form of electricity (1 MJ = 0.278 kWh), the *primary energy* with predominantly thermal power generation is around a factor 2–3 higher, because, according to the second principle of thermodynamics, the maximum efficiency $\eta$ amounts to

$$\eta = \frac{(T_2 - T_1)}{T_2} \quad (3.2)$$

Here $T_2$ is the upper and $T_1$ the lower temperature (K) of the thermal engine (Carnot cycle).

Whereas the physical conversion results in 1 kWh = 3.6 MJ, the conversion under consideration of the second principle (thermal) as rough approximation amounts to

$$1 \text{ kWh (electrical final energy)} \approx 10 \text{ MJ (primary energy)}$$

This is valid only for the average European electricity mix with predominantly thermal electricity generation, not, however, for countries with a large portion of hydropower (e.g. Norway, Austria and Switzerland), see also Section 3.2.4. Table 3.2 lists the efficiencies of electricity generation in relation to the assigned primary energy carriers for some countries.

Today in technology, predominantly 'concentrated' forms of energy are still employed (chemical energy and nuclear energy) which undergo some type of degradation during conversion into other useful forms of energy, for example, kinetic energy. This is because the forms of energy with lower concentration cannot endlessly be converted into those with higher concentration. This is the practical consequence of the second principle of thermodynamics which denotes the limits of conversion.

The most important applications of the second principle concern the conversion of thermal energy into other forms of energy (see above) and the specification

**Table 3.2** Energy carriers and efficiencies of electricity production in different European countries (1999).[41]

| Country | Water power (%) | Nuclear power (%) | Coal (%) | Oil (%) | Gas (%) | Other (%) | Average efficiency (%) |
|---|---|---|---|---|---|---|---|
| Austria | 68.44 | 0.00 | 9.14 | 4.65 | 14.72 | 3.04 | 64.83 |
| Switzerland | 58.37 | 37.69 | 0.00 | 0.25 | 1.46 | 2.23 | 61.52 |
| Germany | 3.53 | 30.84 | 51.87 | 1.06 | 9.99 | 2.72 | 33.85 |
| France | 13.76 | 75.99 | 6.17 | 1.96 | 1.45 | 0.67 | 40.82 |
| Norway | 99.33 | 0.00 | 0.18 | 0.01 | 0.23 | 0.25 | 79.71 |

41) Boustead, 2003.

of free energy or free enthalpy during chemical processes. Only if the latter is negative – thus delivered by the system – a reaction can take place voluntarily. Only the free energy can be converted, for example, into electricity, the remainder is dissipated as heat.

Energy dissipated as 'waste heat' can be economically used, within an appropriate infrastructure, for heating purposes (also for cooling).[42] The overall efficiency including the use of heat can be much higher (approximately 0.8) than the efficiency that refers to electrical or mechanical energy (work) alone.

## 3.2.2
## Cumulative Energy Demand (CED)

### 3.2.2.1 Definition
The determination of the cumulative energy demand[43] (CED[44]) is used in comparative assessments of the primary energy demand of technical processes and product systems. Prior to standardisation, this value was called *energy equivalent value*.[45] Following a somehow clumsy, but exact nomenclature, the VDI recommends according to guideline 4600 (see also[46]):

> *The Cumulative Energy Demand KEA[47] (CED) states the entire demand, valued as primary energy, which arises in connection with production, use and disposal of an economic good or which may be attributed to it in a causal relation. This energy demand represents the sum of Cumulative Energy Demands for production, use and disposal of the economic good.*

In VDI 4600, the common expression 'product' in LCA, according to ISO 14040, which includes goods and services, is replaced by 'economic good'. As a formula, the definition can be presented as follows:

$$KEA = KEA_H + KEA_N + KEA_E \tag{3.3}$$

whereby the subscripts refer to H = production (Herstellung), N = use including maintenance (Nutzung), and E = disposal (Entsorgung). Because of this formula, life cycle thinking can be perceived. The transportation is not assessed separately, but included in the definitions of concepts and sub-concepts. The same applies for 'production and auxiliary materials, consumables and production facilities'.

### 3.2.2.2 Partial Amounts
The CED consists of different amounts that include energy consumption in the narrow sense, and the content of energy resources and other materials with calorific

---
42) Depending on the perspective, a refrigerator is always also an oven; this technique is applied in so-called passive houses.
43) VDI, 1997; ecoinvent 1, 2004a (Chapter 2).
44) Klöpffer, 1997.
45) Kindler and Nikles, 1980.
46) Mauch and Schaefer, 1996.
47) KEA = Kumulierter Energieaufwand. In the following, the German acronyms KEA, and so on are used, as in the English version of VDI 4600 VDI (1997).

value in the products:

$$KEA(CED) = KPA + KNA \tag{3.4}$$

KPA = cumulative process energy demand
KNA = cumulative non-energy demand

KPA encompasses all traded Final Energies (EE) for heat, power, light, and the generation of other useful electricity valued as primary energy through overall efficiencies of energy supply.

The cumulative non-energy demand is the sum of the energy content of all energy carriers employed for non-energy purposes (non-energetic consumption, NEV)[48] and the *inherent energy* of working materials (SEI)[49] valued as primary energy.

The KNA, in the USA is often set equal to the energy content of non-energetically used energetic resources.[50] This implies an allocation problem because, for example, wood in many countries is rated as an important energy source; in the USA, however, it is as such marginalised: it is not part of the *feedstock energy* and not included into statistics as a resource. VDI-Guideline 4600 quotes:

$$KNA = NEV + SEI \tag{3.5}$$

NEV = non-energetic consumption (of sources of energy)
SEI = inherent energy (of materials used)

Both subgroups are valuated as primary energy. It is differentiated – as in the USA – between sources of energy (listed as such in the statistics) and inflammable materials generally not rated as energy source.

In contrast to the US practice, both subgroups are added. To us, this distinction seems artificial as each inflammable material can become an energy source. Examples of SEI-materials are starch, cellulose, vegetable and animal fats and oils, most food, and so on.

For the determination of the primary energy consumption, the final energy consumption (EEV[51]) must therefore be complemented by the NEV and the material-inherent energy content (SEI) content. The evaluation of the primary energy has to consider the overall efficiency of supply of fuels (g) referring to the respective energy contribution (for definition see Section 3.2.3.2). Therefore, the entire CED is the *sum* of all weighted ($i, j$ and $k$) partial contributions to the final energy ($EE_i$) and to the material-bound amounts of energy ($NEV_j$ and $SEI_k$):

$$KEA(CED) = \sum_i \frac{EEV_i}{g_i} + \sum_j \frac{NEV_j}{g_j} + \sum_k \frac{\Sigma_k SEI_k}{g_k} \tag{3.6}$$

---

48) Nicht Energetischer Verbrauch.
49) Stoffgebundener Energieinhalt.
50) Boguski et al., 1996.
51) Endenergieverbrauch.

The efficiency thus serves as weighting factor. During data procurement of single processes in the enterprise, mainly the partial amounts of the final energy consumption will be determinable. As the energy is usually purchased, invoices can be evaluated, and on the basis of machine data and running times the final energy consumption can be calculated or be measured by electric meters. For LCAs, however, the primary energy demand is essential. For instance, $g = 0.2$ will result in a fivefold higher primary energy demand (because 80% of the primary energy, usually in the form of waste heat, are lost).

With the above stated formulas, the CED can also be determined without a complete life cycle inventory. The latter would require the consideration of all input and output categories. The determination of CED is therefore often regarded as a methodology of its own,[52] although VDI 4600 refers to a fU and LCA in general. The CED determination within the inventory analysis is done using the sum of energy sources and is often split into a $CED_{fossil}$ and a $CED_{renewable}$.

### 3.2.2.3 Balancing Boundaries

The CED is a very useful aggregated quantity which provides a good overview on the integrated primary energy demand of a product system.

It has been argued that the determination of CED in case of certain forms of primary energy (nuclear energy, solar energy, and wind energy) is ambiguous.[53]

A convention is needed on how to define primary energy, in particular, with respect to solar energy. Solar radiation hitting the earth's surface has so far not been assessed; therefore wood and similar renewable energy carriers and materials have not been evaluated in regard of their primary energy potential. This is also true for hydropower, which is an indirect form of solar energy and is quantified in the form of potential energy. Because of the high (approximately 85%) efficiency in the conversion of potential energy of the water into electrical energy, the electricity itself (starting from the power station) can be defined as primary energy, as a first approximation. For the energy supply to the consumer, corrections due to varying losses depending upon distance and voltage and wire losses must be made in the calculations. The available generic data records[54] provide various options.

For nuclear energy the CED is calculated using the thermal efficiency of electricity generation: the primary energy is thus defined as the energy stored in the fissile atomic nuclei. Here and in the case of solar energy a convention is certainly needed, which should be specified in the form of a standard. VDI 4600 does not propose rules on how to deal with these forms of primary energy in the context of LCA. In ISO 14044 an aggregation of energy to CED is not expressly mentioned contrary to a year-long practice. The Dutch guideline of 2001 provides CED as an option.[55]

---

52) Wrisberg et al., 2002.
53) Frischknecht, 1997.
54) Fritsche et al., 1997 and updates : Globales Emissions-Modell Integrierter Systeme (GEMIS 4.3) (downloaded 2 May, 2007) http://www.oeko.de/service/gemis/de/index.htm; Version of 2010 is Gemis 4.7; ecoinvent 1, 2004a; Frischknecht et al., 2005.
55) Guinée et al., 2002.

In CED guideline VDI 4600 it is pointed out that the evaluation of nuclear energy and renewable energies is 'not clearly definable', and the following proposals are made:

1. *Hydropower*: The system boundary is the intake structure of the power plant. The efficiency of the energy supply is, according to this definition, the relationship of the net energy production (electric current) to the processable energy of the water, thus the potential energy that results from the usable gross height of fall.
2. *Wind power*: Analogous to hydropower, the system boundary is equal to the rotor disc of the power plant. The supply level, according to this definition, is the relationship of the net energy production (electric current) to the kinetic energy of the wind that passes the rotor blades.
3. *Photovoltaic energy*: The system boundary is the gross module surface. The supply level is, according to this definition, the relationship of the net energy production (electric current) to the solar energy irradiated on the gross surface.
4. *Nuclear energy*: Evaluation of the primary energy is done with the thermal efficiency of the nuclear power stations and the efficiency of utilisation for nuclear fuels. For Germany this results in an average value of 0.33.
5. *Fuels and biomass*: For fuels used to generate energy (including also garbage, etc.) the low heat value is inserted, in the case of biomass related to the harvested plants.

The use of these definitions and specifications is recommended until an International Standard is provided. Such definitions can never be completely satisfying. Thus Frischknecht has correctly pointed out[56] that water craft is also based on solar energy, which induces evaporation. Because photovoltaic energy can only attain 20% efficiency (compared to 80% by water craft) with respect to primary energy and electricity production, the determination of the system boundary seems to be inequitable at first sight. A closer look reveals that $100 - 20 = 80\%$ (solar) energy is not lost and can be applied for thermal use as in the case of the conversion of fossil into electrical energy. A photovoltaic system that uses waste heat (e.g. for the supply of industrial water) will fare better in the analysis than a system that only provides electricity! In addition, it is to be considered that these specifications produce higher overall efficiencies for solar cells, and this will yield a lower CED which will fare better in comparative assessments.

Wood as biomass according to (5) is introduced with low heat value, that is, the efficiency of wood production as related to solar energy is avoided. This efficiency is low and, as it appears in the denominator in Equation 3.6, would lead to a dominance 'of CED wood' in all wooden products. This approach would only be justified if solar energy were a scarce source (as in case of the fossil energy sources).

For renewable energies, Frischknecht and co-workers[57] propose to use the energy extractable with today's technology as weighting factors consistently. This is an

---

56) Frischknecht, 1997.
57) Frischknecht *et al.*, 2007b; SIA, 2010.

**Table 3.3** Worldwide extraction of fossil energy sources.[58]

| Source of energy | $(10^9 \, t \, a^{-1})^a$ (min–max) | Lower heating value (LHV) |
|---|---|---|
| Lignite | 0.9–1.3 (1980–1993) | 10.9 MJ kg$^{-1}$ |
| Hard coal | 3.2–3.6 (1985–1993) | 29.3 (26.8–35.4) MJ kg$^{-1 b}$ |
| Oil (crude oil) | 2.7–3.0 (1980–1992) | 42.5 (38–46) MJ kg$^{-1}$ |
| Natural gas | Approximately 1.4 | 36.0 (32–38) MJ m$^{-3}$ |
| Sum | 8.2–9.3 | |

$^a$The tonne [t] is according to ISO 1000 a recognized designation for the megagram (Mg); $10^9$ Mg = $10^6$ Gg = $10^3$ Tg = 1 Pg Mills *et al.* (1988) and ISO (1981).
$^b$Water and ash-free; by definition, 1 kgce = 29.3 MJ kg$^{-1}$.

*alternative* concept to the one proposed in VDI 4600. It is consistent, but eliminates most efficiencies of energy transformation (biomass, photovoltaics). There are open questions related to the use of hard coal and lignite. Most questions with regard to efficiency would vanish, if the waste heat was used consistently. Because this is *not* the case, the more efficient processes should have an advantage, as scarce raw materials should be protected (fossil fuels) or less area should be used (photovoltaic, wind).

From the point of view of LCA, every specification must keep an eye on system-thinking and the awareness that the goal is not 'art pour l'art' but aims to achieve an economic management and a lifestyle closer to the required sustainability.

### 3.2.3
### Energy Content of Inflammable Materials

#### 3.2.3.1 Fossil Fuels
Fossil fuels are still the most important primary energy carriers. Estimated annual extractions are of the order of magnitude of some $10^9$ metric tonnes (1 t = 1 Mg), compared to the annual production of mass chemicals which 'only' amounts to some million ($10^6$) tonnes. Table 3.3 lists annual extractions of the most important fossil energy sources.

#### 3.2.3.2 Quantification
Energy carriers are quantified either as mass or volume (standard cubic meter in the case of gases). For energy assessment and also from the practical view of quality, energy units are, however, more meaningful. Therefore a measure is necessary for the chemical energy content (more precisely: enthalpy) of the energy source. For this purpose, the *lower heating value* (LHV)[59] is usually chosen in technology. In reactions of hydrocarbons of the general formula $C_n H_m$, it is related to the energy

---

58) Hulpke, Koch and Wagner, 1993; Falbe and Reglitz, 1995; Österreichisches Statistisches Zentralamt, 1995.
59) Synonyms are: net caloriferic and lower caloriferic value.

$E_{therm}$ supplied by combustion with oxygen according to Equation 3.7:

$$C_nH_m + \left(n + \frac{m}{4}\right)O_2 \rightarrow nCO_2 + \frac{m}{2}H_2O + E_{therm} \quad (3.7)$$

for example, for methane ($n = 1$, $m = 4$):

$$CH_4 + 2O_2 \rightarrow 1CO_2 + 2H_2O + \Delta H \quad (3.8)$$

$\Delta H = -857$ kJ mol$^{-1}$ [60]

The reaction enthalpy ($\Delta H$) on the right side of Equation 3.8 refers to 25 °C and liquid water as final product. The minus sign of the reaction enthalpy in Equation 3.8 corresponds to a convention for exothermic reactions, which thus deliver energy (enthalpy). In endothermic reactions (taking up energy from the surroundings), this enthalpy has a positive sign. This convention is valid in physical chemistry. In technology this rule is not applied.

Coals and crude oils, the most important non-gaseous fossil energy carriers, are chemically badly defined mixtures that may also contain, besides C and H, different elements. This explains the ranges of heating values in Table 3.3. Furthermore, the use of molar units does not make sense. Technical heat values of solid and liquid energy carriers usually refer to a mass unit.

For the determination of the LHV, the reactants must be present before and after combustion at 25 °C (298.1 K). The water formed during this combustion process is considered to be in the vapour state, which is usual in technical processes (in spite of water being liquid at 25 °C, the final temperature in the definition of the heat value).

Thermodynamically more meaningful is the *higher heating value* (HHV),[61] which is also defined at 25 °C as starting and final temperature; the water formed here is, however, in the liquid state. The HHV is usually (as the name suggests) higher than the LHV, because during condensation the heat of vapourisation is set free as condensation heat and adds to the total measured enthalpy.

For LCAs, according to Boustead,[62] the thermodynamically more correct HHV value is preferable. In praxis, however, the LHV values as provided in technology are more easily available. The chemical composition of the energy carrier (the amount of hydrogen must be known), which varies according to the origin of the fossil fuel, or the amount of water formed during incineration has to be known for conversion to occur. The difference between the numerical values of LHV and HHV amounts to a maximum of 10% if the fuel is rich in hydrogen like methane; and it nearly disappears for fuels poor in hydrogen, for example, (hard) coal (Table 3.4). The vapourisation enthalpy of water needed for the conversion amounts to 2.45 kJ g$^{-1}$ H$_2$O. The sign is positive for evaporation and negative for condensation. The latter implies that the input energy during evaporation (with constant pressure it is called *enthalpy*) will again be released during condensation.

---

60) Per mol means per formula (i.e. not per kilogram or per standard cubic metre).
61) Other names include gross energy, upper heating value or gross caloriferic or higher caloriferic value (*HCV*).
62) Boustead, 1992.

**Table 3.4** Specific heat values of some fuels Boustead, 1992.

| Fuel | Higher heating value (HHV) (MJ kg$^{-1}$) | Lower heating value (LHV) (MJ kg$^{-1}$) |
| --- | --- | --- |
| Gasoline | 45.85 | 42.95 |
| Propane | 50.00 | 46.95 |
| Methane | 53.42 | 48.16 |
| Natural gas | Approximately 51.50 | 46.10[a] |
| Diesel (fuel oil) | 42.90 | 40.50 |
| Coal | 30.60 | 29.65 |

[a] Average value from Table 3.3 converted at density of 0.78 kg m$^{-3}$.

Because HHV and LHV values of fossil energy carriers vary with deposits, if the exact values are not known, averages are used for a mass-to-energy conversion. The best-known average is

$$< LHV(Coal) > = 29.3 \, MJ \, kg^{-1}$$

This defines the 'kilograms of coal equivalent (kgce),' which is not approved officially but is often used in statistics. It signifies an *energy* unit or LHV of 1 kg of average coal. This was formerly approximated as 7000 kcal kg$^{-1}$, thus (1 cal = 4.184 J)[63)]

$$1 \, kgce = 7000 \, [kcal \, kg^{-1}] \times 4.184 \, [kJ \, kcal^{-1}] = 29\,288 \, [kJ \, kg^{-1}]$$
$$\approx 29.3 \, MJ \, kg^{-1}$$

Thus the accuracy of this specification is only apparent and is caused by the conversion of the obsolete unit Calorie to the SI unit Joule.

In analogy to kgce, the unit 'tonnes of coal equivalent (tce)' is used: 1 tce = 29.3 GJ. The advantage of these units is their *descriptiveness* by providing a mass equivalent in (kg) or (t) for energy.

Concluding the discussion on lower versus higher heating values, it should be pointed out that the guideline to CED (VDI 4600), contrary to the Boustead recommendation, uses the LHV as data for the energy content of fuels. Additional use of HHV is not excluded but should explicitly be noted.

Corresponding to VDI 4600, the overall efficiency of supply of fuels ($g_{OES}$) is defined according to the following Equation 3.9:

$$g_{OES} = \frac{LHV}{CED_s} \quad (3.9)$$

$CED_s$ = cumulative energy demand for the supply of an energy carrier, in an LCA, calculated per functional unit

---

63) Calorie is a badly defined unit and should not be used any more. The conversion factor 4.184 (precise) refers to a 'thermochemical' or 'defined' calorie.

> **Exercise: Calculation of emissions based on final energy**
>
> Coal is declared as '$CO_2$-loaded' and natural gas as 'less $CO_2$-loaded'. Justify this statement by showing in an arithmetical example of how much $CO_2$ emissions result from the production of 100 MJ usable energy from coal and natural gas. (Basis LHV; with an implied efficiency in both cases of 35%.) Use the data in Table 3.4 as well as the following additional information:
>
> - C-content natural gas: 75% (w/w)
> - C-content hard coal: 80% (w/w).

### 3.2.3.3 Infrastructure

The inclusion of the construction of plants (infrastructure, capital goods, etc.) with energy-intensive goods and processes is not uniformly handled in LCA. Usually these CEDs relating to construction of power plants, factories, and roads are cut-off because they contribute less than 1% to the total energy. The energy necessary to build a power plant is of the order of 1 ppm of the energy the plant will provide during its 30–50 years' life time. This can be illustrated by the following rough calculation:

The production of electricity of a large 1000 MW ($10^6$ kW) power plant with a 60% average capacity factor and 50 years working time results in (1 kWh ≡ 3.6 MJ):

$$\textbf{Electrical energy} = 10^6 [\text{kW}] \times 50 [\text{a}] \times 365 [\text{d a}^{-1}] \times 24 [\text{h d}^{-1}] \times 0.6 [-]$$
$$= 2.628 \times 10^{11} \text{ kWh} = 9.46 \times 10^{11} \text{ MJ} = 9.46 \times 10^{17} \text{ J}$$
$$\approx 1 \, \textbf{EJ (exajoule)}$$

This energy is not yet the final usable energy, but corresponds to the electric energy provided without transmission and distribution losses (thus starting from the power station). If the energy for the construction of a power plant is, for simplification, set equal to the steel and concrete applied, the following power supply is required (average value for a nuclear power station, or a coal power station of comparable size):

$$\text{CED (plant)} \approx 10^{15} \text{ J} (= 0.001 \text{ EJ})$$

Comparison of these rough estimations shows that only *1 per mil* of the provided energy is necessary for the two most important building materials. Mauch and Schaefer[64] indicate a range of 0.1–0.2%. Although in this case the neglect is not relevant for the final result, this will not always be the case. For the first time in the history of LCA, the so-called ETH data[65] included infrastructure data. On the basis of the ecoinvent data that can be used either with or without capital goods, it was shown that a neglect of these data is not always acceptable.[66]

---

64) Mauch and Schaefer, 1996.
65) See query Section 1: this tradition is maintained in the Swiss data base ecoinvent, ecoinvent 1, 2004a; Frischknecht et al., 2005.
66) Frischknecht et al., 2007a.

There are impact categories (aquatic ecotoxicity, human toxicity and land use) and product groups (photovoltaics and wind energy), which strongly depend on capital goods. In any case it should be carefully examined whether by use of infrastructure data consistency and symmetry problems occur, particularly in comparative LCAs.

### 3.2.4
### Supply of Electricity

Electricity as a special form of energy plays an important role in LCA. It has already been pointed out that the evaluation of the primary energy within this form of energy is particularly urgent (see Section 3.2.2). On the basis of the second principle of thermodynamics, conversion losses have been identified: only around 30–40% of thermal energy can be converted into electrical energy in conventional thermal power stations. The remainder *could* largely be used as low temperature input for heating; contrary to the practice in Sweden, this is not yet common practice in Germany.

The notion 'electricity mix' is of fundamental importance for a specification of average supply levels – and thus for calculating the primary energies – of national grids. They are rarely supplied with one kind of primary energy alone (exceptions: 90–100% hydropower in Norway and Brazil; about 80% nuclear power in France) (see also Table 3.2). Rather, a mixture of the following sources is typical:

- Fossil energy sources (hard coal and lignite, natural gas and oil),
- nuclear energy,
- hydropower,
- renewable energy sources without hydropower (biomass, wind energy, solar energy, etc.) and
- import (weighted mix of countries exporting into the country under examination).

If the manufacturing plants of the examined product are scattered all over Europe, a European electricity-mix is often applied. For these average values, the Western European Electricity Network is often used as the basis. The Union for the Coordination of Transmission of Electricity (UCTE)[67] is, with 2530 TWh (2006), one of the largest electricity networks of the world. The transnational energy flow within the UCTE amounted to approximately 297 TWh in 2006, amounting to 12% of the produced quantity of electricity. The Scandinavian states (except Denmark west), Great Britain and Ireland as well as the Baltic States and the GUS states have their own networks. For Continental Europe the UCTE is a good approximation. The publicly accessible statistics of the UCTE, however, show small resolution with respect to the primary energies used. An improved resolution is provided by the country specific statistics of the International Energy Agency (IEA) for EU-27. Table 3.5 presents the electricity mix data of 2005 for the European Union and

---

67) www.ucte.org, see also BUWAL, 1991 (UCTE since 1999; 1951–1998: 'Union pour la Coordination de la Production et du Transport de l'Electricité' (UCPTE).

**Table 3.5** Electricity mix EU-27 2005, Germany, France, Switzerland and Austria.[68]

| Production | EU-27 (GWh) (%) | Germany (GWh) (%) | France (GWh) (%) | Switzerland (GWh) (%) | Austria (GWh) (%) |
|---|---|---|---|---|---|
| Nuclear energy | 997 699 (30.1) | 163 055 (26.3) | 451 529 (78.5) | 23 341 (39.1) | 0 |
| Coal | 1 000 829 (30.2) | 305 547 (49.3) | 30 641 (5.3) | 0 | 8 482 (12.9) |
| Oil | 138 503 (4.2) | 10 583 (1.7) | 7 227 (1.3) | 191 (0.3) | 1 641 (2.5) |
| Natural gas | 663 744 (20.0) | 69 398 (11.2) | 22 961 (4.0) | 869 (1.5) | 13.036 (19.8) |
| Σ Fossil fuels | 1 803 076 (54.5) | 385 528 (62.1) | 60 829 (10.6) | 1 060 (1.8) | 23 159 (35.2) |
| Waste | 27 086 (0.8) | 6 094 (1.0) | 3 260 (0.6) | 1 872 (3.1) | 546 (0.8) |
| Hydro power | 340 846 (10.3) | 26 717 (4.3) | 56 404 (9.8) | 33 086 (55.5) | 38 612 (58.8) |
| Biomass | 57 332 (1.7) | 10 495 (1.7) | 1 821 (0.3) | 226 (0.4) | 2 039 (3.1) |
| Geothermal | 5 397 (0.2) | 0 | 0 | 0 | 2 (0.003) |
| Solar-photovoltaics | 1 491 (0.05) | 1 282 (0.2) | 15 (0.003) | 19 (0.03) | 14 (0.02) |
| Wind power | 70 496 (2.1) | 27 229 (4.4) | 959 (0.2) | 8 (0.01) | 1 328 (2.0) |
| Tidal power | 534 (0.02) | 0 | 534 (0.1) | 0 | 0 |
| Σ Renewable energy sources | 476 096 (14.4) | 65 723 (10.6) | 59 733 (10.4) | 33 339 (55.9) | 41 995 (63.9) |
| Other sources | 7043 (0.2) | 0 | 0 | 0 | 18 (0.03) |
| Σ Production | 3 311 000 | 620 300 | 575 351 | 59 612 | 62 990 |
| Σ Percent | 99.9 | 100 | 100.1 | 99.9 | 100 |

three selected member countries (Germany, France and Austria) as well as for Switzerland.

Unfortunately in Table 3.5 coal as an important energy resource is not split into brown coal (lignite) and hard coal. According to Eurostat[69] the proportion of hard coal (EU-27 2005) amounts to 19.1% and that of of brown coal to 9.3%.

There are also similar statistics from OECD[70] (OECD world, OECD Europe, single countries) and single non-OECD countries. The tables can also be downloaded

---

68) Energy statistics of the International Energy Agency (IEA), Paris.
69) Eurostat 2008: European Commission: Europe in figures, Eurostat yearbook 2008; ISSN 1681-4789; download at *http://epp.eurostat.ec.europa.eu*.
70) Organisation for Economic Co-operation and Development.

from the web site of IEA (*www.iea.org*). These statistics also provide the exports and imports, not classified by countries though.

Apart from the smaller countries that produce either a surplus or are dependent on import, imports and exports are usually the same and together make up about 10% of the total amount of electricity produced in the country. The numbers of OECD (world) of 2007 depict an overall electricity production of 10 000 TWh including a startling high proportion of more than 85% of nuclear energy and fossil energy sources (the same as in Europe). Apart from hydropower (13%), renewable energies do not yet contribute significantly.

For the energy calculation in the context of LCI, the final electricity consumption must be extrapolated to the primary energy needed to produce the usable electricity. For this purpose, the mix of electricity production serves as the basis. The efficiencies with regard to specific energy sources have to consider the average operating state of the power plant (conversion losses), transmission and distribution losses as well as the typical upstream processes like transportation of energy sources, and so on. These processes are considered by good generic data bases or can be deduced from the sub processes.

In the context of renewable energy contributing to the electricity mix, the primary energy it is related to has to be defined (see Section 3.2.2). As for hydropower, the primary energy is set equal to the potential energy of the water by using an efficiency of 90% or 85%. For thermal processes, the corresponding average efficiency is approximately 35%. The CED, for example, per kilowatt hour, is equal to a weighted and averaged primary energy necessary for the supply of 1 kWh (see Section 3.2.2).

In LCA the primary energy should be split into renewable/non-renewable and fossil. This is necessary for the conducting an impact assessment later (see Chapter 4), in particular for the calculation of the global warming potential – GWP. High electricity consumption in countries with a high proportion of fossil primary energy always implies a high GWP whereas hydropower and other renewable forms of energy contribute far less.

The production of electricity does not only cause the emission of greenhouse gases (mostly $CO_2$ and $CH_4$) but also other emissions that have other impacts on the environment (e.g. radioactive emissions and acid forming gases). For an LCI, those data must be procured or extracted from a generic data base and then be attributed to the unit processes. Apparently friendly types of production like hydropower can have serious effects on natural ecological systems (dams and artificial lakes), which are difficult to quantify (impact category land use). Obvious damage occurs with the extraction of coal and lignite in surface mining (coal pits), which relates to the same impact category. Transmission of electric current can also have environmental impacts, for example, by the use of the isolation gas $SF_6$, which is an extremely strong greenhouse gas that may be released to the atmosphere by leakages. However, on the other hand it has to be assessed whether its use does not save more energy and thus reduce more greenhouse gas emissions than caused by $SF_6$.

## 3.2.5
**Transports**

Transportations occur in every LCA and are often not expressly referred to as *individual unit processes* but added to other unit processes whereby double counting must be avoided (the output of an 'upper' unit process can be the input of a 'lower' one). For reasons of transparency, the transportation processes must be modelled just as carefully as all other unit processes.

Transportation data are nearly always taken from (external and/or internal) generic data bases except when the transport itself is the system to be studied, in which case relevant primary data is required. Many systems that have been optimised by re-use and recycling (e.g. reusable packaging) are substantially affected or even determined in the final result by transportation, in particular due to:

- Distance,
- Means of transport and
- Extent of utilisation/logistics.

The transportation of raw materials, energy sources, materials, products, and waste is usually at the centre of attention, whereas the passenger transportation plays a minor role. Things of course change if transportation is the topic of the study. With regard to passenger transportation, the comparison between rail, car and air plane may be under investigation; with respect to transportation of goods, different transportation variants such as rail, truck, and ship may be compared.

Usually two key figures are employed for quantification in LCI:

- passenger kilometre (Pkm).
  One person is transported over a distance of 1 km.
- tonne-kilometre (tkm).
  A mass of 1 ton (Mg) is transported over a distance of 1 km.

For calculating passenger kilometres, the number of persons is multiplied with the distance in (km). Relating environmental loads of the transport (e.g. $CO_2$ releases per 100 km) to passenger kilometres provides a meaningful measure for comparing different transport types: If a person travels 100 km in a car, the environmental loads per passenger kilometre are about four times as high compared to a car occupied by four passengers. Likewise, the degree of utilisation (actual number of passengers/maximum capacity) for transports by bus, train and air plane plays an important part in the quantification of environmental load per passenger kilometre which, however, is not always easy to obtain.

For the determination of ton kilometres, the transported mass (tons) is multiplied with the distance covered (km) and the environmental load of the vehicle (e.g. its fuel consumption) is related to the ton kilometres. In contrast to the calculation of the environmental load per passenger kilometre, it is considered that the energy consumption is split into a load-independent portion of the empty vehicle and a load-dependent portion. As the energy consumption of the empty vehicle is allocated to the loaded goods, the specific energy consumption (in (l fuel tkm$^{-1}$) or (MJ tkm$^{-1}$)) decreases with increasing degree utilisation (real load/maximal load capacity) (seeexercise in this chapter).

The degree of utilisation is thus of major importance for the calculation of the environmental loads per person kilometre and per ton kilometre: A fully loaded passenger car or truck will be more favourable per person or per ton than a partly loaded one, despite larger fuel consumption per vehicle and kilometre. As the specific degree of utilisation for the transport of persons or goods cannot easily be determined, averages are often used.

Even if for most cases no complete data are available for modelling the transportation processes, the transportation distances and vehicles should be determined as specifically as possible in order to select meaningful generic data records. In case generic data is resorted to for environmental loads of the means of transport, which are the basis for calculations related to passenger kilometre or tonne-kilometre, their acquisition date must be considered: fuel (or electricity) consumption has decreased over the past years as also emissions, as a result of slow but intensified change of legislation.[71]

Not all transportation means depend on motion (railways, motor vehicles, ships, air planes, etc.); they can also be based on tubing systems (pipelines) whose energy requirements and maintenance must be obtained and related to the transported mass/volume/energy. If such data are present in good quality, the conversion to a fU is straightforward.

Here, transportation is discussed under the aspect of 'energy'; without any question, energy and the correlated resource consumption represent a major problem in environmental politics. In addition, as in the case of electricity production, the emissions of harmful gases and particles have to be considered as output. These data are necessary for the determination of several impact categories and indicators (see Chapter 4). In particular these are (the most important emissions are in parentheses):

- climate change ($CO_2$, $CH_4$, $N_2O$, Freon substitutes),
- formation of photo oxidants (volatile organic compounds/VOCs, CO, $NO_x$),
- terrestrial eutrophication ($NO_x$),
- acidification ($NO_x$, $SO_2$) and
- human toxicity (VOC, $NO_x$, fine dust, PAH).

Road traffic in particular substantially contributes to these emissions. However, shipping traffic also contributes essentially to emissions of sulphur dioxide by the use of fuel oil (bunker oil).The Federal Environmental Agency (UBA, Berlin and Dessau) points out that the $SO_2$-load in ports is predominantly due to open sea vessels (e.g. Hamburg 80%). The same is valid for offshore regions and the North Sea.

Tanker accidents imply severe regional loads but can, only with difficulty, be attributed to a specific product group (see also Section 4.5.5.1).

---

71) Institut für Energie- und Umweltforschung Heidelberg, 2006.

Data for traffic-dependent energy supply and traffic-dependent emissions are included in the Transport Emission Model TREMOD 4.0.[72] Data based on the real situation (until 2003) are complemented by scenarios of future development (until 2030). TREMOD is, however, not publicly accessible (because of its extent and complexity). Data for road traffic are coordinated with the handbook on traffic emissions HBEFA[73] which in its version 2.1 contains emission factors for Germany, Switzerland and Austria.

Noise emissions should be procured for the impact category 'noise', but comprehensive data can hardly be collected at justifiable expenditure (see Section 4.5.4.3). A calculation of noise emissions in proximity to roads is accomplished by an expansion of the computational emissions programme MOBILEV.[74]

> **Exercise: Calculation of environmental loads by transport (without supply chain of the fuel)**
>
> For the calculation of the environmental loads of transportation not only are distances and means of transportation relevant, but also the transport capacity related to the fU.
>
> In an LCA the packaging system 'carton packaging for beverages' is examined with the following fU:
>
> 'Supply of 1000 l filling good at the point of sale'
>
> Transports must, therefore, also be assessed.
>
> Products are transported by a long-distance truck (40 t permitted total weight, 25 t maximum payload). The truck houses a maximum of 34 loading positions for Euro-pallets, of which 24 are used.
>
> The following table shows the energy consumption of the truck (on the average: motorway, highway, built-up areas).[75] The truck is supplied with Diesel fuel (LHV: 42.96 MJ kg$^{-1}$; density: 0.832 kg l$^{-1}$).
>
> *Empty trip*          9.29 MJ km$^{-1}$
> *50% utilisation rate*   0.87 MJ tkm$^{-1}$
> *100% utilisation rate*  0.50 MJ tkm$^{-1}$.
>
> Since the energy consumption of the empty vehicle is allocated to uploaded goods, the specific energy consumption (in (MJ tkm$^{-1}$) related to the transportation weight) decreases with increasing utilisation rate.
>
> Fuel consumption as a function of the degree of utilisation divides into a load-independent part (B_empty[76]), needed by an empty truck, and a load-dependent

---

[72] German: Transport Emissions Modell; Institut für Energie- und Umweltforschung Heidelberg, 2006 (The summary can be downloaded at *www.ifeu.de*); INFRAS, 2004b.
[73] German: Handbuch Emissionsfaktoren des Straßenverkehres.
[74] Fige GmbH (quoted according to UBA Dessau, Verkehr, Daten and Modelle, 2008; http://www.uba.de).
[75] Forty-tonne truck average value in Germany in 2005; personal communication IFEU, 2008.
[76] B signifies 'burdens', more precisely fuel consumption and emissions of the truck.

part (B_loaded) which increases linearly with the transport weight and the utilisation degree (Figure 3.8).

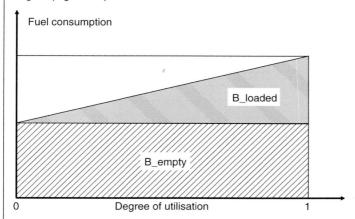

**Figure 3.8** Fuel consumption as a function of the degree of utilisation.

The following table shows packaging weights and pallet configuration (for practical example see Table 3.9 in Section 3.7)

| Packaging (1-l beverage carton) | Weight (g) |
|---|---|
| Weight primary packaging (carton) | 31.5 |
| Secondary packaging (corrugated cardboard trays) | 128 |
| Transportation packaging | 24 000 |
| Euro-pallets (wood) | |
| Pallet foil | 280 |
| **Pallet pattern** | |
| Carton per tray | 12 pieces |
| Trays per layer | 12 pieces |
| Layers per pallet | 5 pieces |
| Cartons per pallet | 720 pieces |

1. Calculate the degree of utilisation of the truck (for filling good simplified: 1 l = 1 kg)
2. Calculate the fuel consumption (in litre Diesel) for a distance of 100 km for a truck loaded with 24 pallets. Assume a linear dependence between the fuel consumption and the utilisation rate. Finally, derive the specific energy consumption from the fuel consumption.
3. Calculate the fuel consumption (in litre Diesel) for a distance of 100 km for a truck loaded with 24 pallets related to the fU.

## 3.3
## Allocation

### 3.3.1
### Fundamentals of Allocation

Allocation means the attribution of environmental burdens during the life cycle, for co-production, recycling and disposal. In doing so, however, fundamental problems of science theory that have not yet been addressed occur, because the focus until now has been the firm scientific and technical methodology of LCI (for an exceptionally clear presentation of the fundamentals of LCI see Boguski[77]):

- validity of the basic laws of physics and chemistry;
- efficiency parameters of technical plants, agricultural processes, and so on;
- clear and unambiguous cut-off criteria.

Limits to a strict scientific-technical analysis are reached for the first time when attributing environmental burdens in the 'upper' part of the product tree for a *simultaneous production of several products in one unit process*.[78] This is best demonstrated in the case of co-production.

### 3.3.2
### Allocation by the Example of Co-production

#### 3.3.2.1 Definition of Co-production
In Figure 3.9 a procedure without co-products is shown:

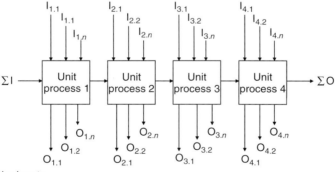

I = Inputs
O = Outputs

**Figure 3.9** Simple chain (section of a life cycle) without co-production.

---

77) Boguski et al., 1996.
78) Heintz and Baisnée, 1992; Boustead, 1994b; Huppes and Schneider, 1994; Klöpffer and Volkwein, 1995; Klöpffer, 1996a; Grahl and Schmincke, 1996; Heijungs and Frischknecht, 1998; Tukker, 1998; Heijungs, 1997, 2001; Curran, 2007.

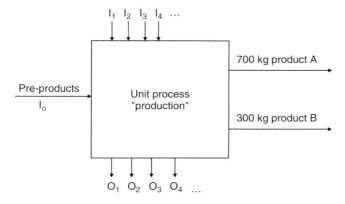

**Figure 3.10** Unit process with co-products.

The simple chain with $i=4$ unit processes shows the principle of a procedure in an LCI without co-production: The sum of the inputs ($I$) and outputs ($O$) of the life cycle is the *sum of individual amounts* whereby inputs can also be side chains (branches in the picture of the product tree) or partial LCIs as, for example, the aggregated data by APME (Plastics Europe) or by ECOSOL (for surfactants). According to Figure 3.9, for the conduction of this simplest LCI, only the following additions have to be made in order to get the total input and the total output:

$$I_{tot} = \sum_i I_i$$

$$O_{tot} = \sum_i O_i$$

It goes without saying that an adding of non matching inputs and outputs does not make sense.

Co-production implies that at least two products are formed in one unit process (Figure 3.10). This is particularly frequent in the chemical industry,[79] agriculture, the mining industry, oil refining and extractive metallurgy, less frequent in machine and tool-making. In chemistry the formation of several substances in one reaction is more of a rule than an exception.

### 3.3.2.2 'Fair' Allocation?

The major task is to fairly allocate the environmental load, that is, inputs and outputs, to the products A and B (more generally A, B, C, …). The choice of the attribute 'fair' indicates that a strict scientific solution cannot be provided. In the science of economics the problem of allocation has been known for over 150 a. It concerns the allocation of fair costs to the individual products. Costs for individual products must be derived from the total costs. The British political economist John

---

79) Riebel, 1955.

## 3 Life Cycle Inventory Analysis

Stuart Mill[80] is said to have been the first to recognise the problem of allocation, and illustrated it by the following example:

Chicken (→ meat)/Eggs

Similarly descriptive allocations are, for example, with cattle:

Meat/tallow(→ soap)/skin(→ leathers)

Before proposals for a solution of the allocation problem are presented in Section 3.3.2.3, the following two important strategies are discussed: 'allocation per mass', the oldest and still the method of choice for many multi-output processes and 'system expansion', recommended by ISO 14044.

**3.3.2.2.1 Allocation per Mass**  The allocation per mass requires that all inputs and all outputs are partitioned according to the mass ratio of the co-products formed. If, for example a unit process with two co-products A and B (see Figure 3.10) results in 700 kg A and 300 kg B per fU, according to this rule $700/(700+300) = 0.7$ or 70% of all emissions, energy consumption, ancillary materials, and so on, are attributed to A and 30% to B. It is important to note that, for a consideration of multiple unit processes, the allocation per mass has to be made for all upstream unit processes. A simplified example is illustrated in Figure 3.11. Co-products 1, 2.1 and 2.2 leave the system and are employed in other product systems.

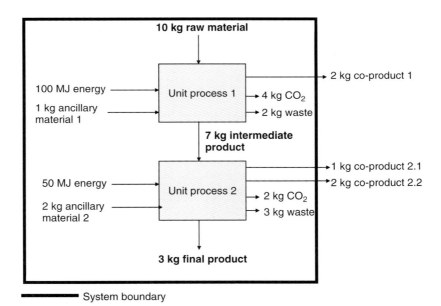

**Figure 3.11**  Allocation of interconnected multi-output processes; example for an allocation per mass.

---

80) Mill, 1848.

Allocation usually starts with the process that exhibits the final product as output (in Figure 3.11 unit process 2). The production of this unit process amounts to 6 kg overall, with the following constituents: 3 kg of the final product (50%), 1 kg of the co-product 2.1 (16.67%) and 2 kg of the co-product 2.2 (33.33%). All inputs (here: energy, ancillary materials and intermediate products) and outputs (here: $CO_2$ and waste) are distributed ('allocated') according to the mass ratio of the product output. The following loads are therefore allocated to 3 kg of the final product in process 2: 25 MJ (50%) energy input, 1 kg (50%) of ancillary material 2 input, 1 kg (50%) $CO_2$ output, 1.5 kg (50%) waste output, and 3.5 kg (50%) of intermediate products (input). Only those loads of process 1 can be allocated to the final product that are related to the production of 3.5 kg (50%) of the intermediate product input into unit process 2.

In addition, co-product 1 is formed in unit process 1. The allocation per mass in process 1 is primarily done exactly as described in process 2. All in all, the total sum of 9 kg is produced with 7 kg (77.78%) intermediate product and 2 kg co-product 1 (22.22%). 7 kg of intermediate product are therefore loaded with 7.78 kg raw material, 77.78 MJ energy, 0.78 kg of ancillary materials 1, 1.56 kg waste and 3.11 kg $CO_2$.

However, according to the allocation per mass for unit process 2, only 50% of the intermediate product can be allocated to the final product. Hence the loads in unit process 1 must be allocated correspondingly. If the allocation in unit process 2 were not done, the co-products in unit process 2 would never be loaded with consumptions and emissions of upstream processes. This would surely not be fair.

To 3 kg final product 50% of load from process 1 has to be allocated which has to be added to the load from process 2. The energy, for example, results are

25 MJ/3 kg final product from process 2

+38.89 MJ/3 kg final product from process 1

= 63.89 MJ/3 kg final product

This corresponds to 21.3 MJ $kg^{-1}$ of the final product.

An indication per (kg final product) or a reference to the reference flow that results from the choice of the fU is common practice.

Another example of an allocation per mass is illustrated in Figure 3.7 (see Section 3.1.4). The mass flow of raw material and intermediate products per 1000 kg LAS are shown in the flow chart according to the ECOSOL study. Inputs of the final process 'sulfonation and neutralisation' are 127 kg of NaOH ($3.18 \times 10^3$ mol). The assessment of the production of 127 kg NaOH includes the 'extraction of 99 kg NaCl' ($1.71 \times 10^3$ mol). Because of the reaction equation of chlorine alkali electrolysis, however, the molar ratio 1:1 of NaOH:NaCl can be assumed:

$$2NaCl + 2H_2O\ (+e^-) \rightarrow 2NaOH + Cl_2 + H_2$$

117 g  36 g    80 g    71 g  2 g

             52.3%   46.4% 1.3%

(stoichiometric conversion with rounded mol masses).

The reason for the mass flow as indicated is owing to the fact that $Cl_2$ leaves the system as co-product. Thus only 52.3% of the environmental loads of the NaCl production can be allocated to NaOH production. If for the production of LAS 127 kg of NaOH are necessary, a stoichiometric 186 kg of NaCl are required for the process of chlorine alkali electrolysis. Of these, however, only 97.3 kg (52.3%) are allocated to sodium hydroxide according to an allocation per mass. If $H_2$ is not considered for allocation (remains in the system) the indicated 99 kg of the ECOSOL study results.

**Exercise: Allocation per mass in a process chain (anonymised case example)**

A product is made of crude oil. The process chain is represented as a flow chart. For each process step data for the energy consumption and the mass of resulting co-products are available. Calculate the energy consumption of the final product in (MJ kg$^{-1}$).

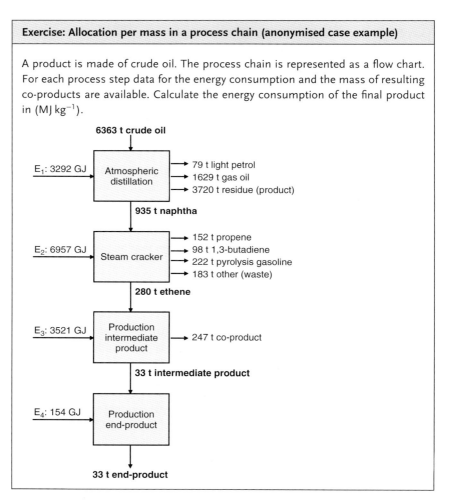

**3.3.2.2.2 System Expansion** In Figure 3.11 the co-products leave the system. In contrast, with 'system expansion' the co-products remain in the system (Figure 3.12). Consequently these have to be analysed and downstream assessed in their life cycle including all unit processes until disposal. By an 'allocation per

3.3 Allocation | 97

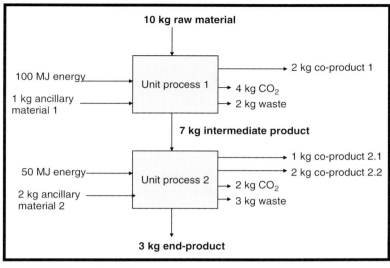

Figure 3.12  System expansion, compared to allocation per mass in Figure 3.11.

mass' loads to be allocated to the final product were calculated. In the case of system expansion, calculated loads are, however, allocated to a product-mix of the total system including final product, co-product 1, co-product 2.1, and co-product 2.2.

With this method very large systems may be studied. This can make sense for an overall representation of environmental loads of large industrial systems.[81] Usually, however, in accordance with the definition of the goal and the specification of an appropriate fU an allocation of load to specific products is required. Thus in the case of system expansion the following additional problems have to be solved:

- The fU must be revised, since the performance of assessed co-products must be included. This is an added value since it occurs in addition to the performance of the examined product. If two products are compared with each other, identical co-products rarely ever occur and thereby different added values have to be considered. In order to compare the systems, the system symmetry has to be restored again. In Section 2.2.5.3 (Figure 2.5) it has already been discussed in an example of the waste industry as to how the added value by system expansion can be included.
- System expansion can result in intransparent large systems, especially for a yield of multiple co-products, in their turn used for multiple applications.
- Since all co-products must be analysed and downstream assessed, a substantially increased effort at investigation is necessary.

---

81) Tiedemann, 2000.

As the allocation requires assumptions, which cannot be strictly and scientifically deduced, for example, the decision as to whether an allocation per mass, energy or price is to be made (see Section 3.3.2.3 'Diamond paradox'), ISO 14044[82] recommends the *avoidance of allocation* by system expansion as the most scientific solution. Because of the substantially higher effort in practice and depending on the respective goal definition, multi-output processes are mostly modelled using defined allocation rules. Most important applications of system expansion can be found in the assessment of options for the waste industry (see Section 3.3.5) as well as in the context of open-loop recycling (see Section 3.3.4).

### 3.3.2.3 Proposed Solutions

For the solution of the problem of allocation different strategies were developed, of which none, however, completely satisfies.[83] A unique solution for all objectives and goal definitions is probably impossible.

The rules to be found in ISO 14041 and 14044 and the results of the fundamental debates at the allocation workshop in Leiden, 1994[84] are of main concern. Allocation in LCA has been the topic of several reports that did not always distinguish between co-products and the related problem of OLR[85] (see Section 3.3.4).

The list below, which can also be used as some sort of a check list, intends to show the strategies based on ISO 14044, but in a less dogmatical way. The actual product under investigation by LCA is indicated by A, the co-products by B, C, ...

1. Statement whether product B (C, ...) has a performance and is merchantable, thus represents an economic good. If this is not the case, B is *waste* (more precisely: waste for disposal) and *no* environmental loads are attributed to it. The underlying logic implies that nobody will conduct a technical process only in order to produce waste.
2. Examination whether *system expansion* is possible with justifiable effort. If so, a scientific solution is possible.
3. Examination whether *system reduction* is possible: if the unit process is too large, for example, a whole factory was selected that produces several products (A, B, C, ...); If so, the large unit process can be divided into smaller ones considering, for example, a production line, a reactor, a field, and so on, with only one product under investigation.

    This restriction can shift the problem: The separation of a large unit process into smaller ones implies that the data requirement is much larger and more differentiated; this means an allocation problem can result in a data problem. A carefully applied system reduction can also be valued as strictly scientific if its application does not imply subjective assumptions.

---

82) ISO, 2006b.
83) Fava et al., 1991; Heintz and Baisnée, 1992; Society of Environmental Toxicology and Chemistry – Europe, 1992; Society of Environmental Toxicology and Chemistry (SETAC), 1993; Boustead, 1994b; Huppes and Schneider, 1994; Ekvall and Tillman, 1997; International Standard Organization (ISO), 1998a; ISO, 2006b; Curran, 2007, 2008.
84) Huppes and Schneider, 1994; International Standard Organization (ISO), 1998a; ISO, 2006b.
85) Klöpffer, 1996a; Ekvall and Tillman, 1997; Curran, 2007, 2008.

A very good example of system reduction is the NaOH production in the context of chlorine alkali electrolysis according to the amalgam process,[86] a typical coupled production (Equation 3.10):

$$2NaCl + E_{el} \rightarrow 2Na + Cl_2(gas) \qquad (3.10a)$$

$$2Na + 2H_2O \rightarrow 2NaOH + H_2(gas) \qquad (3.10b)$$

-----------------------

$$2NaCl + 2H_2O + E_{el} \rightarrow 2NaOH + Cl_2 + H_2 \qquad (3.10)$$

Commercially useful products are primarily chlorine gas ($Cl_2$) and sodium hydroxide (NaOH). Besides, hydrogen gas is formed which is mostly used thermally in the factory. The sodium is not isolated as Na-metal but dissolves to sodium amalgam in the liquid mercury electrode, which in a second stage decomposes into caustic soda solution (Equation 3.10b). This solution must be *concentrated* for transport and sale for which thermal energy is necessary, which can be unambiguously attributed to this production step. As the remaining energy demand is supplied by electricity, it is justified to assign the entire thermal energy (primary energy carriers are usually fossil fuels) to the production of the caustic soda solution alone. Electricity, however, must be assigned to the three products (NaOH, $Cl_2$ and $H_2$). The allocation problem is thus only partly solved by system reduction.

4. *Physical causation*: Scientific-technical arguments can be a reason for the allocation of environmental loads in defined sub processes. A frequent application is the assessment of emissions of an incineration plant, which are to be allocated to the assessed product to be burnt as waste. If the ingredients of the waste are known by type and quantity, its oxidation products can be ideally calculated according to chemical stoichiometry, and thereby a justified estimation of emissions in the exhaust air can be made. Because of different conditions in the incineration processes in reality compared to controlled oxidation processes of single substances, the calculation of toxic emissions as traces can only with great difficulty be allocated to individual wastes.[87] A detailed description of the difficulties with allocation of emissions to single waste materials as well as possible solutions can be found in (UBA, 2000, p. 81 ff., loc. cit).

Boundaries between (3) and (4) are fluent. The allocation of the thermal energy to NaOH as illustrated above has physical causes; it can, however, only be calculated following system reduction.

Only if steps (1) to (4) fail, a need for allocation rules is required according to ISO 14044, which finally can only be provided by agreement (convention) and whose application, case by case, must be justified and secured by sensitivity analysis (see Chapter 5).

---

86) Boustead, 1994b.
87) Tiedemann, 2000.

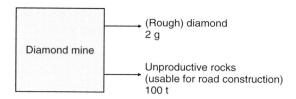

**Figure 3.13** Mass ratios in the unit process diamond mine.

5. Allocation per *mass*.

    This is the oldest allocation rule and most commonly used with multi-output processes (see Section 3.3.2.1). It appears to be scientific without being so in reality.

    As LCA is primarily based on mass flow analysis, the allocation proportional to mass lends itself as an arbitrary though simple and universal rule. Has the allocation problem thereby been solved? Unfortunately not. Limitations of the allocation per mass can best be demonstrated by the so-called *diamond paradox* (Figure 3.13). Typical exploitable concentrations for diamonds are around 10 ct (carat) per 100 t rocks. With 1 ct = 200 mg this results to 2 g diamond per 100 t rocks (0.02 ppm). There are also higher values, but lower ones are also workable.[88] In this fictitious example it is assumed that for the rocks a (small) revenue is made, for example, as material for road construction. In an allocation per mass nearly all of the environmental load would be allocated to rock material (unproductive rocks), which is obviously unreasonable. Nobody would operate complex mines just to extract material used as underground for roads! Therefore, the loads have to be allocated, largely or even exclusively, to the product diamond, for the exploitation of which the technical effort was made.

    A pragmatic solution to the problem would be to allocate the unproductive rocks as *waste* not disposed but with market value providing some revenue. Waste is not considered as co-product and thereby not allocated with loads. The question arises as to where the border line between waste and co-product is.

    An alternative and to date preferred[89] solution in such cases is the allocation on economic value:

6. Allocation on the *economic value* of the products, approximated by *price*. This allocation is also primarily an allocation per mass although averaged by economic value (measured by price). An allocation per price that can be obtained for the products A, B, C, ... certainly provides a solution to the diamond problem because the prices per mass differs by orders of magnitude. If in the example of Figure 3.10 product A costs 20 € kg$^{-1}$ (700 kg thus 14 000 €) and B costs 5 € kg$^{-1}$ (300 kg thus 1500 €) cost, the following weighting of masses per price results:

$$\text{Weighting factor A} = 700\,[\text{kg}] \times \frac{20\,[\text{€}\,\text{kg}^{-1}]}{15\,500\,[\text{€}]} = 0.903$$

---

88) Pohl, 1992.
89) Guinée et al., 2002; Guinée, Heijungs and Huppes, 2004.

$$\text{Weighting factor B} = 300\,[\text{kg}] \times \frac{5\,[\text{€}\,\text{kg}^{-1}]}{15\,500\,[\text{€}]} = 0.097$$

With this allocation thus approximately 90% instead of 70% (= unweighted mass-proportional allocation) is allocated to the product system A and only approximately 10% instead of 30% to product system B. The choice of allocation method thus changes the result of the LCA.

Problems with price-proportional allocations are often considerable with market-dependent price fluctuations. In order to adjust these, average values for longer periods (e.g. 10 a) or, more simply, for the reference year or reference period of the LCA (defined in Goal and Scope) should be formed. Furthermore, the geographical basis has to be defined (World market? European Union? OECD? On the basis of which currency?) This allocation is complicated by the fact that many prices are based on secret arrangements. Despite these difficulties the price-weighted allocation per mass represents the only universally applicable 'subjective rule'. It is, even though not scientifically verifiable, by no means arbitrary, because economic activities are usually accomplished in view of the most valuable product (extreme case: see diamond example) and not in order to produce by-products or waste. In modern economy the latter is avoided as far as possible, or again included into production by recycling.

7. Further proposals for allocation:

As further reference units for allocation, *molecular mass and calorific value* have been suggested. Both units are not universally applicable. They are, however, occasionally used for special applications. The calorific (heating) value is applied for an allocation of refinery products but because of very comparable calorific values of the co-products this leads to similar results as with an allocation per mass. The mol mass as basis for allocations is not suited for chemically badly defined mixtures. It was, for example, used in ecoprofiles of Plastics Europe (APME studies) for an allocation of chlorine alkali electrolysis[90] (see above allocation per mass). For a comparison of approximately 15 (!) further possibilities of allocations for the same process (some probably humorously meant) see Boustead (1994b).

Considering the fact that allocations cannot be conducted with total objectivity, in ISO 14044 a transparent reasoning is required for a deviation from scientific methods; this always results in an *avoidance* of allocations. In case of using subjective allocation methods, the conduct and discussion of at least one sensitivity analysis in the interpretation phase is mandatory (see Chapter 5). This is to evaluate the impact of the choice of the allocation method on the final result.

### 3.3.2.4 Further Approaches to the Allocation of Co-products

The above discussion shows a historically grown argumentation line which has predominantly developed from common practice in assessment and by International

---

90) Boustead, 1994b.

Standards.[91] Price-weighted allocation, uncommon in the early period of LCA, has been discussed, in particular, by Huppes.[92]

A completely different approach going back to matrix calculations for quantitative descriptions of product systems as developed in economic theory has been introduced by Heijungs.[93] The most important result is the occurrence of allocation problems in all cases where the matrix equations describing the system cannot be solved.[94] Using an example of CLR it can be shown by formal derivation that an allocation problem does not exist, in accordance with common practice and experience.

An overview by Curran of the allocation of co-products[95] that also refers to recycling and disposal, starts with the inventory and the distinction between foreground and background processes.[96] The former relate to the scope of the examined product system that can be directly influenced by the decision maker. For this, specific data are usually available or can easily be procured. Raw material acquisition, material production, and the supply of energy, transportations, and so on, are termed *background process*. For the latter, mostly generic data (see Section 3.4.3.1) that represent averages of many single processes are used. Foreground data can be investigated by their reaction to (small) changes in technology. A special case represents the modelling of a profound change of technology-mix called *discrete change*, which may follow fundamental changes in society.

### 3.3.2.5 System Expansion

The basic idea of system expansion and also the subsequent problems of very large systems have already been described in Section 3.3.2.1 (see Figures 3.11 and 3.12). The following example shows possible solutions for a product comparison if system expansion is applied. For reasons of clarity Figures 3.14–3.16 illustrate the production alone. Use and disposal are not integrated but of course have to be considered in the modelling of a system expansion.

System expansion is illustrated by a comparison of products A and C, where A is formed together with co-product B (Figure 3.14). The benefit of the systems 1 and 2 is not identical because in system 1 two useful products (A and B) are formed, whereas in system 2 only one product (C) is formed – the one to be compared with A. In order to achieve comparability, the fU is changed by an expansion to A + B and C + B. Due to the co-production of B in system 1, the entire and *separate* production of B must be added to C in system 2, yielding the same amount B as in system 1 (Figure 3.15). A production system for B has to be modelled as an equivalent system. The system boundary is the same as the one that has been

---

91) Heintz and Baisnée, 1992; Boustead, 1994b; International Standard Organization (ISO), 1998a; Frischknecht, 2000; Kim and Overcash, 2000; Werner and Richter, 2000; Ekvall and Finnveden, 2001; Guinée et al., 2002.
92) Huppes and Schneider, 1994; Guinée et al., 2002.
93) Heijungs, 1997, 2001; Heijungs and Suh, 2002.
94) Heijungs and Frischknecht, 1998.
95) Curran, 2007, 2008.
96) First in: SETAC-Europe, 1996.

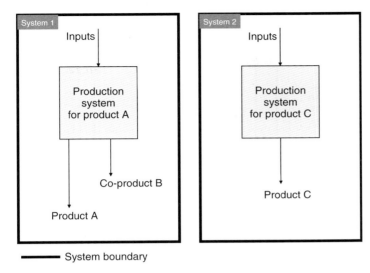

**Figure 3.14** System expansion with product comparison – both systems do not have an equal benefit.

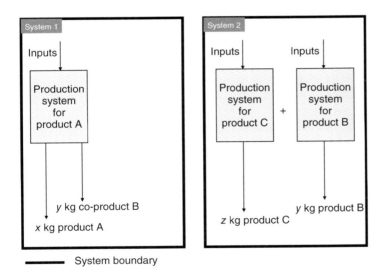

**Figure 3.15** Providing equal benefit by addition of an equivalent system.

assigned for A and C. If the full life cycle of A and C including use and disposal has to be considered this is also true for the equivalent system B.

If several production routes exist for B, the most common route, which may include some arbitrariness, should be applied for the assessment. If B can *only* be produced as co-product of A: bad luck! Should other co-products be formed in those

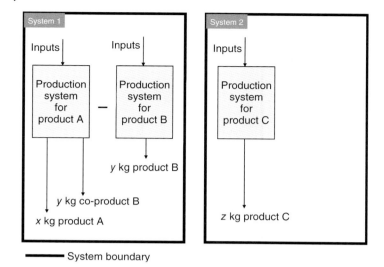

**Figure 3.16** Providing equal benefit by subtraction of an equivalent system.

other production systems of B, the system gets more and more complex. Thus even system expansions do not always provide unambiguous results and require substantially more data. If supplementary data cannot be procured they have to be estimated, another issue of uncertainty in the assessment.

As an alternative for an expansion of the fU, a comparison between A and C can be maintained if the environmental loads related to the production of B are *subtracted* from A (credit).

This so-called *avoided burden approach* can only figure as system expansion if the considered system boundaries of system 1 and system 2 are symmetrical. The advantage of this approach consists in the simple fU. Arbitrariness remains, however, concerning the choice of an equivalent production system for B and the data demand increases.

Ekvall and Weidema[97] distinguish between retrospective or attributional (classical)[98] and prospective or consequential LCAs where system modelling is based on market driven future scenarios. Some reasons are provided on why system expansions are more difficult in retrospective LCAs: in addition to the difficulties already mentioned it is a fact that in retrospective LCAs a representation of the status quo is given raising doubts on system expansions with alternative production lines. For prospective LCAs system expansions is exemplified as to be always recommendable and declared as the method of choice. Only the development of

---

97) Ekvall, 1999; Weidema, 2000; Weidema, Frees and Nielsen, 1999.
98) The designation retrospective for traditional or classical LCA with constant economical background is misleading, as in this form of assessment, by far the most predominant world-wide, comparisons with new products or products under development can be made.

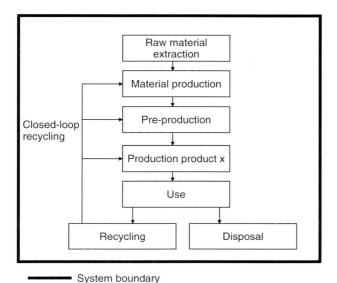

**Figure 3.17** Simplified production chain with closed-loop recycling (CLR).

the methodology and its common acceptance[99] will however show whether the implicit assumptions and related uncertainties as well as an increased data demand are justified. In any case the choice of LCA variant has to be justified by goal and scope of the investigation (see Chapter 2) and its uncertainties have to be discussed in the interpretation (Chapter 5).

### 3.3.3
### Allocation and Recycling in Closed-Loops and Re-use

Compared to the allocation of co-products, CLR is straightforward. In the simplest case a product enters the production chain of the same product again after use (Figure 3.17).

It is evident from Figure 3.17 that a 100% effective CLR of the final product

- makes its disposal unnecessary (ideal cycle);
- means that less raw material (not necessarily less energy!) is required.

In reality, however, a complete recycling is not possible. An extensively examined example concerns the refillable and returnable (strictly speaking: reused) bottles (refillable bottle, RB); the obtained savings depend on the *trippage rate* (*TR*), which indicates the average number of re-uses of the bottle. Although this figure is not easily procured good estimations are available. This is especially true for systems in steady state. The TR can amount to approximately 50 with well established multi-use systems (e.g. 0.5-l RB Euro beer bottle). Thereby, material dependent key

---

[99] So far this methodology seems to be used particularly in Scandinavia.

figures of the bottles decrease by 1/TR, for example, assuming an effective mass of 800 g (bottle weight) to 20 g per filling for a TR of 40! This is, of course, not true for expenditures that are incurred for each filling:

- Cleaning,
- Transport,
- Label and
- Closure.

In good multi-use systems or those with high TR, these factors have a much higher relative environmental impact. An optimum multi-use system is characterised by:

- standardised, stable containers,
- sale in crates,
- deposit system,
- decentralised supply, short transport distances and
- minimum environmental impact by the filling process, closure, label, and so on.

The first three measures aim at an increase of the TR, the latter two at a decrease of the impacts of the remaining processes. LCAs have contributed a lot to an understanding of these production systems and processes.[100] An often vigorous discussion on single versus multi-use packaging is thus not focussed on general principles but aims at the generalisation of results that were obtained for individual systems. For comparison, the system boundaries of the compared product systems and the data quality must be strictly examined.

The use of collected one-way containers in a new production after melting is also a part of CLR. Other examples of CLR include production scraps, for example, thermoplastics, glass and metals, which by melting can frequently be funnelled back into the production process (e.g. extrusion).

The treatment of CLR in LCI can be summarised as follows: in case of sufficient data no assumptions have to made, the processes can be derived by scientific-technical means. An allocation problem does not occur, because all processes take place within the system boundary. This is also the result of a formal deduction which has been briefly appreciated in Section 3.3.1.[101]

> **Exercise: Closed-loop recycling of production scraps**
>
> The figure shows the production of a formed steel sheet as a highly simplified flow chart. The main mass flow and the energy consumption are indicated. Calculate the amount of energy and iron ore per kg product that can be saved by a feedback of scraps into the converter. As simplification, treat slag as waste.

---

100) BUWAL, 1991; Hunt *et al.*, 1992; Günther and Holley, 1995; Schmitz, Oels and Tiedemann, 1995; Curran, 1996; BUWAL (1996, 1998); UBA, 2000; UBA, 2002.
101) Heijungs and Frischknecht, 1998.

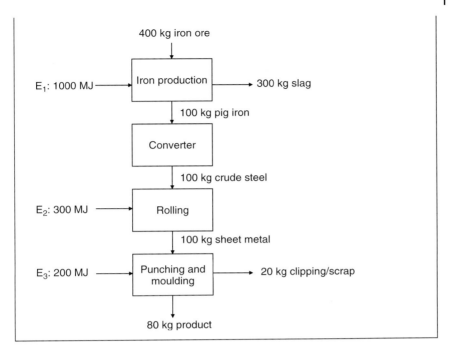

### 3.3.4
### Allocation and Recycling for Open-Loop Recycling (COLR)

#### 3.3.4.1 Definition of the Problem

OLR, contrary to CLR, represents a difficult case for allocations comparable to coupled production.[102] We first have a look at two systems A and B. Note that the originally separated systems A and B are now contained in a common system boundary and connected by a box 'collection, transport, upgrade' (Figure 3.18). The reason is that for the production in system B, the product in system A after its use phase can be (fully or partially) used as *secondary raw material*. A and B now form one system and the problem is on how environmental burdens and resource uses to the subsystems are to be allocated – one of which is being investigated.

By the small boxes (waste disposal of A) in subsystem A and (raw material extraction for B) in subsystem B it is suggested that a certain fraction of A despite the recycling has to be disposed as waste and a fraction from B has to be produced from primary raw material. Since product B is not recycled here – according to the model assumption – the process chain of B ends with the disposal. Generally of course, product B can also be employed for other products after use, collection, and so on.

---

102) SETAC, 1991; Hunt et al., 1992; Society of Environmental Toxicology and Chemistry – Europe, 1992; Society of Environmental Toxicology and Chemistry (SETAC), 1993; Curran, 1996, 2007, 2008; Klöpffer, 1996a; Ekvall and Tillman, 1997; International Standard Organization (ISO), 1998a; UBA, 2000; UBA, 2002; ISO, 2006b.

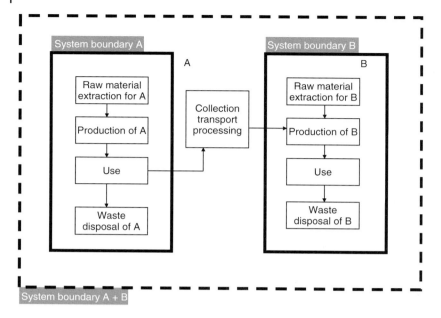

**Figure 3.18** Simplified presentation of two product systems with recycling in open-loop recycling – OLR.

The question concerning the 'right' allocation is: How are environmental advantages and disadvantages to be 'fairly' or 'suitably' allocated to the subsystems A and B (generally +C, D, ... )? The following environmental advantages apparently occur in this simple example:

1. Less waste accumulation in A (extreme case: no waste at all by use of product A).
2. Less primary raw material (resource) consumption in B (extreme case: no material resource consumption for production of product B).

A scientifically strict solution to the problem (indicated by the dotted framework in Figure 3.18) is *system expansion*, which within a simple A/B system still seems possible with justifiable effort. This system expansion is also recommended by ISO 14044.[103] It assumes however that system B is known in detail and data are available for an analysis of B. This for OLR is often just not the case! Besides, the benefit of the expanded system and thereby its fU must be newly defined. System expansion avoids allocation, but the price is often too high.

Typical secondary raw materials for OLR are:

- Waste paper and carton,
- Waste glass,
- Metal scrap and
- Thermoplastic polymers.

---

103) International Standard Organization (ISO), 1998a; ISO, 2006b; Curran, 2007, 2008.

For these secondary raw materials well-developed collecting systems and applications exist (but are not applied in all countries). A market for such materials has formed and is actually quite old (e.g. scrap). Some examples: Today in Germany newsprint, almost exclusively, and cardboard, to a large extent, are produced from waste paper or waste carton. Waste glass is added in a certain proportion for bottle production, and a high portion of scrap iron has for a long time been employed with some steel grades. It is nevertheless rarely known into which new product (B) or products (B, C, D, … ) the secondary raw material from the examined system is integrated. System expansion in these cases is thus *not feasible or only with very uncertain assumptions*. Usually only the product *group* is known, whereby the secondary raw material is preferably applied.

Allocation rules[104] present an alternative to system expansion.

### 3.3.4.2 Allocation per Equal Parts

The seemingly simplest and oldest rule is the so-called 1 : 1 – or 50 : 50 – Rule[105]

1. Waste avoidance in A gets a credit entry to even parts in systems A and B.
2. Raw material saving in B likewise gets a credit entry to even parts in systems A and B.

However, this arbitrary rule – but nevertheless regarded as just – also presupposes knowledge of both systems. The advantage to system expansion consists of the fact that allocation can be limited to specific process steps (see Figures 3.19–3.21). For a system expansion, however, the complete system B must be assessed.

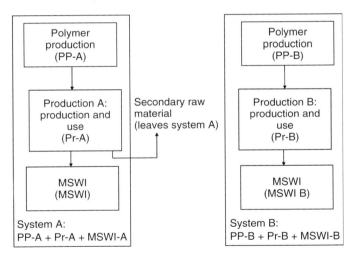

**Figure 3.19** Process pattern of noncoupled systems[106] (disposal here: municipal solid waste incineration – MSWI).

---

104) Huppes and Schneider, 1994; Klöpffer, 1996a; Ekvall and Tillman, 1997; Curran, 2007, 2008.
105) SETAC, 1991; EPA, 1993; Klöpffer, 1996a; UBA, 2002; EPA, 2006.
106) IFEU, 2006.

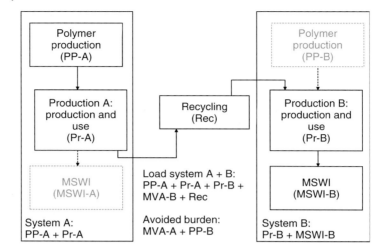

**Figure 3.20** Process pattern for coupled systems with system expansion[107] (disposal, here: MSWI).

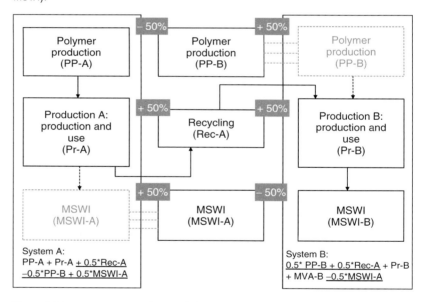

**Figure 3.21** Process pattern for coupled systems with allocation according to the 50:50 method[108] (MSWI disposal).

The 50:50-rule is 'fair' in that it rewards both the secondary raw material deliverer as well as the receiver. Reason: a recycling economy requires the cooperation of all people involved, a correct behaviour should thus also be apparent in LCAs and lead

---

107) IFEU, 2006.
108) IFEU, 2006.

to a behaviour which benefits the environment. In addition a distribution of loads and advantages in equal parts is intuitively judged as fair.

A 50:50 rule may, however, be less justified if an already flourishing market for the special secondary raw material of the examined system has been established, and no stimulation is needed for the provider of used material of A.

The sample study (IFEU, 2006, loc. cit) specifies the following allocation rules for an OLR in the context of production systems employing polymers.

In a first step the authors consider two independent systems (Figure 3.19).

If system expansion is made, all processes necessary for recycling (e.g. collection, transportation, sorting, processing) must also be assessed for the system A + B. Figure 3.20 shows the process pattern for coupled systems with system expansion. The fU must now refer to the complete system (system A + B).

Allocation rules have to be defined if the fU should refer to the individual systems. In the sample study the 50:50 rule was applied as illustrated in Figure 3.21. The following segments were defined:

- *Recycling*: Collection, transport and processing of product A, so that the material can be inserted into production of B: 50% of the environmental loads connected with the recycling are added to both systems.
- *Waste incineration of A*: If after use product A were not prepared as secondary raw material a waste incineration would follow the production and use phase of A. Since the environmental loads of the MSWI in system A can only be avoided by inserting secondary raw material into system B, system A is burdened with 50% of the MSWI environmental loads (+50%) while system B will be 'rewarded' by 50% (−50%) of the MSWI environmental loads as credit entry.
- *Polymer production (PP-B)*: If system B would not take in the secondary raw material of System A, the polymer for the production of product B would have to be manufactured from primary raw material. As this primary production of the polymer was fully assessed in system A (Figure 3.20) for 50% of the polymer production necessary in system B without recycling a credit entry is inserted into system A (−50% PP-B) and a debit entry of 50% is inserted into system B (+50%).

For an assessment of system A via 50:50 allocation, system B need not be completely considered. Therefore less data are required compared to system expansion. Besides, fUs can be defined separately for A and B which is required for most goal definitions in LCA.

In the 100:0 allocation, credit entries for secondary material are completely allocated to the delivering system. This variant is often used to examine the relevance of results of the allocation method (see Section 3.3.4.4 and Chapter 5).

### 3.3.4.3 Cut-off Rule

A further important rule (rule 2, *cut-off rule*)[109] provides exactly defined separation boundaries between systems A and B, for instance at garbage collection or separation, and the loads of the two systems are assessed independently. Based on the

---

109) Not to be confused with cut-off-criteria.

system expansion in Figure 3.20, the cut can be defined in the process 'recycling'. This has as a consequence:

1. In system A all loads related to waste disposal are avoided. (The recycled portion is not assigned as waste.) There are no environmental impacts in connection with waste disposal.
2. In system B all loads related to raw materials are avoided (the portion used as secondary raw material has no loads owing to raw material extraction, no resource consumption, etc.). These loads are completely allocated to system A.
3. The environmental loads for the recycling are allocated to the systems A and B according to defined rules.

In the sense of a cycle economy[110] the rule is fair, it rewards *both* systems. The avoidance of loads are different by nature, but do abide by logic:

- for A, recycling is actually waste avoidance;
- for B, the use of secondary raw material implies saving of primary raw material.

The behaviour is environmentally friendly if the special material supplied by A is scarce and the upgrading process always included in the recycling step (Rec.), see Figure 3.20, is not more demanding (energy use, emissions) than the production of virgin raw material for B.

The greatest advantage of the rule is the ability to make an analysis based on the knowledge of one system only (the one to be analysed). This can best be explained by the following:

After use product A is collected and up-graded; the transport to a waste separation plant is still part of A, as well as returned parts plus useless scraps, which can be rated as waste of A. All further sub processes are allocated to System B (transport, cleansing, upgrading if necessary, melting or granulation). This method was successfully applied by Holley and co-workers in an inventory on beverage packaging commissioned by UBA Berlin.[111] The secondary raw material is shifted beyond the system boundary of A, gets temporarily stored and is shifted back from there into system boundary B. Sorting and cleaning are divided between the two systems as already indicated. An exact demarcation of systems has to be specified for the individual case.

The application of the cut-off rule can be critical in cases of the carbon assessments or determination of *carbon footprints*: the entire $CO_2$ release from incineration, for example, plastics, will be allocated to system B. At the same time the carbon assessment of system A is no longer closed.

The resulting consequences of this simplification on the final results of the study have to be examined in the context of the sensitivity analysis (see Chapter 5).

---

110) There is no really good English equivalent for the German word 'Kreislaufwirtschaft'; a similar concept is called *Industrial ecology* which stresses the similarity of an environmental friendly business with natural ecosystems.
111) Günther and Holley, 1995; Schmitz *et al.*, 1995.

The treatment of secondary raw materials often implies environmental loads and must be examined individually. By no means is recycling a priori more environmentally friendly than an adequate waste disposal, for example, by incineration where the energy is used for steam production, heating, and/or power generation. Also *down cycling* usually relates to recycling as secondary raw materials (B, C, …) often prove to be of lower quality than the appropriate primary ones or they may have to be cleaned or otherwise improved with a surplus of energy and material.[112] It often makes sense to assess quality losses directly. Thus for the production of a defined cardboard quality often more waste paper fibre, compared to primary fibre, is needed. All these processes are accessible to LCA, which therefore can serve as a decision guideline.

#### 3.3.4.4 Overall Load to System B

A further rule (rule 3)[113] defines as load of B the environmental loads of primary raw materials, if B were exclusively produced from these. The delivering system will get the same amount as bonus, so that all loads will remain in B. In the sense of a cycle economy this appears unfair, because only the manufacturer of secondary material will be 'rewarded', not, however, the customer. Here again system B must be known in detail. If B also supplies secondary raw material it can be subtracted. There is no double counting. All in all: though rule 3 is mathematically correct, it is 'unfair' in the sense of circle economy, for the user of secondary raw materials. In the case of metals as a classical group of recyclable materials, there is already a healthy market of high quality secondary raw materials. It has been argued that in this case the actors of system A have to be motivated to provide their waste (scrap) for recycling[114] rather than the users. Rule 3 can contribute to such an incentive, if justified.

An optimised recycle factor regarding environmental aspects can be calculated according to an ecological *break-even point* by Fleischer.[115] According to advancement by Schmidt[116] it can also be dynamically calculated by the inclusion of increasing environmental burdens (in exceptional cases decreasing) due to increasing recycling rates into the calculation.

### 3.3.5
### Allocation within Waste-LCAs

A complete LCA is conducted 'cradle to grave', where the 'grave' is called *end-of-life phase (EOL)*. This phase can be implemented as recycling (CLR or OLR) or by the different conventional garbage disposal procedures mostly by waste incineration and disposal sites. An EOL phase is part of most LCAs except the product enters during use into an environmental medium. Thus, detergents reach the purification

---

112) Huppes and Schneider, 1994.
113) Fleischer, 1993; Klöpffer, 1996a.
114) Atherton, 2007.
115) Fleischer, 1993.
116) Schmidt, 1997.

plant with waste water and incineration gases of fuels or evaporated propellants of sprays are released into the air. In these cases that are better classified with consumption (rather than use) disposal does either not exist or will be delegated to the waste water sewage plant. Environmental loads due to use and consumption are within the system boundary.[117]

Two substantial questions need to be answered in the context of disposal in LCA.

1. *Modelling of waste disposal of a product* (Section 3.3.5.1). If a product becomes waste, depending upon the country, waste flows enter a disposal system. According to the determined average mass flows, the waste disposal is analysed and modelled by an LCA.
2. *Comparison of different waste disposal options* (Section 3.3.5.2). Different possibilities of waste disposal are to be compared with one another.

Both questions and their handling by system modelling are discussed in the following example 'Disposal of Cardboard Packaging'.

### 3.3.5.1 Modelling of Waste Disposal of a Product

Figure 3.22 presents schematically two disposal options for cardboard packaging: In this example 80% of the cardboard packaging is assumed to be recycled and 20% to be burnt in a waste incineration plant. In a specific LCA the different utilisation pathways common to the country are to be determined in the inventory phase. The primary benefit of both options is the disposal of waste cartons. Both variants have, however, an added value. In the first case (recycling) the added value means a saving of 70 kg cardboard production from raw materials, in the second case (waste incineration) an extra 14 MJ electricity and 80 MJ heat are obtained.

Added values must be considered in the assessment. For example, in the comparison of cardboard versus aluminium packaging different disposal pathways exist and thus different added values.

Concerning the system modelling there are two variants to what can occur to the 70 kg secondary material, 14 MJ electricity and 80 MJ heat (Figure 3.22):

- They can be reused for carton production in the same system. In this case it is a CLR (see Section 3.3.3).
- They can be used in other systems. Then it is an OLR (see Section 3.3.4). All decisions concerning the allocation of environmental loads of the distributing and the receiving system, which were discussed there, are to be made. If multiple receiving systems are involved, the examined complete system might become very large.

For a simplification of the examined system it is often treated as a closed-loop system without the secondary material, electricity or heat being really used by the same system. For a valuation of saved environmental loads equivalence systems are assessed which are treated as credit entry in the examined system. Hereby the technical equivalence has nevertheless to be verified carefully (see Figure 3.23).

---

117) Klöpffer, 1996b.

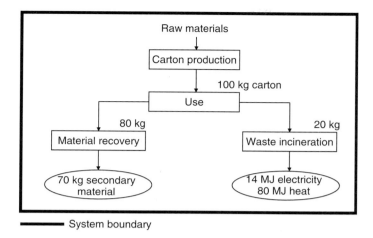

**Figure 3.22** Process pattern for waste carton disposal.

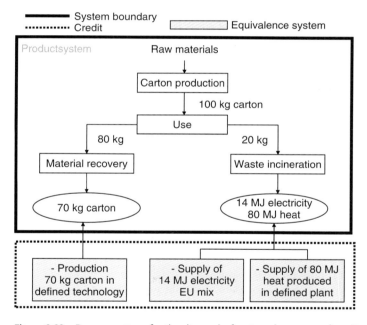

**Figure 3.23** Process pattern for the disposal of cartons by means of credit entries via equivalence systems.

- Since in the base system 70 kg cardboard are made available due to material recovery, production of 70 kg cardboard from raw materials can be avoided. Thus, the loads of the production of 70 kg cardboard from raw materials are accounted and subtracted from the loads of the base system (credit entry).

- Since the base system generates 14 MJ electricity by incineration, this amount does not need to be made available otherwise. Therefore the environmental loads for an (EU-mix) electricity supply of 14 MJ are balanced and credited to the base system.
- Likewise, 80 MJ heat result from incineration. Again an equivalent system has to be modelled, for example, the heat supply by incineration of light fuel oil in a defined plant. These environmental loads will be balanced and also credited to the base system.

The treatment of the outputs from waste treatment of products under consideration of OLR often leads to large systems that imply an elaborate modelling. Therefore a credit entry by means of system equivalence is widely applied.

In many LCAs the system modelling under consideration of equivalence systems (credit-entry method) is used in the same sense as 'system expansion' avoiding allocation. This is, however, not true; the system is rather treated as a 'quasi-closed system', as such simplified, and all links to other systems, unavoidable in OLR, are not considered.

### 3.3.5.2 Comparison of Different Options of Waste Disposal

LCAs are frequently employed to determine the environmentally most favourable option of waste disposal.[118] In many waste disposal technologies usable energy that can be used for the production of electricity and/or steam and hot water for district heating is generated.[119] These added values have to be considered in an LCA by credit entry or system expansion.[120] Added values may also occur for landfill by capturing of landfill gas and for mechanical-biological waste treatment, for example, by the production of agrarian gas.

The main idea of this system expansion ('basket of benefit approach') is the equal benefit of the compared disposal routes. It is a prerequisite to a comparison of systems, and by consequence the fU is mostly modified (inclusion of added values) (see Section 2.2.5.3).

Figure 3.24 shows the examined system if the goal definition of the LCA aims to find out which of the two systems, material recovery or incineration of waste cardboard, is environmentally more friendly. The fU of the system in this case is: 'Disposal of 100 kg cardboard'.

The system expansion conducted here is done according to the same logic as in the previously discussed case of co-products. While system expansion is seldom used for co-products, its application in the assessment of waste disposal systems is common place. Here, no reasonable alternatives exist and the systems mostly remain feasible.

Since a waste incineration with energy generation (system B) implies an added value (14 MJ electricity, 80 MJ thermal energy per functional unit), these amounts of energy have to be *added* to system A where no energy is formed. Assumptions

---

118) White, Franke and Hindle, 1995; Giegrich et al., 1999.
119) In summer spare energy can also be used for cooling, which unfortunately is rarely applied.
120) Fleischer, 1995; Fleischer and Schmidt, 1996.

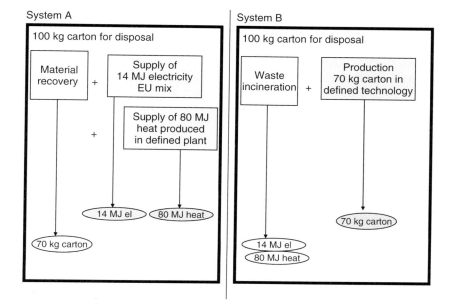

**Figure 3.24** Process pattern for a comparison of two disposal variants for cardboard ('basket of benefit').

must be made on the complementary processes. The fU can now no more be defined as the disposal of a specific mass waste but by disposal + supply of energy $x$, $y$, and so on, for example, per tonne waste. The 'winner' will be the system which provides the waste disposal *and* additional energy supply with the smallest environmental loads.[121]

### 3.3.6
### Summary on Allocation

Allocation can only partly be determined in a strictly objective, scientific mode, with the exception of CLR which is clearly defined. This can already be seen in the choice of words 'fair, just, unjust', which refers to scientifically not definable issues.

Science and the international standards[122] require: Avoid allocations, expand or reduce the system boundary, look for physical causes. That is it. These demands are often unrealistic and in the case of system expansion may result in unmanageable, extremely complex systems that can only be handled in large LCAs with national

---

121) With mixed waste, for example urban waste, the prehistory of the waste is not assessed. This is permitted, if only the ecologically most favourable kind of disposal is concerned.
122) International Standard Organization (ISO), 1998a; ISO, 2006b.

objectives and by accordingly generous means. The LCA on graphic papers[123] and on plastic recycling within Dual System Germany (DSD)[124],[125] in Germany can serve as examples for such 'large' LCAs. System reduction if applicable is undoubtedly preferable.

A solution for 'small' LCAs, which strive for a comparison of product systems, alternatives of disposal, or a system optimisation by classical means, can only be achieved by an application of allocation rules by *convention*. International standards are too vague to be as such regarded as convention. They primarily recommend transparency, comprehensibility, and sensitivity analyses; this means in the interpretation phase (see Chapter 5) the question shall be answered on how a specific choice of allocation rule impacts the final result.

At present, a solution by an accepted convention is not yet in view. Our recommendations in the case where system expansion or reduction for reasons either of price or practicability and usefulness are not possible are:

- mass-proportional allocation, if necessary including a weighting by means of prices of co-products;
- 50:50-rule (rule 1) or cut-off rule (rule 2) with OLR;
- other allocations with OLR following the goal definition and if necessary in coordination with the most important 'actors';
- credits in case of consideration of the waste management in product comparisons;
- basket of benefit method within the comparison of waste management technologies.

Examples of allocations according to ISO 14044 are listed in Chapter 7 of the technical report ISO TR 14049.[126] The avoidance of allocations by system expansion is also discussed on the basis of examples.

## 3.4
**Procurement, Origin and Quality of Data**

### 3.4.1
**Refining the System Flow Chart and Preparing Data Procurement**

Data are 'the alpha and omega' of an LCI. They concern in principle all inputs and outputs of unit processes that have been identified in phase 1 ('Goal & Scope') as necessary for an adequate description of the system(s). The starting point of the product system description is a careful analysis of the production processes starting with the extraction of the raw materials. An analysis of the transport processes and waste flows within selected geographical and temporal boundaries

---

123) Institut für Energie- und Umweltforschung Heidelberg GmbH (IFEU), 2000.
124) Part of the Environmental Recycle Act in Germany, 'Duales System Deutschland'.
125) Heyde and Kremer, 1999.
126) International Organization for Standardization (ISO), 2000.

should not be underestimated in relation to effort and significance to the final result.

How can it be done practically? With regard to 'material' products (goods), the product description is the first step; with respect to an 'immaterial' product (service), it is the flow of activities. In both cases upstream material flows and downstream waste flows are described.

For material products (goods), a *product description* indicates which materials in which quantities are used, referred, for example, to piece, mass or another meaningful unit. These data have to be procured in a way that is easy to convert to the reference flow according to the fU. To better understand the product system, both the inspection of the product as well as sketches and plans on the functioning of materials and manufacturing processes are useful and should be documented.

Knowledge of the materials and their share in the product may in some cases already allow a *first rough estimation* of the order of magnitude of environmental loads: should inventory data record for all included materials be available as generic data in a data base these can be aggregated according to their share in the product. An aggregation in the simplest case is the addition. For this first estimation the CED is suited as a parameter (see Section 3.2.2). As none of the life cycle phases of production, use, and disposal are considered in detail, and no specific data on transport are available, these calculations serve as an orientation for the orders of magnitude which can be expected, nothing more.

The next step would be a *system analysis with a differentiated elaboration of the system flow chart* and the cut-off of ancillary branches according to the cut-off criteria chosen. The unit processes identified are the basis for data procurement. All inputs and outputs to every unit process have to be identified. If a preliminary system analysis has already been conducted during the goal and scope definition, it has now to be expanded and refined. The result will be a differentiated system flow chart (see Section 3.7.3).

At this state of work a screening-LCA[127] might be worthwhile if a comprehensive LCA is intended.

### 3.4.2
### Procurement of Specific Data

It is rarely ever possible to procure all data as primary data, that means to gather specific data at specific plants for specific processes. Therefore a real LCI always consists of primary data, generic data and, where the one or other is not available, of estimations[128] (see Section 3.4.3.1). For all these data sets, the *documentation* of their origin and quality is essential, because comprehensibility and transparency are central requirements according to ISO 14040/44.[129]

---

127) Christiansen, 1997.
128) Bretz and Frankhauser, 1996.
129) European Commission, 2010; Sonnemann and Vigon, 2011.

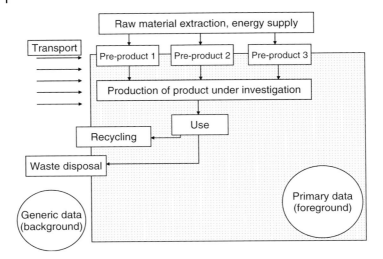

**Figure 3.25** Schematical data procurement.

To which extent primary data are available or can be procured depends substantially on whether or not the manufacturer of the examined product is integrated into the conduct of the LCA. If the manufacturer is the commissioner of the LCA, the data base of processes of the respective manufacturing plant will be very good. Missing primary data in this case can be accounted for with justifiable effort. Based on solid contacts to upstream suppliers of the manufacturer, the latter can usually persuade them to make data available. Close upstream processes therefore often imply good (foreground) data quality (see Section 3.7).

The same is true for CLR if it concerns processes within the manufacturing plant or the supplier. For waste disposal usually non-specific data are available and can be used unless the product requires specific techniques developed by the manufacturer (see Figure 3.25).

The availability of primary data often depends on the willingness of companies to procure data and make them available. Generally it is true that the data situation gets unspecified to the same extent as the unit process in the production chain is remote from the commissioner.

If an LCA is not commissioned by an enterprise but by an authority like an Environmental Agency, closest possible cooperation with enterprises (or trade associations) concerned with the production and disposal of the examined product results in improved data records.

The hatched part in Figure 3.25 corresponds to the 'foreground' of the inventory according to a SETAC Europe working group on LCI.[130] In literature several terms for data categorisation are used. Here, the terms foreground data and primary data are used synonymously. The same applies for background data and generic data (see Section 3.4.3.1).

---

130) SETAC-Europe, 1996.

> **Example: Determination of primary data 'collector's passion'**
>
> They still exist; the leading production operators checking electric meters and measuring instruments before entering the office; the environmental representatives collecting data before it became common practice to publish ecological reports (recently called *sustainability reports*). The development of LCAs, especially of LCIs, owes these passionate data collectors a lot. Here, within the enterprises, the interface between LCA and operational environmental assessment can be found. Primary data procured here are at the centre of every LCA. Without them LCAs would be a rather useless venture.
>
> Interestingly enough, the first 'proto-LCAs' were in the rarest cases commissioned by the upper management, but as a result of the activities of engaged employees, they served as the starting point of a broader commitment to the subject. Examples were listed in a broad based study on the implementation of LCA in industry.Frankl and Rubik, 2000. In the long run, however, the conduct of LCA can only be implemented if it will be part of the corporate identity and management.

Specific data records generally allow an improved spatial and temporal assignment of emissions and resource consumption, which also may have a stronger influence on future LCIAs (see Section 4.5).

Fairly easily procured specific data in an enterprise are the following:

- Demand and nature of material
- Used energy and forms of energy (heat, electricity, fuels)
- Co-products
- Production and nature of wastes
- Operating and ancillary materials
- Transportation, to or from and within the examined enterprise.

High quality data procured with a higher effort include:

- Emissions into the air (after filter)
- Emissions into water (after waste water purification)
- Contamination of soil and groundwater
- Use of pesticides and fertilisers (which substances? how much?)
- Data concerning ionising radiation, biological emissions and nuisances (noise, odour).

The emissions are usually measured and documented for other purposes (environmental legislations in most developed countries). Therefore, often only the sum or group parameters are measured and collected, for example, chemical oxygen demand (COD) or biological oxygen demand (BOD), sum of volatile organic compounds, adsorbable organic halogen compounds (AOX), and so on. Depending on the country of origin, procurable substances can be differently defined or measured, or vice versa, as parameters often directly depend on the measuring procedure.

SETAC's 'code of practice' provides extensive recommendations on how to proceed with these difficult cases.[132]

The required data are often present within a company or at suppliers but in a form useless for LCI. They have to be transformed or newly gathered in a way applicable for LCIs which implies a substantial effort.

Things improve if an operational input–output analysis of the plant or the production site has already been performed,[133] for example, as part of the operational environmental assessment within an environmental management system. Both methods complement each other, particularly if producer and supplier(s) as part of the supply chain co-operate (*chain management*).[134] Exaggerated secrecy, however, often prevents the confident cooperation of producer, supplier and client out of a fear of revealing production secrets and of financial cheating by a – though limited – disclosure of production costs and price structuring.

---

**Example: Data formats in the unit process 'Punching of Steel Sheet' (fictitious data)**

In an enterprise steel sheets are punched as moulded parts from a Coil. The simplest case is assumed: A company in Germany punches only one type, which in this form is transported to only one customer, who builds two of these sheet metals into one of his products P. The Coils originate from one supplier only. Available data could, for example, be conveyed as follows:

- *Input*
  a. electric energy: $5 \times 10^5$ kWh a$^{-1}$
  b. coils: 1000 t a$^{-1}$
  c. transportation distance: 100 km.
- *Output*
  a. product (P): $1.2 \times 10^6$ pieces a$^{-1}$
  b. scrap iron (blend): 40 t a$^{-1}$ (back to supplier)
  c. transport distance: 50 km.

The example shows that data of this format cannot be used directly for a calculation of an LCA of product P with the fU '1 piece of P'.

---

The transformation of operational data into unit process data which are applicable in an LCA will be shown by the electricity consumption of 'Punching of Steel Sheets' in the above example (see also Figure 3.27).

The following must be considered in detail:

---

132) Beaufort-Langeveld et al., 2003.
133) Braunschweig and Müller-Wenk, 1993; see the European environmental management system for enterprises and organizations EMAS and the international standard ISO 14001 (ISO, 2004); Finkbeiner et al., 1998.
134) Udo de Haes and De Snoo, 1996; a co-operation within the supply chain is also requested by the European Union's chemicals legislation REACH (2006).

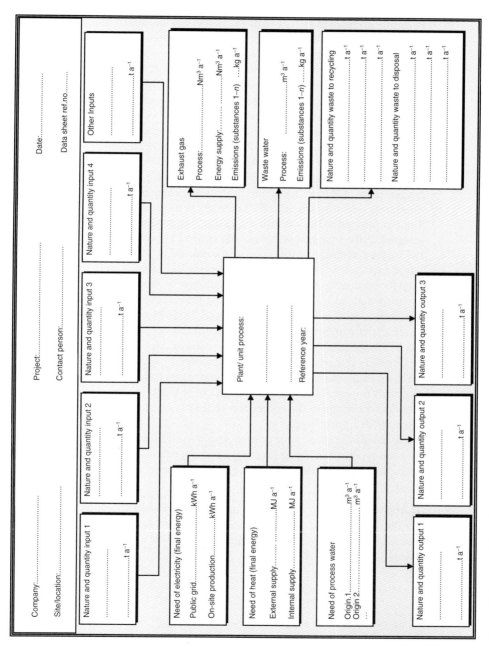

**Figure 3.26** General form for the structuring of the data collection (IFEU).

**Table 3.6** Operational data for the unit process 'Punching of Steel Sheets' (fictitious data).

| INPUT | Quantity | Unit | Factor | Quantity | Unit | Per piece | Unit |
|---|---|---|---|---|---|---|---|
| Electric energy | $5 \times 10^5$ | $kWh\,a^{-1}$ | $3.6\,MJ\,kWh^{-1}$ | $1.8 \times 10^6$ | $MJ_{el}\,a^{-1}$ | 1.5 | $MJ_{el}$ |

- For a procurement of primary data to defined unit processes, usually a questionnaire will be handed out to the company where inputs and output are registered (see Figure 3.26; also appendix A in ISO 14044 lists examples of data acquisition sheets). If the processes are well known to the data surveyors, the questionnaire can be refined more specifically, which usually increases the data. An important step in LCI is consequently a description as precise as possible of all processes for which primary data shall be gathered.
- Even if the data situation is better than those in the example above, the conveyed data usually have to be converted. For further use of the data in LCA it is common practice to convert those data to 1 kg of the product of the appropriate unit process or to a multiple of the reference flow according to the fU. In the above example not the mass of the product P but the number of pieces is indicated. Therefore the energy consumption in Table 3.6 relates to the number of pieces.
- The environmental loads due to the energy supply of the 1.5 $MJ_{el}$/piece must be assigned to the unit process 'Punching of Steel Sheets'. It is important to know the country of origin or provider of the electricity. In Germany, for example, 'Strommix Deutschland[135]' ('electricity mix': the average primary energy mix used for producing electricity) can be used (see Section 3.2.4).
- In many data bases and publications the generic data record to the electricity mix is available. It lists all environmental loads from raw material extraction to electricity supply at the customers site. Table 3.7 shows this data record on the left. Appendix B lists the complete standard report sheet to current mixes in Germany related to 1 kJ of electrical energy.[136] In some countries, as in the USA, no single electricity net exists. In such cases it should be known where the electricity is used so that the regional net can be identified and used; alternatively, a weighted average of the energy mix in such countries can be used. In Europe often the EU-mix is used.
- For a calculation of the unit process data 'Punching of Steel Sheets' the energy demand procured in the operational inventory (transformed in units of $MJ_{el}$) is multiplied with the electricity mix generic data record. The result is shown in Table 3.7 on the right.

If appropriate data are available upstream processes can thus be included, according to the principle of inclusion of an upstream electricity mix generic dataset. Unit processes can be seamlessly linked to one another with suitable software (see Section 3.4.3.3).

---

135) See Section 3.2.4.
136) UBA, 2000, Materialsammlung S. 179 ff.

## 3.4 Procurement, Origin and Quality of Data

**Table 3.7** Consideration of generic data record of the electricity grid in Germany in the unit process 'Punching of Steel Sheets'.

| Energy supply electricity grid Germany Functional unit: 1 kJ electrical energy | | | | | |
|---|---|---|---|---|---|
| Multiplier: 1.5 MJ electrical energy per piece | | | | Data for unit process | |
| Input | Quantity | Unit | Factor | Quantity | Unit |
| **Cumulated energy demand (CED)** | | | | | |
| CED (nuclear energy) | 1.08E+00 | kJ | 1.5 | 1.62E+00 | MJ per piece |
| CED (hydropower) | 6.07E–02 | kJ | 1.5 | 9.11E–02 | MJ per piece |
| CED (fossil total) | 2.28E+00 | kJ | 1.5 | 3.42E+00 | MJ per piece |
| CED (non-specific) | 6.48E–07 | kJ | 1.5 | 9.72E–07 | MJ per piece |
| **Raw materials in stores** | | | | | |
| **Source of energy** | | | | | |
| Natural gas | 5.57E–06 | kg | 1500 | 8.36E–03 | kg per piece |
| Oil | 1.45E–06 | kg | 1500 | 2.18E–03 | kg per piece |
| Brown coal | 1.08E–04 | kg | 1500 | 1.62E–01 | kg per piece |
| Hard coal | 3.70E–05 | kg | 1500 | 5.55E–02 | kg per piece |
| **Non-energy carriers** | | | | | |
| Limestone | 1.46E–06 | kg | 1500 | 2.19E–03 | kg per piece |
| **Water** | | | | | kg per piece |
| Cooling water | 6.93E–03 | kg | 1500 | 1.04E+01 | kg per piece |
| Water (process) | 1.49E–05 | kg | 1500 | 2.24E–02 | kg per piece |
| Sum CED | 3.42E+00 | kJ | | 5.13E+00 | MJ per piece |
| Sum mass flow | 7.10E–03 | kg | | 1.06E+01 | kg per piece |

| | | | | Data for unit process | |
|---|---|---|---|---|---|
| Output | Quantity | Unit | Factor | Quantity | Unit |
| **Wastes** | | | | | |
| **Wastes for disposal** | | | | | |
| Ash and slag | 7.42E–06 | kg | 1500 | 1.09E–02 | kg per piece |
| Sewage sludge | 1.10E–09 | kg | 1500 | 1.65E–06 | kg per piece |
| Hazardous waste | 9.40E–09 | kg | 1500 | 1.41E–05 | kg per piece |
| **Waste destined for recovery** | | | | | |
| Ash and slag | 4.12E–06 | kg | 1500 | 6.18E–03 | kg per piece |
| Wastes unspecified | 1.87E–08 | kg | 1500 | 2.81E–05 | kg per piece |
| **Emissions into air** | | | | | |
| Dust | 1.25E–07 | kg | 1500 | 1.88E–04 | kg per piece |
| **Inorganic compounds** | | | | | |
| Ammonia | 1.14E–09 | kg | 1500 | 1.71E–06 | kg per piece |
| Hydrogen chloride | 3.37E–08 | kg | 1500 | 5.06E–05 | kg per piece |
| Dinitrogen monoxide | 1.34E–09 | kg | 1500 | 2.01E–06 | kg per piece |
| Hydrogen fluoride | 4.65E–09 | kg | 1500 | 6.98E–06 | kg per piece |

*(continued overleaf)*

Table 3.7 (continued)

| Energy supply electricity grid Germany Functional unit: 1 kJ electrical energy  Multiplier: 1.5 MJ electrical energy per piece | | | | Data for unit process | |
|---|---|---|---|---|---|
| Output | Quantity | Unit | Factor | Quantity | Unit |
| Carbon dioxide, fossil | 2.11E–04 | kg | 1500 | 3.17E–01 | kg per piece |
| Carbon monoxide | 2.48E–08 | kg | 1500 | 3.72E–05 | kg per piece |
| NO$_x$ | 2.52E–07 | kg | 1500 | 3.78E–04 | kg per piece |
| Sulphur dioxide | 8.98E–07 | kg | 1500 | 1.35E–03 | kg per piece |
| **Metals** | | | | | |
| Arsenic | 9.52E–13 | kg | 1500 | 1.43E–09 | kg per piece |
| Cadmium | 2.76E–13 | kg | 1500 | 4.14E–10 | kg per piece |
| Chromium | 1.67E–12 | kg | 1500 | 2.51E–09 | kg per piece |
| Nickel | 1.63E–11 | kg | 1500 | 2.45E–08 | kg per piece |
| **VOC** | | | | | |
| Methane, fossil | 5.61E–07 | kg | 1500 | 8.42E–04 | kg per piece |
| Benzene | 6.33E–11 | kg | 1500 | 9.50E–08 | kg per piece |
| PCDD/PCDF | 8.90E–18 | kg | 1500 | 1.34E–14 | kg per piece |
| Non methane volatile organic compound (NMVOC), unspecific | 6.04E–09 | kg | 1500 | 9.06E–06 | kg per piece |
| **PAH** | | | | | |
| Benzo(a)pyrene | 3.56E–16 | kg | 1500 | 5.34E–13 | kg per piece |
| PAH without B(a)p | 1.78E–12 | kg | 1500 | 2.67E–09 | kg per piece |
| PAH, unspecific | 9.72E–14 | kg | 1500 | 1.46E–10 | kg per piece |
| VOC, unspecific | 3.18E–13 | kg | 1500 | 4.77E–10 | kg per piece |
| *Emissions into water* | | | | | |
| Inorganic salts | 1.37E–14 | kg | 1500 | 2.06E–11 | kg per piece |
| Nitrogen compounds as N | 2.47E–15 | kg | 1500 | 3.71E–12 | kg per piece |
| **Indicator parameters** | | | | | |
| AOX | 2.75E–18 | kg | 1500 | 4.13E–15 | kg per piece |
| BOD-5 | 5.49E–17 | kg | 1500 | 8.24E–14 | kg per piece |
| COD | 1.18E–15 | kg | 1500 | 1.77E–12 | kg per piece |
| *Secondary sources of energy* | | | | | |
| Electrical energy | 1.00E+00 | kJ | 1.5 | 1.50E+00 | MJ per piece |
| **Minerals** | | | | | |
| Gypsum (FGD derived from flue gas desulphurisation) | 2.64E–06 | kg | 1500 | 3.96E–03 | kg per piece |
| *Water* | | | | | |
| Waste water (cooling water) | 6.63E–03 | kg | 1500 | 9.95E+00 | kg per piece |
| Waste water (process) | 3.34E–06 | kg | 1500 | 5.01E–03 | kg per piece |
| Sum **electrical energy** | 1.00E–00 | kJ | | 1.50E+00 | MJ per piece |
| Sum mass flow | 6.86E–03 | kg | | 1.03E+01 | kg per piece |

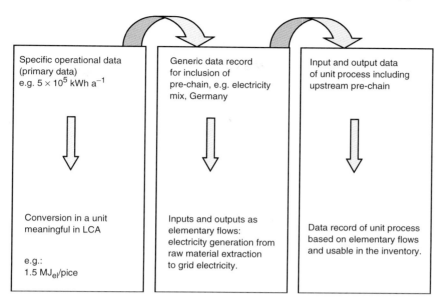

**Figure 3.27** From operational data to the unit process data record.

## 3.4.3
### Generic Data and Partial LCIs

#### 3.4.3.1  Which Data are 'Generic'?

The translation 'generic', a word of Greek origin, 'of general concern', indicates that specifics are not meant. They are averages or representative single values. An example: For each plastic production, for example, polyethene, crude oil is used, which is processed in a refinery. It would be completely inadequate if for each product made of plastic the primary data would have to be procured from the refineries again and again. Besides, because of trade flows, the refinery from which the molecules that were polymerised to specific kilograms of plastic originated is usually not traceable. Therefore it makes sense to determine average values of environmental loads for a type of refinery which manufactures pre-products for the plastics industry allocated to the appropriate monomer, for example, on ethene. The resulting data record would then be 'a generic' data record. For an estimate of its usefulness it is important to include an indication of the geographical region of the examined refineries and the time horizon of the data. In other words, the indication of the temporal and geographical system boundary of such data records is very important.

Similarly, the data record for the production of electrical energy (electricity mix Germany) listed in Table 3.7, is generic. It is based on the analysis of the technology mix of power plants applied for the production of electricity in 1996 (see data sheet in the Appendix B).

Thus, generic or background data are always based on the specific analysis of material and energy flows in defined plants. By building meaningful averages the data is processed to be usable in unit processes of an LCA. If average values cannot be calculated, carefully selected, representative single values can also be defined as generic data. In any case a careful inquiry is necessary on whether they are actually useful for the objective formulated in the LCA under study.

The use of generic data makes sense – even if some specific data were available – if the origin of a special raw material is not known or if it is not known in which factory a material or intermediate product was manufactured, and so on. Even if the information were available by a certain cut-off date it could have changed or be incomplete the next day, because one producer in the supply chain would have changed his suppliers or the oil would have another origin. These considerations are to show the fact that generic data are not a necessary evil but the only meaningful alternative for 'background' data. It is thus astonishing that the Dutch 'Handbook on Life Cycle Assessment – Operational Guide to the ISO Standards' recommends to avoid the use of generic data for detailed LCAs.[137]

Transports also are mostly calculated with generic data on fuel consumption and emissions. Distances, mode of transport and moved material, capacity utilisation, logistics, should on the other hand be specifically dealt with in the foreground (see Section 3.2.5).

Beside the actual raw materials there is a wide range of mass products, called *commodities*, which are bought on the market, if no long-term contracts cause permanent links of the producer to a few suppliers. Among others, the most important commodities are the most important metals, plastics, building materials and base chemicals. The long-term contracts mentioned move the suppliers to the foreground, whereas in case of rapidly changing suppliers, commodities and background processes have to be discussed and procured with appropriate generic data.

Generic data are indispensable for the conduct of LCIs. These data can be averaged unit process data or results of partial inventories (cradle-to- (factory-) gate) LCIs, which should be representative for a specific technology or region. The expression 'cradle to factory gate' indicates that these data represent no genuine LCIs ('from cradle to grave') but only a part of the life cycle. Partial LCIs as generic data are indispensable for a conduct of complete inventories and LCAs.

An approximate match of system boundaries, especially geographical ones, is a prerequisite for the correct application of generic data and their respective partial inventories. Raw materials, materials and chemicals sold worldwide should also exhibit world-mix data for a calculation of averages. Electricity is mostly supplied by the national grid (exports and imports are nearly balanced). If on the other hand production occurs somewhere in Europe, (without precise data), a European average is preferable (Section 3.2.4). The same is true for other large regions.

---

137) Guinée et al., 2002; Klöpffer, 2002.

### 3.4 Procurement, Origin and Quality of Data | 129

The most important applications for the use of generic data are

1. *Energy*
   a. Sources of fossil origin (natural gas, Diesel/light oil, fuel oil, gasoline, hard coal, brown coal (lignite)) starting from deposits
   b. Uranium ore starting from deposits, enrichment
   c. Primary energy-mix for electricity generation and transmission
2. *Transportation*
   a. Rail (electric, Diesel, mixed)
   b. Truck (different sizes if necessary)
   c. Passenger car
   d. Ship (open sea vessel, river boat)
   e. Air plane
   f. Pipeline
3. *Commodities*
   a. Metals (iron/steel, aluminium, copper, zinc, tin, ... )
   b. Plastics (low density polyethylene (LDPE), high density polyethylene (HDPE), polystyrene (PS), PVC, polyethylene terephthalate (PET), polyamide (PA), polyurethane (PU), ... )
   c. Building materials and – products (concrete, brick, insulating materials, floor covering, roof covering, ... )
   d. Packaging materials (paper and cardboard, glass, plastics, metals, ... )
4. *Chemicals*
   a. Tensides and builders
   b. Agrarian chemicals (pesticides, above all herbicides, fertilisers, ... )
   c. Large volume (base) chemicals such as sulphuric acid
   d. Solvents
   e. Plasticisers, for example, for PVC

Depending on financial funds, the references quoted in Sections 3.4.3.2 and 3.4.3.3 can be applied as *sources* for generic data.

#### 3.4.3.2  Reports, Publications, Web Sites
The most well-known are:

BUWAL (Swiss Federal Office for Environment, Forest and Landscape)[138] comprises inventories for packaging material (metals, plastics, glass, paper cardboard) and energy (UCPTE). This data base conceptualised for LCA of packagings have had an altogether big influence on the development of LCA. The same is true for the data base *Inventories for Energy Systems* of ETH Zurich (Chapter 1). An update of the BUWAL data has been accomplished[139] by ecoinvent (see below) and is therefore no longer free of charge.

---

138) Now Federal Agency for Environment Protection, Berne (BAFU); BUWAL, 1991; BUWAL, 1996, 1998.
139) Roland Hischier at *http://www.ecoinvent.org/fileadmin/documents/en/presentation˙papers/packaging˙DF˙eng.pdf.*

*APME*[140] (Association of Plastic Manufacturers in Europe): The leading European plastic data collection for the most important bulk plastics. Unfortunately those reports are only available in the form of summaries (highly aggregated data, small transparency). Important emissions, like, for example, monomer styrene and vinyl chloride, are not listed separately.

The data represent weighted average values from the most important European manufacturing plants and -procedures. For strongly varying production methods (e.g. emulsion and bulk polymerisation) data to individual procedures are separately available in short form.

*ECOSOL* (European LCI Surfactant Study Group with administrative support of the CEFIC/ECOSOL Sector Group). The complete study[141] – see also Section 1.5 published 1995 in the magazine 'Tenside, Surfactants, Detergents' – covers the most important tensides manufactured in Europe (most important operational area: surface-active content of detergents) and some important intermediate products, like alcohols, sulphur, soda and materials based on vegetable oils. The study was commissioned by US-American company Franklin Associates Ltd. (FAL) and submitted to a peer review Klöpffer, Grießhammer and Sundström (1995) and Klöpffer, Sundström and Grießhammer (1996). The data among others were integrated into a detergent study at the Öko-Institut e.V. Freiburg Grießhammer, Bunke and Gensch (1997) financed by the German Federal Environmental Agency. The study is supplemented by similar work on zeolites (phosphate replacement in detergents) and water glass Fawer (1996, 1997), which are other materials used in detergents. A new version of the ECOSOL data collection is being prepared by PE International for CEFIC (project ERASM-SLE).[142]

*ProBas* (process orientated base data for environmental management instruments) a web portal of the German Federal Environmental Agency, Dessau,[143] linked to a library of LCA data which is also, even primarily, suited for operational environmental assessment. However for the latter, self procured location-specific data should be used. Generic data, for example, for electricity, transportation, upstream materials, and so on, may only be used as supplement. ProBas has been updated 2012.[144] The CED-data base[145] is also accessible by ProBas.

*GEMIS* (total release model of integrated systems)[146] is an important data source for energy data with emphasis on the European Union. GEMIS was originally developed by the Ökoinstitut Darmstadt commissioned by the

---

140) Now: Plastic Europe Association. Boustead, 1992, 1993a,b, 1994a,b, 1995a,b, 1996a, 1997a,b,c; Boustead and Fawer, 1994; updated versions at *http://www.plasticseurope.org*.
141) Janzen, 1995; Stalmans *et al.*, 1995.
142) To be published 2014.
143) *http://www.probas.umweltbundesamt.de*.
144) *http://www.probas.umweltbundesamt.de* (13 July, 2012).
145) *http://www.oeko.de/service/CED*.
146) Fritsche *et al.* (1997); available at ProBas; GEMIS Austria: *http://www.umweltbundesamt.at/ueberus/products/gemis*.

Hessian Ministry for Environment, Youth, Family and Health Fritsche et al. (1997).

Newer data sources are the *network on life cycle data*[147] of the Research Centre Karlsruhe[148] and of the *European Platform on Life Cycle Assessment* at the Joint Research Centre Ispra of the European Union.[149] A goal of these programmes is to collect existing data records, which are already available in different organisations into a useful and uniform data format applied in LCA and material flow analysis and to provide these free of charge to professionals.

Further important data sources are technical encyclopaedias and product specifications, various Internet sites, and so on. The International Journal of Life Cycle Assessment (Springer) publishes regularly edited versions and case histories of LCAs and inventories, with data for special systems and regions. Further scientific magazines applicable as data sources are the Journal of Industrial Ecology (Wiley), Cleaner Production (Elsevier) and Integrated Environmental Assessment and Management – IEAM (SETAC press); see also Section 1.5.

### 3.4.3.3 Purchasable Data Bases and Software Systems

A by now unfortunately outdated compilation of available data sources by SPOLD (Society for the Promotion of LCA Development)[150] comprises, in addition to reports, also commercial databases. Further critical comparative discussions of altogether approximately 50 products (worldwide) are listed.[151] Many databases are integrated into LCA software systems. These are, however, often not original databases but are supplied by original data surveys, like, for example, ecoinvent. No effort is made here to clarify the complex context.

The most well-known European products of this kind are

- The *Boustead Model* (UK), developed and licensed by Boustead Consulting Ltd., is the result of a collecting activity for many decades by an LCA pioneer[152] and is considered as an extensive LCI data collection. Critical attention must be paid to the age of the data, even if current updates are presumed.
- *ecoinvent* (CH) is the result of a national effort for many years in Switzerland, which led to a qualitatively leading and (possibly worldwide) outstanding LCI database at present in Europe.[153] The data consider both Switzerland and Europe, which makes them applicable both nationally as well as internationally provided it is put to correct use. The software integrated into the purchasable product also permits the conduct of impact assessments according to different standard methods (see also Chapter 4). Detailed additional information which

---

147) German: Netzwerk Lebenszyklusdaten.
148) Bauer, Buchgeister and Schebek, 2004; http://www.netzwerk-lebenszyklusdaten.de.
149) http://lca.jrc.ec.europa.eu.
150) Hemming, 1995.
151) Vigon, 1996; Rice et al., 1997; Siegenthaler et al., 1997.
152) Boustead and Hancock, 1979; Boustead, 1996b; http://www.boustead-consulting.co.uk.
153) ecoinvent 1, 2004a; ecoinvent 2, 2004b; ecoinvent 3, 2004c; Frischknecht et al., 2005; http://www.ecoinvent.ch (Swiss Centre for Life Cycle Inventories); http://www.ecoinvent.org/database; version 3 (2013).

(unfortunately) cannot be found within the freely accessible reports also belongs to the product. The most recent version appeared in 2013.[154]

- *GaBi* (University of Stuttgart and PE International, DE) is based on fundamental work on LCA at the University of Stuttgart.[155] This database is particularly highly recommended by the engineering sector and has, from the beginning, considered the interests of the auto mobile industry, its suppliers and its producers of raw material. GaBi is both a database as well as an LCA software.
- *SimaPro* (Pré Consultants, NL)[156] is the most distributed LCA software worldwide. It also implements extensive databases. Which reached a new dimension by the cooperation with ecoinvent. SimaPro is also known for a specific impact assessment ('eco indicator'),[157] which, however, according to ISO 14040, can only be applied for internal use if comparative assertions are concerned.
- *Umberto* (Ifu, DE)[158] was developed by 'ifu Hamburg – material flows and software' in cooperation with the Institute for energy and environmental research in Heidelberg (IFEU), and is primarily a software for flows of material, energy and mass analyses plus price calculations and is also well suited for LCAs. Umberto also implements a database for standard unit processes.

### 3.4.4
**Estimations**

The most unpleasant experience for an LCA practitioner is the absence of any data for a material, a construction element, a chemical, or an agricultural process, and if the information needed cannot or will not be provided by the industry. In this case there are two possibilities:

- omission of the process or
- introduction of estimations.

The latter is preferable despite of all uncertainties[159], because otherwise – if not an insignificant sub process is concerned – the less examined system would always do better (the missing cooperation of manufacturers would be rewarded; offence of the symmetry principle). Possible ways of estimation are with:

- Older data or data from other geographical areas (other system boundaries),
- Data of chemically similar compounds, materials, and so on,
- Estimations based on information in technical manuals.[160]

---

154) A supplement issue of International Journal of Life Cycle Assessment about the most recent version 3 of ecoinvent is in preparation (2013).
155) University Stuttgart, IKB and PE International; Eyrer, 1996; Spatari et al., 2001; http://www.gabi-software.de.
156) http://www.pre.nl/software.
157) Goedkoop, 1995; Goedkoop et al., 1998.
158) http://www.umberto.de.
159) Fleischer, 1993; Hunt et al., 1998.
160) Bretz and Frankhauser, 1996.

In particular, chemicals are candidates for estimations[161] because of their very large number (in the European Union more than 100 000 substances are registered according to EINECS and REACH). Data procured by estimation have to be tagged in the inventory and have to be discussed during the interpretation phase. ISO standards 14040 and 14044 contain several strict regulations for the documentation of data quality and possible effects of estimations on the final results (see Chapter 5).

### 3.4.5
### Data Quality and Documentation

One danger in the use of electronic data bases is the difficulty to reconstruct underlying assumptions of the data generation. In other words, a judgement of their quality,[162] (original data, average values, estimation, etc) and on their suitability for a specific study is difficult. Even original data may contain measuring and allocation errors that have to be documented. Therefore, commissioned by SPOLD, a uniform data format was developed, to facilitate the electronic data exchange on the Internet.[163] It was conceived as a network of users and providers of databases for the exchange of data in the same format. It has been pointed out that, as a start, the SPOLD format is a transfer format, which may not yet necessarily represent the best structure for databases. Recent versions of SPOLD are being used in modern data collections (see below).

LCA data records can only with difficulty be evaluated statistically. This was already shown on a SETAC Workshop on data quality in Life Cycle Assessment in Wintergreen, Virginia, USA. As previously illustrated, the procurement of suitable data is a central problem for LCA. Different options for the procurement such as those discussed in Section 3.4.4, imply highly varying data quality, which usually cannot be represented by indicating average values and mean deviations. These representations are practically only suited for original measurements on single unit processes. Many of the data used in the inventory can be generic data or have already been weighted, averaged, aggregated, and include allocations and cut-off rules applied for their procurement with little transparency to the user.

As the reliability of LCA-results depends considerably on the quality of input data, questions of quality have been frequently discussed in recent years.[164] It cannot yet be foreseen, which data quality model will be generally accepted (if a one-fits-all solution is possible at all). It seems to be certain that a transparent description of data origin and certain quality criteria remain an important issue in data quality management. This requires a uniform data format, which was first postulated by the SPOLD workgroup 'Promoting Sound Practices'. The paper format converted

---

161) Bretz and Frankhauser, 1996; Geisler, Hofstetter and Hungerbühler, 2004.
162) Fava et al., 1994.
163) Singhofen et al., 1996; Hindle and de Oude, 1996; Bretz, 1998; *http://www.spold.org*.
164) Fava et al., 1994; Chevalier and Le Téno, 1996; De Smet and Stalmans, 1996; Kennedy, Montgomery and Quay, 1996; Coulon et al., 1997; Fernandez and Le Téno, 1997; Kennedy et al., 1997; Huijbregts et al., 2001; Ross, Evans and Webber, 2002; Beaufort-Langeveld et al., 2003; Ciroth, Fleischer and Steinbach, 2004; Sonnemann and Vigon, 2011.

into an electronic format can be downloaded from the Internet and may be used for developments within the framework *general public licence*.[165] Meanwhile SPOLD[166] has terminated its activities, the name SPOLD, however, is used in SPOLD data formats. Documentation and update of these is done by the consulting firm 2.-0 LCA Consultants (DK). At present downloadable 'SPOLD Data Exchange Software' is a development of the original 'SPOLD Format Software' in 1997.[167] It also uses nomenclature[168] recommended by SETAC in its 'Code of Life Cycle Practice'. The SPOLD format has also been adapted by ecoinvent as 'ecoSpold data format'.[169] Data collection and handling has been dealt with in the 'Shonan guidelines' by UNEP/SETAC.[170]

At around the same time as SPOLD the Swedish data format SPINE was developed.[171] In contrast to SPOLD as data communication format, SPINE was conceived as a database format, as pointed out by Weidema.[172] By consequence several attempts were made to merge the two concepts into one. In the Technical Guideline ISO 14048[173] an attempt was made to develop a general data documentation format on the basis of existing approaches as recommendation to LCA practitioners and data providers. However, the problem could not be completely solved. An *Open Source Software*[174] provides a converter for compatibility issues of different data formats.

## 3.5
### Data Aggregation and Units

The simplest data aggregation is the addition of homogeneous inputs and outputs. In this case the data are standardised in such a way that all unit processes refer to the selected fU and the reference flow. This is very simply in principle and is done by the PC more or less automatically (e.g. with a spread-sheet programme like Excel or with the help of commercial software systems; see Section 3.4.3.3). Difficulties usually arise if data from different sources are to be used in one LCI; it has to be carefully considered which data, for example, different waste categories, are equivalent. In cases of doubt, it is urgently recommended to consult the SETAC Code of LCI Practice[175] and the global 'Shonan Guiding Principles'.[176]

It appears trivial to point out that only data with the same unit have to be added. There is, however, no international convention concerning units to be used in

---

165) http://lca-net.com/spold/download/index.
166) Hindle and de Oude, 1996; Bretz, 1998.
167) Weidema, Bo: SPOLD '99 format – at electronics DATA format for exchange of LCI DATA (1999.06.24), 35 sides with four appendices, download at http://www.spold.org.
168) Beaufort-Langeveld et al., 2003.
169) http://www.ecoinvent.org/de/ecospold-data-format/.
170) Sonnemann and Vigon, 2011.
171) Carlson, Löfgren and Steen, 1995; Arvidsson, Carlson and Pålsson, 1999.
172) Weidema, 1998.
173) ISO/TC 207/SC 5, 2002; Beaufort-Langeveld et al., 2003.
174) Ciroth, 2007.
175) Beaufort-Langeveld et al., 2003.
176) Sonnemann and Vigon, 2011.

LCIs. The praxis in The International Journal of Life Cycle Assessment allows only SI (Système International) units as well as their multiples and a few other units authorized by SI, so kilowatt hour, day, hour, minute (ISO 1000,[177] DIN 1301).[178] As such, besides the use of Joule and its multiples (in LCI mostly as MJ), also kilowatt hour (1 kWh ≡ 3.6 MJ) is allowed to distinguish electrical energy (mostly given in kilowatt hour) from other forms of energy. Inadmissible for this journal is the use of obsolete US-American units, for example, the BTU (British Thermal Unit), calorie, pound, gallon and miles.[179]

Heijungs[180] shows by examples how important it is to comply with the rules, in LCA also. SI also regulates the designations of units and the prefixes. Heijungs further points to the problem of dimensionless quantity (physicists say, such a quantity is of dimension 1). These are usually quotients, for example, the 'dimensionless Henry coefficient'; here, an additional note or a clear designation helps, like 'air/water-distribution coefficient'.

---

**Box 3.1 Units**

According to The International Journal of Life Cycle Assessment, Instructions for authors[181].

Only use metric units (SI) and some other units specified in ISO 1000. The Système International d'unités, the modern variant of the meter convention, comprises the following fundamental units: meter, kilograms, second, ampere, Kelvin, mol and candela and a set of derived units (e.g. Newton for force, Joule for energy, Watt for power and Pascal for pressure). A complete list is part of ISO 1000/ISO 80 000.

Examples of permitted additional units, which do not belong to SI, are hectare (= $10^4$ m$^2$), symbol ha; litre (= dm$^3$), symbol: l or L; day (= 86 400 s), symbol d; hour (= 3600 s), symbol h; minute (= 60 s), symbol min; kilometre per hour (km h$^{-1}$); metric ton (= 1000 kg, Mg), symbol t; Watt hour and multiples, like kWh, MWh and GWh (in LCIs *only* to differentiate electricity from other forms of energy). Furthermore, decibel (dB); mol per litre (mol dm$^{-3}$), symbol mol l$^{-1}$ or mol L$^{-1}$; electron volt (eV) and its multiples, like keV, MeV and GeV (only common for the energy of elementary particles and photons).

However, degree Celsius, symbol °C, defined by 0 °C ≡ 273.15 Kelvin (K) *belongs to SI*.

---

The data sets related to a unit process or a phase of the life cycle *are not to be lost* due to the aggregation of data over the entire life cycle. This is important for the detection – and possibly improvement of – those unit processes or life cycle

---

177) ISO 1000 is now part of ISO 80 000 (2009).
178) Deutsche Normen, 1978a,b; International Standard Organization (ISO), 1998b.
179) Such units can, of course, be used in LCA studies of regional importance and for audiences unaware of metric/SI units.
180) Heijungs, 2005.
181) *http://www.springer.com/environment/journal/11367*.

phases that cause high burdens. This is done in the context of the so-called sectoral analysis in the interpretation phase (see Chapter 5).

In the LCI, weighting or valuating aggregations are to be avoided. Examples were the 'critical volumes' according to BUWAL[182] (see Section 4.2). These aggregation types should rather serve as early forms of the LCIA.[183] To the same category belongs the controversial aggregation per mass (sum of all mass movements, inclusive of spoils) called *Mass Intensity per Service Unit* (MIPS). This method proposed by Friedrich Schmidt-Bleek[184] is based, similar to the CED, on the search for a simple, universal measure for the ecological loads (similar to money in economics) emanating from a product. Given the allowed equation 'service unit = functional unit', the MIPS method can essentially be seen (apart from aggregation) as a LCI. Besides, MIPS cut off small mass flows and therefore the common trace emissions, according to the motto 'megatons, not nanograms'. It is, however, erroneous to assume that MIPS can be determined easier than, for example, CED, because a sound inventory (even without trace emissions that are not included in MIPS) is always necessary. Of course, all problems discussed above as those of allocation, data quality, and so on, also apply to the MIPS method, just as with an LCA-inventory. 'MIPS' is only easier if the LCI is conducted with low quality.

In comparison to CED, however, MIPS has the advantage of a higher clarity (mass vs energy) captured in the colloquial expression 'ecological backpack' (= MIPS). It designates, for example, how many tons (rock) have to be moved to produce 1 g catalyst from noble metals. This clarity cannot usually be achieved with energy units unless by use of (non-standard) coal equivalents (ce)[185] in tons (tce) or kg (kgce) standard coal; this is an energy unit, disguised as mass (see Section 3.2.3.1).

Neither the MIPS method nor a mere calculation of CED provides LCA-results according to ISO 14040/44.

## 3.6
### Presentation of Inventory Results

The representation of an inventory analysis with its plethora of single data is often a tightrope walk between transparency and legibility. To ensure both, a main part with few tables and pictures can be supplemented by an appendix or a material volume (or a CD), containing all original data and intermediate results.

In an *LCI study*,[186] the Goal and Scope definition including all assumptions, fUs, system boundaries, and so on, must not be missing. Furthermore, in this case an interpretation with a discussion must follow. Above all, it should be illustrated whether data quality and statements show a reasonable relationship. This is achieved by sensitivity analyses (see Chapter 5).

---

182) BUWAL, 1991.
183) Klöpffer, 1994, 1995; Klöpffer and Renner, 1995.
184) Schmidt-Bleek, 1993, 1994; remember the famous motto: 'Megatons, not nanograms'.
185) 1 kgce = 29.3 MJ.
186) The acronym LCI stands according to ISO 14040 (§ 3.3) for 'Life Cycle Inventory analysis', the second phase of a full LCA. It is also used sometimes for 'Life Cycle Inventory study', a truncated LCA without the third phase, LCIA.

## 3.7
## Illustration of the Inventory Phase by an Example

Many of the following procedures are integrated into LCA software tools in one form or other. Graphical user interfaces for the modelling of systems and definition of the arithmetic rules are common practice to date. This fact does not, however, release the practitioners from carefully conducting all steps, as otherwise faults occurring during software operation may not be identified. With regard to the content, the following items must be considered:

1. Detailed description of the examined product system (Section 3.7.1): Which materials in the product system in which quantities and related to the fU are to be considered? Which waste flows and which transportation? The fU and the respective reference flows that were determined in the phase 'definition of goal and scope', should again be critically checked and adapted if necessary.
2. Analysis of the manufacturing processes, waste utilisation and disposal technologies and other relevant processes of the product system (Section 3.7.2): Procurement of primary data and ensuring the availability of generic data. Transfer of the operationally procured primary data into data records of unit processes. Characterisation of the data records concerning origin, quality and system boundaries (technically, geographically and temporarily). Ensuring that the data relevant for impact assessment have been procured as inputs and outputs of the unit processes: impact categories and associated indicators have already been defined in the phase 'definition of goal and scope' (they will be extensively discussed in Chapter 4).
3. Elaboration of a differentiated system flow chart including reference flows (Section 3.7.3): In this step the sizes of the unit processes can be specified. If generic data for complete life cycle sections are available in appropriate quality, for example, the production of LDPE from raw material to LDPE pellets, the production of LDPE pellets can be treated as a single unit process (see Section 3.1.3).
4. Allocation rules (Section 3.7.4):
   In an LCA, allocation rules, according to ISO 14040/44, are already defined in the phase 'definition of goal and scope'. As, for a better understanding, 'allocation' is discussed in Section 3.3, the specifications of the sample study are presented here. They are:
   a. Definition of allocation rules on process level for multi-output and multi-input processes;
   b. Definition of allocation rules on system level for OLR;
   c. Definition of the rules for obtaining equality of benefit during waste treatment.
5. Modelling of the system (Section 3.7.5):
   Definition of the arithmetic rules with consideration of allocation rules by which the unit processes are to be interlaced.
6. Calculation of the inventory (Section 3.7.6):

The phases of the inventory will be illustrated as follows based on the system variants: 1-l-beverage carton and 1-l-PET bottle for juice and juice nectar of the sample study.[187]

### 3.7.1
### Differentiated Description of the Examined Product Systems

On the basis of the first system analysis in the phase 'definition of goal and scope', differentiated product descriptions are provided in the inventory. The following questions have to be examined:

1. Which materials in which quantities related to the fU are to be considered in the product system?
2. Which data are available concerning mass flows with regard to disposal or recycling and waste treatment after use of the product?
3. Which transportations have to be considered?

#### 3.7.1.1  Materials in the Product System

The composition of the packaging systems selected for illustration purposes is listed in Table 3.8. From these material listings the reference flows for all materials result relative to the defined fU. These reference flows are the basis to model the production from raw material to disposal, and to finally quantify the elementary flows. Material listings can be quite elaborate. This is demonstrated by the following descriptions (IFEU, 2006, loc. cit):

> *Beverage Carton*
>
> For the German market in the year 2005, the average material composition of beverage cartons was separately deduced for the three examined filling goods and the individual box sizes. For this, packaging data were procured from three manufacturers (Tetra Pak, SIG Combibloc and Elopak), and a weighted average of market shares of the individual packaging types was formed. The packaging tares listed in the tables are thus generic values and do not (necessarily) represent real packagings in trade.
>
> Data on weight share of the dry printing ink of the composite material are available only for one of the three manufacturers. Because of an overall proportion of only approximately 0.5%, the cut-off criterion is effective for this material flow. The production of the printing ink is not considered in this study.

---

187) IFEU, 2006 (Quotations from the study).

## PET Bottle

For a deduction of representative packaging specifications, contrary to cartons, no reference to market data for an average calculation was possible as access to the necessary data was missing. The multiplicity of preform and bottle manufacturers makes an estimate of the total market even more difficult. The selection of PET one-way bottles examined here therefore relates to bottle types which were used in the year 2005 in Germany and their market share was regarded as representative in the respective range of application.

Bottle weights and packaging specifications listed in Table 3.8 are based on information of beverage manufacturers as well as on market knowledge of companies represented in the project advisory board. The validity of the selected calculations was verified additionally by test purchases.

**Table 3.8** Specifications of the assessed packaging systems for juice and juice nectar: 1-l-cartons with closure and 1-l-PET-bottle IFEU (2006).

| Storage packaging: beverage carton with aluminium layer | Juice/nectar 1 l with closure | Storage packaging: PET bottle | Juice/nectar 1 l |
|---|---|---|---|
| **Primary packaging** | **31.50 g** | **Primary packaging** | **43.1 g** |
| Composite, thereof: | 28.84 g | Bottle (95% PET; 5% PA) | 38.0 g |
| Raw cardboard | 21.37 g | | |
| LDPE | 5.89 g | Closure (HDPE) | 3.3 g |
| Aluminium | 1.45 g | Label (paper) | 1.8 g |
| Imprint | 0.13 g | Bottle type | Multilayer/PA fraction 5% |
| | | Bottle colour | Clear/brown |
| Closure (HDPE) | 2.66 g | | |
| *Secondary packaging* | | *Secondary packaging* | |
| Tray (corrugated cardboard) | 128 g | Shrink foil (LDPE) | 10.03 g |
| *Transportation packaging* | | *Transportation packaging* | |
| Pallet (Euro-pallet, wood) | 24 000 g | Pallet (Euro-pallet, wood) | 24 000 g |
| Pallet foil (LDPE) | 280 g | Pallet foil (LDPE) | 480 g |
| | | Intermediate layer per pallet (corrugated cardboard) | 4 × 475 g |
| *Pallet configuration* | | *Pallet configuration* | |
| Cartons per tray | 12 | Bottles per shrink pack | 6 |
| Tray per layer | 12 | Packs per layer | 26 |
| Layers per pallet | 5 | Layers per pallet | 5 |
| Cartons per pallet | 720 | Bottles per pallet | 780 |

For the fillings of juice and juice nectar in 1-l-bottles multilayer bottles were examined. Typically one plastic layer with higher barrier characteristics (hindering gases like oxygen to pass the bottle wall) is enclosed by two PET layers. Injection of the barrier layers occurs during injection moulding of the bottle preform (co-injection). PA is mostly used as barrier material for multilayer bottles, other substances like ethyl-vinyl alcohol (EVOH) are also possible. Clear transparent, brown dyed bottles are employed. No information on type and quantity of colouring materials is available. For this study it is assumed that the bottles consist of PET and PA only. Furthermore it is not known whether scavenger materials are contained in the screw-type cap, therefore only polyethylene (HDPE) is considered here. The label adhesive abiding the cut-off criterion is not regarded here.

On the basis of the fU (packaging necessary for the supply of 1000 l filling material to the point-of-sale; see Section 2.3.2) and with the help of the data in Table 3.8, reference flows for all materials are known. Thus, for the variant carton packaging, for example 1.45 kg aluminium must be traced back to the raw material and considered in the phase of disposal. On the basis of the pallet pattern, the transportation expenses of the variants can be calculated (see Section 3.2.5).

### 3.7.1.2 Mass Flows of the Product after Use Phase

For the modelling of the *EOL* phase, the detailed knowledge of waste flows of a product after its use is necessary. Both utilisation ratios and the type of utilisation – material or thermal – must be known. In the sample study the following statements for the examined 1-l vessels are made, which also indicate the research routes and the research details (Figure 3.27, Table 3.11):

*Beverage Carton*

After use, beverage cartons are partly retraced by collection and waste recovery structures of the DSD.[188] The goal of material recovery is the production of paper fibres to be used for the production of packagings.

'Official' recycling rates in Germany according to the Packaging Ordinance[189] have been, for the last few years, constantly around 65% in relation to the total carton put into circulation.[190] This ratio has been used as the modelling basis in the study.[191] Moreover, the remaining 35% of the not materially recycled cartons are treated thermally.

In a sorting plant beverage composite cartons are 'positively' sorted and allotted to a target fraction. Positive recognition plus sorting result in a sorting rate of approximately 90%, the remaining 10% are treated thermally

---

188) Duales System Deutschland GmbH: *www.gruener-punkt.de*; waste management company.
189) The Packaging Ordinance (Verpackungsverordnung/VerpackV is intended to avoid the impact of packaging waste on the environment or to reduce it.
190) See FKN homepage at *www.getraenkekarton.de*.
191) IFEU, 2004a (unpublished); Only after conclusion of LCA calculations, an 'official' utilisation ratio was published for the year 2005. According to data of the FKN it its about 66%.

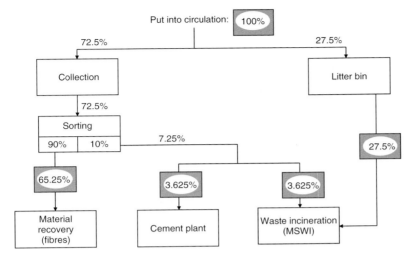

**Figure 3.28** Material flow of carton packaging after use.

together with other remainders. The collection rate of used cartons is thereby the recycling rate divided by sorting rate amounting to around 72.5% of sold packagings (see Table 3.9 and Figure 3.28). Non-retraced beverage cartons (27.5%) are burnt in the waste incineration plant (MSWI).

**Table 3.9** Collection and recycling of used 1-l-beverage cartons.

| Recycling of beverage cartons | 1000 ml |
|---|---|
| Collection rate | 72.5% |
| Deviation from average collection rate[a] | — |
| Sorting rate[b] | 90% |
| **Recycling rate** | **65%**[c] |

[a] Assumptions by IFEU (by consent in the project advisory group).
[b] DSD.
[c] FKN (provisional average utilisation ratio for 2005, as in previous years).

*PET Bottles*

PET bottles regarded here were not pledged in the year of reference 2005 and partially retraced after use by the dual system for recycling. In the year 2005 approximately 58% of collected bottles were recycled as PET flakes (material recovery) which are used for the production of fibres or tapes. The remaining bottles are recovered thermally or as raw material. The collection rate of PET bottles according to DSD is about 80%.

To date, according to DSD only transparent bottles are suited for material recycling.[192] Coloured bottles can, however, be recovered as material if they are transparent. The PA portion of multilayer bottles is generally no obstacle for recycling.

The material recycling rate results from the following material flows: 80% of the PET bottles are collected separately by DSD, 20% of the remaining bottles are dumped in the litter bin of households (these are collected by the municipal waste disposal system). An 80% of DSD collected bottles are distributed in the sorting plants as follows: PET bottles 58%, mixed plastics 27% and other remainders 15%. The PET bottle fraction is subject to re-sorting with a yield of 97%. This results in a recycling rate (referred to the total quantity) of 45% (see Table 3.10 and Figure 3.29).

### 3.7.1.3 Handling of Sorting Residues and Mixed Plastics Fraction

Sorting residues of by DSD collected beverage cartons and PET bottles in 2005 according to information provided by DSD to equal parts have been

- disposed as waste (MSWI) or
- used in cement plants.

Mixed plastics are recovered thermally (incineration with energy recovery) or as raw material. The mix of recovery is as follows:

- Cement plant (70%),
- Secondary raw material recycling plant (gasification to methanol) (~15%),
- In a blast furnace as fuel(~15%).

The unit processes considered for recycling are presented in Section 3.7.2.

**Table 3.10** Collection and recycling of used PET bottles.

| Recycling of PET bottles | Juice 1000 ml |
|---|---|
| | Transparent |
| Collection rate | 80%[a] |
| Deviation from average collection rate[b] | — |
| sorting rate (PET fraction)[a] | 58% |
| **Recycling rate**[c] | 45% |

[a] DSD 2005.
[b] Assumptions by IFEU (by consent of the project advisory group).
[c] Consideration of further preparation losses.

---

192) Report Mrs. Bremerstein (DSD), 29.06.06.

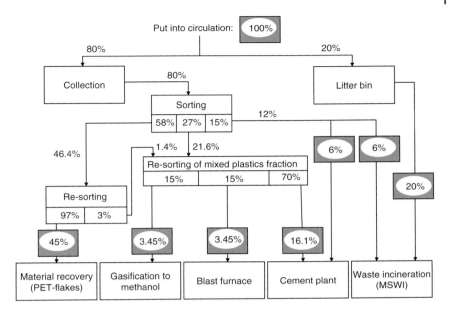

**Figure 3.29** Material flow of PET bottles after use.

### 3.7.1.4 Recovery of Transport Packaging
For transport packaging, recovery is based on:

> LDPE foil: 90% material recovery and 10% thermal recovery (MSWI, Municipal Solid Waste Incineration)
> Corrugated cardboard: 95% material recovery and 5% thermal recovery (Municipal Waste Incineration Plant, MWIP).

## 3.7.2 Analysis of Production, Recovery Technologies and Other Relevant Processes of the Production System

### 3.7.2.1 Production Procedures of the Materials
For the definition of unit processes a differentiated analysis of production procedures starting from the raw materials is necessary. In the course of this analysis it is useful to check the data availability: are useful generic data available or is the gathering of primary data necessary? It is clear from the examples that data records that have been gathered in earlier studies as primary data can be used as generic data in a subsequent study. For the production of materials, no primary data were procured in this study. The quality of available generic data has been judged as being applicable in the study.

In the following the data origins are summarised.

| Material/product | Data basis |
|---|---|
| Raw carton production | For the modelling of the raw carton production *four site-specific data records of the companies Stora Enso, Assi Domän, and Korsnäs are consulted. They represent production at three Swedish and one Finnish location in the year 2002.* The data procurement took place in the framework of an earlier LCA on behalf the FKN IFEU (2004). From the available data an average value for the raw cardboard production is formed, used for all examined beverage carton types |
| | For the calculation of the expenditure for the wood supply information from UBA, 2000a[a]: Material Volume I; Industrial Wood (forest NORTH) 1 is consulted. The data cover forest management, including logging and transport of wood |
| | The transport of raw carton to the composite production in Germany is likewise considered |
| Aluminium foil production | For the production of cardboard composite material exclusively primary aluminium is used. The basis of the consulted LCA data is provided by ecoprofiles which were published in the year 2000 by the European Aluminium Association (EAA), Brussels EAA (2000) |
| | The data record for primary aluminium covers the production of aluminium on the basis of bauxite extraction, aluminium oxide and aluminium production (aluminium ingots) including anode production and electrolysis. *The data are based on collections of the European Aluminium Association (EAA) in the years 1995 and 1998. A representativeness of 92% respective 98% of the European primary aluminium production was obtained* |
| | The data record, according to EAA, was partly updated on the basis of an survey of member firms in the year 2002. This updated data record was provided European Aluminium Association (2006) by the EAA for the use in LCAs. Statements concerning the representativeness of the updated data are not yet available |
| | *The data records for aluminium foils are likewise based on collections of the Association in the year 1998 for the production of semi-finished aluminium. Depending on the product group representativeness between 20 and 70% was obtained* |
| LDPE production | LDPE is manufactured in a high pressure process and contains a high number of long sidechains |
| | The data record covers production of LDPE granulates starting with the extraction of raw materials from deposits including associated processes. *The data refer to a period around 1999. They were procured from a total of 27 polymerisation plants.* The plants considered cover an annual production of 4 480 000 tons. The European total production in 1999 was approx. 4 790 000 tons. *The data thus represent 93.5% of Western European LDPE production* PlasticsEurope (2005a) |
| HDPE production | HDPE is manufactured in different low pressure processes and contains less side chains than the LDPE |

| Material/product | Data basis |
|---|---|
| | The data record covers production of HDPE granulates starting with the extraction of raw materials from deposits including associated processes. *The data refer to a period around 1999. They were procured from a total of 24 polymerisation plants.* The plants considered cover an annual production of 3 870 000 tons. The European total production in 1999 was approximately 4 310 000 tons. *The data thus represent 89.7% of Western European HDPE production* PlasticsEurope (2005b) |
| PP production | Polypropylene results from catalytic polymerisation of unsaturated propylene to long chained polypropylene. The two most important procedures are low pressure-, precipitation- and gas- phase-polymerisation. In a concluding step the polymer powder is processed to granulates by extrusion The data record covers production of PP granulate starting by the extraction of raw materials from deposits including associated processes. *The data refer to a period around 1999. They were procured from a total of 28 polymerisation plants.* The plants considered cover an annual production of 5 690 000 tons. The European total production in 1999 was approximately 7 400 000 tons. *The data thus represent 76.9% of Western European PP production* PlasticsEurope (2005c) |
| PA (Nylon 66) production | Nylon is either produced by direct polymerisation of amino acids or by reaction of a diamine with a two proton acid. Nylon 66 is formed in the reaction of hexamethylenediamine and adipic acid<br>The data record covers production of PA 66 starting with the extraction of raw materials from deposits including associated processes. *The data refer to a period around 1996* PlasticsEurope (2005d). *Information on plants considered and the representativity of the data records is not available* |
| PET production | Data on the primary production of PET are of special importance for this study. *The PET data record used here was compiled on behalf of PETCORE* IFEU (2004) (PET of container recycling Europe) *in the context of an LCA.* Data on the production of PET granulates are so far not yet published; from the point of view of the authors greater transparency of this data record is a strong argument for its use compared to the corresponding data record of PlasticsEurope. (The specific environmental impact for the eligibility of PET would be higher by use of the PlasticEurope data as with application of the newer data record described here.)<br>The following manufacturing process is at its basis: |

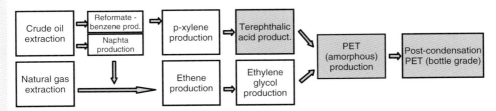

(Another manufacturing process by dimethyl terephthalate (DMT) however is rarely used in Europe, only applied in older, smaller plants and thus not relevant in the context.)

| Material/product | Data basis |
|---|---|
| | To the modelling of data for the production of pre-products naphtha, reformate gasoline, and *p*-xylene, a refinery model was applied, developed *at the IFEU institute*, which reflects the technology used at present in Europe |
| | The data basis for the modelling of ethene production from natural gas and naphtha are data available from the IFEU of *various European crackers in an almost representative distribution*. The data originate from the years 2000 and 2001 |
| | The data on ethyl glycol production characterise a *cross section of European plants*, which refers to the years 1997–2000 |
| | The data on terephthalic acid production are based on data from *European Plants, procured by Professor Rieckmann (FH Cologne) converted by the IFEU according to the requirements of an ISO-conformal LCA* and used |
| | The data record on PET production in bottle-grade is based on the production capacity of five European plants and was arranged by Professor Rieckmann (FH Cologne). The data refer to the period 2002/2003. *Approximately 36% of the European PET production is covered by the PET data record.* In view of the differences concerning selected plants and PET qualities a representative overview of European bottle-PET- production can be concluded |

[a]Umweltbundesamt, 2000a.

### 3.7.2.2 Production by Materials

| Material/product | Data basis |
|---|---|
| Composite cardboard production | The main component of the beverage carton composite is the raw carton. It is sealed by a layer of LDPE on the outside and inside. Depending on production printing takes place before or after this first coating process. For longer durable products, like juice and ice tea, additionally a thin aluminium foil is inserted on the inside for a protection from light and oxygen. It is further separated from the contents by a thin LDPE-layer |
| | The *data on the production of the composite were procured in the context of an LCA on juice.*[a] *They represent production at German Locations of the two largest manufacturers in the period 2002/2003.* These data allow a differentiated view on individual composite types used for different packaging volumes. In each case they contain data on energy and water consumption, emissions, wastes, and packaging material for the dispatch. According to information from the enterprises these data are essentially also valid for the year of reference of this study |

## 3.7 Illustration of the Inventory Phase by an Example

| Material/product | Data basis |
|---|---|
| | The cardboard composite is cut to a given format and transported to the bottler. For transport the cuts are either wrapped or stacked and packed into cartons. Only on the bottling machine the beverage carton is finally formed and sealed |
| PET bottles production | The production of PET bottles usually takes place in two stages, that is, first so-called preforms are manufactured out of dried PET granulates and these in a second step are converted into bottles. Power requirement of the preform and the bottle production is usually limited to the use of electricity for the drying process, the heating and shaping |
| | The energy consumption for the injection moulding of preforms correlates with preform weight. This study assumes a specific constant requirement of electricity of 2.6 kJ $g^{-1}$ |
| | In the context of this study no energy consumption for the production of the examined bottles by stretch blow moulding (SBM) could be procured. *Therefore current data from other projects of the IFEU Institute were taken.* The energy consumption for the SBM process is determined among other things by bottle volume. For the 1000-ml-bottle requirement of electricity of 22.2 kWh/1000 piece is assumed |
| Corrugated cardboard and corrugated cardboard tray production | In this LCA data records of the FEFCO. published in the year 2003 for the production of corrugated base paper and corrugated cardboard packaging were used |
| | The specific data records for the production of 'kraftliner' (predominantly from primary fibres), 'test liners' and 'corrugating medium' (both from waste paper) as well as for the corrugated cardboard packaging were taken. *The data records represent weighted average values in the data procurement of the FEFCO of various European location. They refer to the production of the year 2002. The representativity of the data records ranges from 20% (111 companies) for corrugated cardboard and trays, to >70% for kraftliner* |
| | In corrugated cardboard a portion of fresh fibres are frequently applied for reasons of stability. In the European average this portion amounts to 24% (FEFCO, 2003, loc. cit). For lack of more specific data this split was also presumed in this study |
| Beverage filling | Similar processes can be assigned to the filling of beverage cartons and PET single-use bottles. Cold aseptic filling of fruit juice is the standard technique for both packaging types. Power requirement and the packaging rejects at the appropriate filling lines are also comparable. The filling process of the examined single-use systems is rather subordinated in the LCA as long as the thermal treatment of the beverage is not considered |
| | *For the filling of beverage cartons the data from an LCA on juice, made available by one of the largest German fruit juice manufacturers, were taken* (IFEU 2004, loc. cit). The data on electricity, compressed air, steam, and water consumption were determined by a very high level of detail for individual equipment components, based on measurements of existing plants. Besides the actual filling of the beverage cartons the data also cover shaping of the cartons, sealing, application of the drain and loading of customer pallets |

| Material/product | Data basis |
|---|---|
| | For the filling of PET bottles data for comparable systems were used, which are present at the IFEU institute in context of other projects (IFEU, 2004a, loc. cit)<br>Underlying calculations were transferred to the beverage companies represented in the project advisory group |

[a] IFEU, 2004a.

### 3.7.2.3 Distribution

| Material/product | Data basis |
|---|---|
| Beverage distribution | No information on an average distribution distance is available for the filling materials considered. Collecting and deriving of more up-to-date and representative data for beverage distribution requires a much higher effort and was not the subject of this study<br>As a workaround, the distribution model from UBA-II for the filling type 'Beverages without $CO_2$ of the segment stockpiling purchase (>0.5 l)' was adapted for all barrels considered.[a] There, the distances and applied motor vehicle types (semi-trailer 40 t and 28–32 t, truck with trailer 40 t, truck up to 23 t, truck up to 16.5 t, delivery van) for a two-stage distribution were determined. The average transportation distance is therefore approximately 350 km |

[a] UBA, 2000.

### 3.7.2.4 Collection and Sorting of Used Packaging

The following data basis was used for the collection and sorting of used packaging:

| Material/product | Data basis |
|---|---|
| Collection | The unit process describes the collection and the transport of municipal waste. In this unit process the direct *and* indirect emissions are calculated; that is, the upstream processes of fuel production are included. Calculated consumption and emissions always refer to the weight of the waste, which is regarded<br>The data record for emissions is based on standard emission data, which are arranged, updated and evaluated in the model TREMOD for the Federal Environmental Agency Berlin and the Federal Agency for Environment Protection BUWAL Bern.[a] The original exhaust gas measuring data are |

| Material/product | Data basis |
|---|---|
| | from the TÜV Rhineland. All factors consider appropriate vehicle mix and if necessary the proportion of accountable driven distance |
| Sorting plant | The *data bases for sorting plant is based on DSD researches*. Besides, general prerequisites concerning the type of sorting plant are distinguished if known (e.g. manual sorting, semi-automatic sorting, etc.) |
| | A part of collected transparent PET bottles as well as the majority of coloured, non-transparent bottles are retrieved as mixed plastic in the sorting plant. In a subsequent treatment it is first cut up, cleaned by air classification and melted in a thermal reactor (agglomerator) and pelleted. The energy consumption was set to be 1.188 MJ t$^{-1}$ input The *data go back[b] to those of the[c] and are identical to those of a plant operator which were made available to the contractors* |
| | Since a dry processing is concerned, no direct waste water is released. Data to exhaust air or waste water emissions were not available |

[a] INFRAS, 2004.
[b] UBA, 2001.
[c] **DKR** (DKR:*German*: Deutsche Gesellschaft fur Kunstoff Recycling mbH German society for plastic recycling ltd.).

#### 3.7.2.5 Recovery Technologies (Recycling)

Quite an effort might be necessary for the data acquisition of recovery/recycling processes, if different technologies at different locations with realistic mass flows have to be considered and no generic data are published. In many cases it is necessary to procure data from the industry. These data often must be treated confidentially, that is, they may be used for the inventory calculation of an LCA, but not explicitly be published as data records. The handling of confidential data is a very sensitive area in LCA (see also Section 5.5).

*Recycling of Used Beverage Cartons*

Following sorting, beverage cartons are present as material fraction. The composite material is then processed in two German and one Finnish plants.

The German locations transfer rejects for recovery in cement plants. There the PE portion serves as fuel for the replacement of hard coal, and aluminium serves as replacement for the aggregate bauxite.

Twenty percentage of composite carton collected in Germany was recovered in the Finnish plant in the year 2005. There the reject is used in the nearby Ecogas plant.

The following databases were considered in the unit processes:

| Material/product | Data basis |
|---|---|
| Fibre recovery | The principal purpose of the processing is the recovery of the fibres dissolved from the remaining compound, so-called rejects, by swelling in a pulper. The reject (also contains drains and closures) serves as input to processes cement plant or Ecogas plant. *The fibres are usually processed in the same paper factory into final products. The data basis consists of confidential data of German and Finnish plants* |
| Ecogas plant | In the Ecogas plant the polyethylene separated from the composite is brought into the gaseous phase and burned for energy production. A part is used to fuel the Ecogas plant. The surplus is used in the attached paper plant. According to data of the plant operators about 85% of applied aluminium are recovered and used as material. *The data basis consists of confidential data of the plant operator* |
| Cement plant | The utilisation of rejects in the cement plant was converted as co-incineration process in the system model. It was assumed that the plastics parts replace hard coal during the incineration process and aluminium parts replace the additional material bauxite. *The data basis consists of confidential data of the cement industry* |
| Waste incineration (MSWI) | Used beverage cartons are disposed. Here, a waste incineration is assumed (MSWI). *The waste incineration model applies to a technical standard of the European Union guideline (European Union Incineration Directive – Council Directive 2000/76/EC).* In the model a grate-firing with steam turbine followed by purification of exhaust gas is put at the basis. The energy contained in the waste can partly be recovered (here 10% as electricity and 30% as thermal energy) |

*Recycling of PET Bottles*

By processing PET flakes are retrieved as products. At present the recovery plants for PET bottles in Europe differ by age and procedure details, yet most plants exhibit the same process sequences.

| Material/product | Data basis |
|---|---|
| Material recover: PET flakes | *The process data used in this study were compiled and validated in the context of the study on behalf of PETCORE.*[a] For the processing of multilayer bottles according to information of the DSD a decrease of PET yield around 2% is assumed |

| Material/product | Data basis |
|---|---|
| Cement plant | Recover of remainders and mixed plastics in the cement plant was implemented in the system model as a simultaneous incineration process. It was assumed that the used mixed plastics replace hard coal during the burning process. *The database consists of confidential data from the cement industry* |
| Blast furnace | The blast furnace process was converted in the system model *based on confidential data of the steel industry.* It was assumed that used mixed plastics in the blast furnace replace heavy fuel oil as reducing agents |
| Raw material recovery (plastic gasification to methanol) | A smaller part of the mixed plastics was used as raw material in the secondary raw material utilisation centre SVZ in Cottbus. *The assessment was done on basis of UBA[b] and IVV[c]-Data.* As product methanol is obtained, with a credit entry for the production of methanol from mineral oil |
| Waste incineration (MSWI) | Non-used PET bottles are disposed. Here, a waste incineration is assumed (MSWI). *The waste incineration model applies to a technical standard of the European Union guideline (European Union Incineration Directive – Council Directive 2000/76/EC).* In the model a grate-firing with steam turbine followed by purification of exhaust gas is put as the basis. The energy contained in the waste can partly be recovered (here 10% as electricity and 30% as thermal energy) The database consists of confidential data from the waste incineration plants |

[a] IFEU, 2004b.
[b] UBA, 2000b.
[c] IVV, 2001.

### 3.7.2.6 Recycling of Transport Packagings

The following databases were considered in the unit processes (unit process MWIP, see above):

| Material/product | Data basis |
|---|---|
| Material recovery: PE granulate | Used PE transportation packaging is cleaned and shred/milled afterwards. The PE granulate is reused in plastics industry again and replaces new PE granulates. *The databases of* (Plinke et al., 2000, *loc. cit*) *were taken* |
| Material recovery: corrugated cardboard | Used corrugated cardboard transportation packaging is sorted and reused in the production of waste paper-based corrugated base paper. *The databases for the waste paper sorting of* (Plinke et al., 2000, *loc. cit*) *were taken* |

### 3.7.2.7 Transportation by Truck

No information on an average distribution distance is available for the regarded filling materials. Quite an effort is necessary for the procurement of more up-to-date and more representative data for beverage distribution and this was not the subject of this study. As a substitute for all regarded beverage packagings the distribution model from UBA-II for the filling 'Beverages without $CO_2$ of the stockpile segment (>0.5 l)' was adapted Plinke et al. (2000). The average transportation distance is therefore approximately 350 km.

*The data record is based on standard emission data, which are arranged, validated, updated and evaluated in a 'Handbook for Emission Factors' INFRAS (2004b) for the Federal Environmental Agency Berlin and the Federal Office for Environment Protection BUWAL Bern.*

All factors consider appropriate vehicle mix and the proportion of accountable driven distance.

This handbook is an application of databases and supplies the fuel consumption depending on driving performance plus emissions in the categories *truck classes, road types* and *utilisation ratios* separately.

### 3.7.2.8 Electricity Supply

The balancing of electricity supply (electricity mix, power plant and electricity distribution) by LCA is not only relevant concerning the use of resources but also for the calculation of emissions into the air (see Section 3.2).

The electricity supply for processes within the German reference area were assessed by the German mix of energy sources (Table 3.12). Processes abroad are calculated according to the appropriate regional energy-mix, if the aggregation level of the respective data records allowed a separate modelling of the electricity supply.

The mix of energy sources in the German grid electricity was updated to the year 2003 in accordance with data of (VDEW) (Table 3.11).

**Table 3.11** Power plant split in the model electricity grid, Germany 2003 IFEU (2006).

| Source of energy | Ratio (%) |
|---|---|
| Hard coal | 23.9 |
| Brown coal | 26.1 |
| Mineral oil | 1.1 |
| Natural gas | 12.3 |
| Nuclear energy | 27.8 |
| Water (without pump storage) | 3.6 |
| Wind force | 3.3 |
| Other | 1.8 |

The modelling of the power plants was based on measured values, which were made available to IFEU by operators of German power plants. This data was supplemented by published values.[193] It was also intended to map a regional state of the art of power plants.

## 3.7.3
### Elaboration of a Differentiated System Flow Chart with Reference Flows

As a next step, the system flow chart can be provided based on a thorough analysis of the product system and the structuring of the unit processes identified. Figures 3.30 and 3.31 show such differentiated system flow charts including the mass flows calculated in the LCI and based on the reference flow data (see Table 3.8) for the carton (Figure 3.30) and the PET system (Figure 3.31).

For the calculation of the LCI, the system flow chart must be refined to an extent as to illustrate all considered unit processes and their interdependencies. The degree of refinement depends on the data available (see Section 3.1.3.1); Figures 3.32 and 3.33 show two examples as to how such refinements of system flow chart may look.

## 3.7.4
### Allocation

Since allocations are always based on conventions, the transparent description of allocation rules is of great importance for the credibility and understanding of the study (see Section 3.3). In this exemplary study the allocation rules are described on process level, on system level, and with respect to waste treatment.

### 3.7.4.1  Definition of Allocation Rules on Process Level

**Multi-output Processes**  Beverage raw carton and the necessary fibrous materials are predominantly manufactured by integrated plants of the paper industry. A product range of a set of paper or cardboard products, with different fibre compositions, is usually covered. This and the networking of the energy supply of all fibre lines *make the allocation of energy consumption and energy sources to individual product and fibre lines even more difficult.* In the available data records the appropriate allocations are accomplished by the plant operators themselves. The data records made available are purely inventory data. The original data and applied allocation procedures were not available to the contractors.

---

193) Fritsche *et al.*, 2001; Ecoinvent Centre, 2003.

**154** | *3 Life Cycle Inventory Analysis*

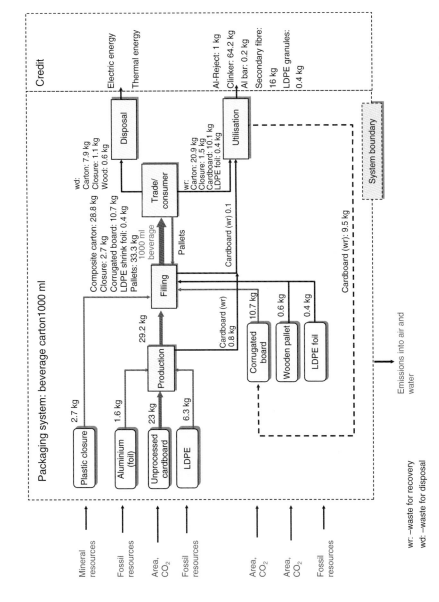

**Figure 3.30** Simplified system flow chart of 1-l-beverage carton covering reference flows. Masses indicated refer to the functional unit: packaging necessary for the supply of 1000 l filling material to the point-of-sale.

## 3.7 Illustration of the Inventory Phase by an Example | 155

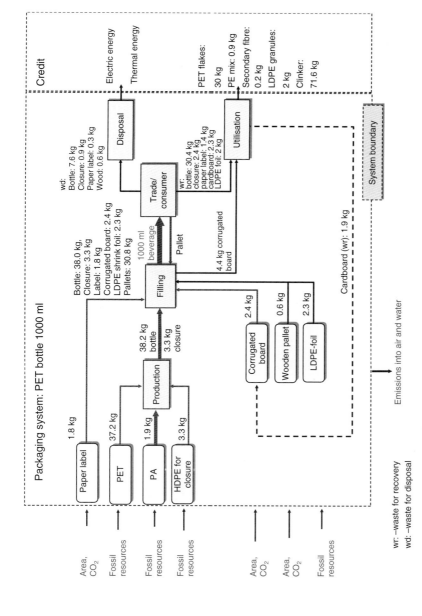

**Figure 3.31** Simplified system flow chart of 1-l-PET bottle covering reference flows. Mass units refer to the functional unit: packaging necessary for the supply of 1000 l filling material to the point-of-sale.

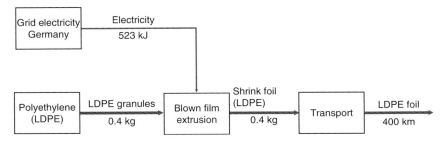

**Figure 3.32** Subsystem LDPE foil Production from Figure 3.30.

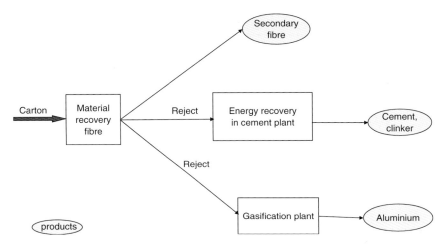

**Figure 3.33** Subsystem beverage carton recycling from Figure 3.28 (see descriptions of unit processes for recycling procedures).

With data records compiled by the authors of the study, the allocation by mass is normally calculated according to the outputs of co-products. Some data records taken from literature also use the heat value or the market value as allocation criteria (e.g. the heat value with PlasticEurope data for plastics).

The respective allocation criteria are documented depending on relevance for individual data records. For published data it is usual to only refer to the appropriate source.

**Multi-input Processes** Multi-input processes predominantly occur within disposal. Appropriate processes are modelled in a way that allows an even and causal allocation of material and energy flows of used packagings to these processes. The modelling of disposal of packaging materials as waste in an incineration plant is a typical example of a multi-input allocation. For an LCA, inputs and outputs are important which can causally be assigned to the incineration of the packagings. According to the introductory remarks on process-related allocations, predominantly physical causations between inputs and output are used.[194]

**Transportation Processes for Distribution**   Distribution of filled packagings requires an allocation of environmental loads between packaging and filling materials, under consideration of the capacity utilisation of the transport vehicle. This approach corresponds to LCA II of the Federal Environmental Agency and is documented (Plinke et al., 2000, loc. cit).

### 3.7.4.2   Definition of Allocation Rules on System Level for Open-Loop Recycling

In this study the allocation for an OLR is made according to the '50 : 50'-method, also a standard technique in UBA-II/2.[195] Here, the use of secondary material is evenly distributed in a 50:50 relationship between the delivering and the receiving system.

In the case of a recycling of beverage cartons, the benefit in the example is a replacement of fresh fibres. This benefit is technically assessed in the beverage carton system in the form of a credit. The height of the credit thereby amounts to 50% of the substituted fresh fibre production proportion due to the employment of secondary fibres.

In this study the original UBA approach is, however, modified. In the allocation method the phase 'disposal' of life cycle 2 (LC 2) of the secondary product is additionally considered in the allocation method.

In Section 3.3.4.2 the explaining illustrations from the sample study were already used for explanation.

### 3.7.5
### Modelling of the System

If all considered unit processes and allocation rules are clearly defined, the calculation rules according to which unit processes are to be connected have to be introduced. This work is comfortably accomplished by the use of software programmes (in this example Umberto) with appropriate input masks and connections to the databases provided.

---

194) For a detailed description of the allocation of input/output by the example of the refuse incineration see Plinke et al., 2000, p. 81.
195) UBA, 2002, p. 14–16.

However, an LCA software cannot provide secure and adequate linkage of the processes of the product system. The conductors of an LCA in general have to correctly realize the logic of the product system *before* the actual calculation can start. The user must not trust a software blindly. This is why the system check is such an important step of the critical review procedure (see Section 5.5).

### 3.7.6
### Calculation of the Life Cycle Inventory

If the system is modelled adequately and all process data records are available in sufficient quality, the LCI of the examined product system can be calculated. Tables 3.12–3.19 show as an example the LCI of the system variant 1-l-beverage carton with closure related to the fU.[196]

In this list both inputs and outputs are indicated as elementary flows. Remember, elementary flows are defined according to ISO 14040 and 14044 Section 3.12 as

> *material or energy entering the system being studied that has been drawn from the environment without previous human transformation or material or energy leaving the system being studied that is released into the environment without subsequent human transformation*

As supplement to these defined elementary flows, the LCI results CED (Section 3.2.2) and use of natural land are included.

Nearly all LCA studies make use of newly procured as well as generic data of diverse origin. Since the designation of inputs and outputs may vary even for the same substance, the same substance may occur with different designations in an LCA, for example: hexafluoroethane or perfluoroethane or $C_2F_6$. If the calculation of an impact assessment is accomplished by means of a software, it must be assured that all variants of the same substance are clearly assigned.

An LCI supplies much more data than used in the next phase, namely LCIA. The reason is that the development of a scientifically justified transfer of LCI data into impact indicators still does not allow a complete use of all LCI data (see Chapter 4). It is thus even more important that the data of the LCI may not be lost, for a retrieval of their information content within the interpretation phase (see Chapter 5).

In Tables 3.12–3.19 the first step from LCI to LCIA is already represented, the classification: The LCI parameters are assigned to the impact categories considered in this specific LCA study serving as a praxis example (see also Section 4.6).

---

196) IFEU (2006) These detailed data have been provided for this book by courtesy of the IFEU.

**Table 3.12** Inventory data energy sources.

| Source of energy (RiD) | Mass/fU (kg) | Impact category |
|---|---|---|
| Natural gas (RiD) | 8.03E+00 | Resource demand |
| Oil (RiD) | 1.22E+01 | Resource demand |
| Brown coal/lignite(RiD) | 5.48E+00 | Resource demand |
| Coal, non-specific (RiD) | 2.67E−03 | Resource demand |
| Hard coal (RiD) | 1.27E+00 | Resource demand |

RiD: raw materials in deposits.

**Table 3.13** Inventory data CED.

| CED | Amount/fU (kJ) | Impact category |
|---|---|---|
| CED (nuclear energy) | 3.17E+05 | — |
| CED (hydropower) | 9.76E+04 | — |
| CED, fossil total | 9.62E+05 | — |
| CED, renewable | 6.58E+05 | — |
| CED, other | 1.23E+03 | — |

The minus sign in this table signifies that the CED is not assigned to an impact category.

### 3.7.6.1 Input

**Energy Carriers** In LCIA, LCI parameters, which in this study are assigned to resource demand, are comprised, according to Table 3.12, in the impact indicator 'crude oil equivalents' (see Section 4.5.1.2).

**Cumulative Energy Demand** The CED is an inventory result of the LCI (Table 3.13). Although it is not transferred into an impact category, it is referred to in the interpretation of nearly all LCAs (see also Section 3.2.2).

**Mineral Raw Materials** In the inventory a multiplicity of mineral raw materials is specified (see Table 3.14).

None of these mineral raw materials is considered in the impact assessment, since exclusively fossil resources were included.

**3.7.6.1.1 Water** Water is used in many processes. The data are listed in the inventory but are not considered in the impact assessment (Table 3.15).

Noticeable is the negative value for 'water, non-specific'. It results from the fact that in the inventory data record, which was used for a credit entry (therefore the

**Table 3.14** Inventory data of mineral raw materials.

| Minerals (RiS) | Mass/fU (kg) | Impact category |
| --- | --- | --- |
| Baryte (RiS) | 2.85E−05 | n. a. |
| Bauxite (RiS) | 5.43E+00 | n. a. |
| Bentonite (RiS) | 4.87E−04 | n. a. |
| Lead (RiS) | 4.57E−06 | n. a. |
| Calcium sulfate (RiS) | 4.92E−05 | n. a. |
| Chromium (CR) (RiS) | 4.88E−08 | n. a. |
| Dolomite (RiS) | 2.58E−05 | n. a. |
| Iron (Fe) (RiS) | 1.68E−04 | n. a. |
| Iron (RiS) | 2.40E−03 | n. a. |
| Iron ore (RiS) | 1.12E−03 | n. a. |
| Feldspar (RiS) | 7.75E−16 | n. a. |
| Ferro-manganese (RiS) | 1.92E−06 | n. a. |
| Fluorspar (RiS) | 5.89E−06 | n. a. |
| Granite (RiS) | 3.24E−10 | n. a. |
| Ilmenite (RiS) | 2.39E−01 | n. a. |
| Limestone (RiS) | 7.94E−01 | n. a. |
| Potassium chloride (RiS) | 1.86E−03 | n. a. |
| Copper (Cu) (RiS) | 4.62E−07 | n. a. |
| Gravel (RiS) | 7.75E−06 | n. a. |
| Chalk (RiS) | 5.81E−27 | n. a. |
| Magnesium (Mg) (RiS) | 2.10E−08 | n. a. |
| Sodium chloride (RiS) | 6.16E−01 | n. a. |
| Sodium nitrate (RiS) | 6.29E−08 | n. a. |
| Nickel (Ni) (RiS) | 5.05E−08 | n. a. |
| Olivin (RiS) | 1.97E−05 | n. a. |
| Quartz ($SiO_2$) (RiS) | 1.82E−27 | n. a. |
| Quartz sand (RiS) | 1.79E−03 | n. a. |
| Mercury (Hg) (RiS) | 1.28E−08 | n. a. |
| Raw phosphate (RiS) | 3.37E−02 | n. a. |
| Raw iron ore (RiS) | 6.39E−03 | n. a. |
| Raw earth (RiS) | 6.47E+00 | n. a. |
| Raw potash (RiS) | 7.97E−02 | n. a. |
| Rutile (RiS) | 8.37E−27 | n. a. |
| Sand (RiS) | 1.19E−03 | n. a. |
| Sulphur (RiS) | 1.88E−01 | n. a. |
| Slate (RiS) | 1.39E−04 | n. a. |
| Talcum powder (RiS) | 2.13E−20 | n. a. |
| Clay/tone (RiS) | 4.81E−07 | n. a. |
| Peat (RiS) | 2.10E−02 | n. a. |
| Zinc (RiS) | 2.20E−03 | n. a. |

n. a.: not assigned.

**Table 3.15** Inventory data water.

| Water | Mass/fU (kg) | Impact category |
|---|---|---|
| Water | | n. a. |
| Cooling water total | 3.70E+02 | n. a. |
| Cooling water public | 6.12E+00 | n. a. |
| Cooling water | 1.14E+04 | n. a. |
| Water (process) total | 2.84E+01 | n. a. |
| Water (process) public | 1.40E+01 | n. a. |
| Water (boiler feed) | 7.99E+02 | n. a. |
| Water (process) | 1.96E+03 | n. a. |
| Water, non-specific | −4.70E+01 | n. a. |
| *Untreated water* | | n. a. |
| Ground water | 1.02E−02 | n. a. |
| Surface water | 2.19E−01 | n. a. |

n. a.: not assigned.

minus sign), neither an indication to the origin nor to the use of water was given. Therefore this entry cannot be balanced with one of the other entries listed.

**3.7.6.1.2 Land Use** Land use as part of area categories II–V is an impact category within the LCIA (see Section 4.5.1.6). The data are documented in the inventory. Land use concerning surfaces of category VII refers to sealed surfaces. These are not considered in the impact assessment.

**Table 3.16** Inventory data natural space.

| Natural space | Area/fU (m$^2$) | Impact category |
|---|---|---|
| Area K2 | 4.74E−01 | Land use |
| Area K2 (FRG) | 3.15E−02 | Land use |
| Area K2 (North) | 1.04E+00 | Land use |
| Area K3 | 4.77E+00 | Land use |
| Area K3 (FRG) | 2.83E−01 | Land use |
| Area K3 (North) | 1.19E+01 | Land use |
| Area K4 | 1.15E+01 | Land use |
| Area K4 (FRG) | 1.36E−01 | Land use |
| Area K4 (North) | 3.16E+01 | Land use |
| Area K5 | 1.79E+00 | Land use |
| Area K5 (FRG) | 2.22E−02 | Land use |
| Area K5 (NORTH) | 7.33E+00 | Land use |
| Area K7 | 2.00E−04 | n. a. |
| Area K7 (FRG) | 5.10E−05 | n. a. |

n. a.: not assigned.

**Table 3.17** Inventory data of emissions into air.

| Emissions (air) | Mass/fU<br>Volume/fU<br>Energy/fU | Impact category |
|---|---|---|
| Exhaust gas (dry standard volume) (A) | 2.40E+02 Nm$^3$ | n. a. |
| Waste heat (A) | 2.58E+05 kJ | n. a. |
| Asbestos (A) | 5.67E−14 kg | n. a. |
| Landfill gas, diffuse (A) | 1.76E−01 m$^3$ | n. a. |
| Fly ash (A) | 2.35E−06 kg | n. a. |
| Carbon (C-total) (A) | 3.02E−04 kg | n. a. |
| Particle (A) | 5.67E−04 kg | n. a. |
| Particle from Diesel releases (A) | 1.00E−04 kg | n. a. |
| Dust (>PM10) (A) | −9.92E−04 kg | n. a. |
| Dust (A) | 2.57E−02 kg | n. a. |
| Dust (PM10) (A) | 5.77E−03 kg | n. a. |
| *Compounds, inorganic (A)* | | |
| Ammonia (A) | 4.04E−03 kg | Acidification<br>Eutrophication (soil) |
| Carbon disulphide (A) | 6.15E−12 kg | Acidification |
| Chlorine (A) | 9.23E−08 kg | n. a. |
| Chloride (A) | 9.54E−06 kg | n. a. |
| Hydrogen chloride (A) | 1.04E−03 kg | Acidification |
| Hydrogen cyanide | 2.41E−09 kh | Acidification |
| Dinitrogen monoxide (A) | 1.13E−03 kg | Greenhouse effect |
| Fluorine (A) | 7.94E−09 kg | n. a. |
| Fluorine, total (A) | 6.67E−04 kg | n. a. |
| Hydrogen fluoride (A) | 1.02E−03 kg | Acidification |
| NOx (A) | 1.54E−01 kg | Summer smog<br>Acidification<br>Eutrophication (soil) |
| Phosphine (A) | 4.43E−08 kg | n. a. |
| Sulphur (A) | 4.16E−08 kg | n. a. |
| Sulphur dioxide (A) | 1.53E−01 kg | Acidification |
| Carbon disulphide (A) | 1.89E−10 kg | Acidification |
| Sulphuric acid (A) | 1.04E−13 kg | Acidification |
| Hydrogen sulphide (A) | 2.11E−04 kg | Acidification |
| Nitrogen (A) | 2.13E−04 kg | n. a. |
| Nitrogen dioxide (A) | 2.10E−02 kg | Summer smog<br>Acidification<br>Eutrophication (soil) |
| Nitrogen oxides, unspecific (A) | 4.18E−03 kg | Summer smog<br>Acidification<br>Eutrophication (soil) |
| TRS; totally reduced sulphur; as S (A) | 7.79E−04 kg | Acidification |
| Hydrogen (A) | 4.60E−04 kg | n. a. |
| *Carbon dioxide (A)* | | |
| Carbon dioxide, fossil (A) | 6.05E+01 kg | Greenhouse effect |

## 3.7 Illustration of the Inventory Phase by an Example

**Table 3.17** (continued)

| Emissions (air) | Mass/fU Volume/fU Energy/fU | Impact category |
|---|---|---|
| Carbon dioxide, renewable (A) | 3.15E+01 kg | n. a. |
| Carbon dioxide, unspecific (A) | 2.21E−01 kg | Greenhouse effect |
| Carbon monoxide (A) | 7.56E−02 kg | n. a. |
| *Metals (A)* | | |
| Antimony (A) | 9.00E−08 kg | n. a. |
| Arsenic (A) | 4.30E−07 kg | n. a. |
| Beryllium (A) | 9.21E−08 kg | n. a. |
| Lead (A) | 1.14E−06 kg | n. a. |
| Cadmium (A) | 4.82E−07 kg | n. a. |
| Chrome (A) | 3.26E−07 kg | n. a. |
| Cobalt (A) | 2.73E−07 kg | n. a. |
| Copper (A) | 8.80E−07 kg | n. a. |
| Manganese (A) | 4.96E−07 kg | n. a. |
| Metals, unspecific (A) | 2.30E−05 kg | n. a. |
| Nickel (A) | 1.20E−05 kg | n. a. |
| Palladium (A) | 1.78E−09 kg | n. a. |
| Platinum (A) | 1.78E−09 kg | n. a. |
| Mercury (A) | 9.93E−07 kg | n. a. |
| Rhodium (A) | 2.49E−09 kg | n. a. |
| Selenium (A) | 2.36E−06 kg | n. a. |
| Silver (A) | 9.95E−16 kg | n. a. |
| Tellurium (A) | 4.91E−07 kg | n. a. |
| Thallium (A) | 3.17E−08 kg | n. a. |
| Uranium (A) | 2.76E−07 kg | n. a. |
| Vanadium (A) | 6.07E−06 kg | n. a. |
| Zinc (A) | 2.75E−06 kg | n. a. |
| Tin (A) | 2.86E−07 kg | n. a. |
| *VOC (A)* | | |
| Methane (A) | 1.01E−01 kg | Greenhouse effect Summer smog |
| Methane, fossil (A) | 5.44E−02 kg | Greenhouse effect Summer smog |
| Methane, regenerative (A) | 6.90E−02 kg | Greenhouse effect Summer smog |
| VOC (hydrocarbons) (A) | 3.35E−03 kg | Summer smog |
| VOC, unspecific (A) | 5.49E−02 kg | Summer smog |
| *NMVOC (A)* | | |
| Ethene (A) | 1.08E−05 kg | Summer smog |
| Hexane (A) | 7.53E−06 kg | Summer smog |
| NMVOC (hydrocarbons) (A) | 1.33E−07 kg | Summer smog |
| NMVOC (HC without benzene) (A) | 1.65E−04 kg | Summer smog |
| NMVOC ( HC without PAH) (A) | 1.20E−02 kg | Summer smog |

*(continued overleaf)*

**Table 3.17** (continued)

| Emissions (air) | Mass/fU<br>Volume/fU<br>Energy/fU | Impact category |
|---|---|---|
| NMVOC from diesel emissions (A) | 2.35E–03 kg | Summer smog |
| NMVOC from diesel emissions (A) | 1.18E–03 kg | Summer smog |
| NMVOC, aromatic, unspecific (A) | 4.69E–04 kg | Summer smog |
| NMVOC, chlorine, unspecific (A) | 2.00E–08 kg | Summer smog |
| NMVOC, fluorine, unspecific (A) | 9.51E–06 kg | Summer smog |
| NMVOC, unspecific (A) | 1.30E–02 kg | Summer smog |
| Propylene (A) | 8.02E–06 kg | Summer smog |
| TOC (A) | 3.11E–04 kg | Summer smog |
| *BTEX(A)* | | |
| Benzene (A) | 7.53E–05 kg | Summer smog |
| Ethyl benzene (A) | 7.21E–11 kg | Summer smog |
| Toluene (A) | 2.35E–07 kg | Summer smog |
| Xylene (A) | 3.17E–07 kg | Summer smog |
| *NMVOC, chlorine, aliphatic (A)* | | |
| Dichloroethane (A) | 1.87E–10 kg | n. a. |
| Dichloroethene (A) | 4.53E–11 kg | n. a. |
| Dichloromethane ($CH_2Cl_2$) (A) | 1.26E–10 kg | n. a. |
| Tetrachloromethane (A) | 1.36E–11 kg | Greenhouse effect |
| Vinyl chloride (A) | 4.06E–09 kg | n. a. |
| *NMVOC, fluorine (A)* | | |
| Hexafluoroethane (A) | 1.21E–09 kg | Greenhouse effect |
| Perfluoroethane (A) | 2.12E–05 kg | Greenhouse effect |
| Perfluoromethane (A) | 2.43E–04 kg | Greenhouse effect |
| Tetrafluoromethane (A) | 9.51E–09 kg | Greenhouse effect |
| *NMVOC, Containing . oxygen (A)* | | |
| Aldehydes, unspecific (A) | 1.27E–07 kg | Summer smog |
| Ethanol (A) | 2.55E–08 kg | Summer smog |
| Formaldehyde (A) | 3.76E–04 kg | Summer smog |
| *HCs, chlorinine, aromatic (A)* | | |
| Chlorbenzene (A) | 1.33E–08 kg | n. a. |
| Chlorodiphenyl (42% Cl) (A) | 6.34E–15 kg | n. a. |
| Chlorophenol (A) | 2.65E–08 kg | n. a. |
| PCB (A) | 3.12E–10 kg | n. a. |
| PCDD, PCDF (A) | 2.22E–11 kg | n. a. |
| *HCs, cont. sulphur (A)* | | |
| Ethane thiol (A) | 1.35E–12 kg | Acidification |
| Mercaptane (A) | 5.77E–08 kg | Acidification |
| *PAH (A)* | | |
| Acenaphtylene (A) | 6.89E–16 kg | n. a. |

**Table 3.17** (continued)

| Emissions (air) | Mass/fU Volume/fU Energy/fU | Impact category |
|---|---|---|
| Benzo (a) pyrene (A) | 2.80E−06 kg | n. a. |
| Dibenzo (a) pyrene (A) | 3.45E−17 kg | n. a. |
| Fluorene (A) | 6.89E−17 kg | n. a. |
| Naphthalene (A) | 3.45E−15 kg | n. a. |
| PAH without B(a)P (A) | 1.04E−04 kg | n. a. |
| PAH, unspecific (A) | 6.45E−09 kg | n. a. |
| Phenanthrene (A) | 6.89E−17 kg | n. a. |
| *HCs, other* | | |
| Biphenyl (A) | 6.89E−17 kg | n. a. |
| Styrene (A) | 3.77E−12 kg | n. a. |
| Tributyl phosphate (A) | 1.14E−08 kg | n. a. |
| *Other* | | |
| Water vapour (A) | 1.29E+01 kg | n. a. |

n. a.: not assigned.

### 3.7.6.2 Output

**3.7.6.2.1 Emissions into Air** Although the emissions into air are listed in four impact categories (climate change (here called *greenhouse effect*), summer smog, acidification and eutrophication of soil) (see Section 4.5.2), this inventory clearly contains substantially more information than is common practice for an impact assessment in many LCAs.

**3.7.6.2.2 Emissions into Water** Emissions into water are considered in this study for the impact category 'aquatic eutrophication' (Table 3.18). As before, the inventory clearly contains substantially more information than is common practice for the impact assessment in many LCAs. This would change if the impact category 'ecotoxicity' was used routinely.

**3.7.6.2.3 Radionuclides** In an LCA the emissions of radionuclides into water and air are listed but not transferred into an impact category (Table 3.19). The data originate from the unit process 'nuclear power plant'.

Tables 3.12–3.19 show the large information provision in an LCI. They, however, also show that it makes sense to bundle and structure data for the interpretation. This bundling of data for the LCIA is discussed in Chapter 4, and the data of these tables are used in Section 4.6 for further explanations.

**Table 3.18** Inventory data of emissions into water.

| Emissions (water) | Mass/fU Energy/fU | Impact category |
|---|---|---|
| Waste heat (W) | 6.63E+04 kJ | n. a. |
| Solids, dissolved (W) | 1.10E−02 kg | n. a. |
| Solids, dispersed (W) | 9.43E−02 kg | n. a. |
| Solids, undissolved (W) | 5.59E−04 kg | n. a. |
| Sand (W) | 9.96E−05 kg | n. a. |
| *Metals (W)* | | |
| Aluminium (W) | 7.97E−06 kg | n. a. |
| Aluminium nitrate (W) | 3.69E−08 kg | n. a. |
| Antimony (W) | 3.98E−10 kg | n. a. |
| Arsenic (W) | 2.01E−07 kg | n. a. |
| Barium (W) | 9.24E−07 kg | n. a. |
| Beryllium (W) | 1.12E−08 kg | n. a. |
| Lead (W) | 1.96E−05 kg | n. a. |
| Cadmium (W) | 9.55E−08 kg | n. a. |
| Calcium (W) | 3.61E−03 kg | n. a. |
| Chrome (W) | 1.40E−06 kg | n. a. |
| Chromium (VI) – oxide (W) | 1.77E−10 kg | n. a. |
| Chrome-III (W) | 1.03E−08 kg | n. a. |
| Chrome-VI (W) | 5.37E−10 kg | n. a. |
| Cobalt (W) | 1.47E−10 kg | n. a. |
| Cyanide (W) | 4.37E−10 kg | n. a. |
| Iron (W) | 8.32E−07 kg | n. a. |
| Fe–Al-oxides (W) | 4.74E−05 kg | n. a. |
| Potassium (W) | 5.69E−05 kg | n. a. |
| Cobalt (W) | 2.55E−09 kg | n. a. |
| Copper (W) | 5.16E−06 kg | n. a. |
| Magnesium (W) | 2.36E−07 kg | n. a. |
| Manganese (W) | 3.75E−05 kg | n. a. |
| Metals, unspecific (W) | 1.29E−04 kg | n. a. |
| Molybdenum (W) | 4.72E−06 kg | n. a. |
| Sodium (W) | 1.60E−02 kg | n. a. |
| Nickel (W) | 1.64E−06 kg | n. a. |
| Mercury (W) | 1.42E−08 kg | n. a. |
| Selenium (W) | 1.40E−06 kg | n. a. |
| Silver (W) | 2.49E−09 kg | n. a. |
| Strontium (W) | 1.93E−10 kg | n. a. |
| Uranium (W) | 6.18E−06 kg | n. a. |
| Vanadium (W) | 3.60E−06 kg | n. a. |
| Zinc (W) | 1.72E−05 kg | n. a. |
| Tin (W) | 6.36E−09 kg | n. a. |
| *Compounds, inorganic (W)* | | |
| Boron (W) | 1.83E−07 kg | n. a. |
| Bromate (W) | 5.83E−09 kg | n. a. |

Table 3.18 (continued)

| Emissions (water) | Mass/fU Energy/fU | Impact category |
|---|---|---|
| Ca-Mg-Hydroxide (W) | 2.47E−04 kg | n. a. |
| Calcium sulphate (W) | 3.07E−05 kg | n. a. |
| Carbonate (W) | 3.60E−04 kg | n. a. |
| Chlorine (W) | 5.62E−05 kg | n. a. |
| Chlorine, dissolved (W) | 1.89E−08 kg | n. a. |
| Chlorates (W) | 7.42E−04 kg | n. a. |
| Chloride (W) | 9.13E−02 kg | n. a. |
| Cyanide (W) | 2.68E−08 kg | n. a. |
| Fluorine (W) | 5.20E−07 kg | n. a. |
| Fluoride (W) | 8.57E−04 kg | n. a. |
| Hydroxide (W) | 4.36E−05 kg | n. a. |
| Limestone (W) | 7.40E−04 kg | n. a. |
| Salts, inorganic (W) | 3.02E−04 kg | n. a. |
| Acids as $H^+$ (W) | 1.71E−04 kg | n. a. |
| Sulphur (W) | 2.03E−08 kg | n. a. |
| Sulphate (W) | 6.84E−02 kg | n. a. |
| Sulphides (W) | 3.21E−08 kg | n. a. |
| *Phosphorus compounds (W)* | | |
| Phosphate (W) | 1.08E−07 kg | Eutrophication water |
| Phosphates (as $P_2O_5$) (W) | 2.94E−08 kg | Eutrophication water |
| Phosphorus (W) | 8.50E−08 kg | Eutrophication water |
| Phosphorus compounds as P (W) | 1.33E−03 kg | Eutrophication water |
| *Nitrogen compounds (W)* | | |
| Amines, tertiary (W) | 1.19E−06 kg | n. a. |
| Ammonia (W) | 1.87E−05 kg | Eutrophication water |
| Ammonium (W) | 1.67E−04 kg | Eutrophication water |
| Ammonium as N (W) | 6.73E−06 kg | Eutrophication water |
| Nitrate (W) | 2.66E−02 kg | Eutrophication water |
| Nitrate as N (W) | 1.25E−08 kg | Eutrophication water |
| Nitric acid (W) | 8.75E−07 kg | Eutrophication water |
| Nitrogen compounds, unspecific (W) | 1.22E−05 kg | Eutrophication water |
| Nitrogen compounds as N (W) | 4.22E−03 kg | Eutrophication water |
| *Compounds, organic (W)* | | |
| Aldehydes, total (W) | 2.79E−04 kg | n. a. |
| Benzene (W) | 2.66E−10 kg | n. a. |
| Detergents, Oil (W) | 9.03E−05 kg | n. a. |
| Dioxins (W) | 2.59E−09 kg | n. a. |
| Fats and oils, total (W) | 9.34E−08 kg | n. a. |
| Isodecanol (as HCs) (W) | 1.59E−06 kg | n. a. |
| Hydrocarbons, aromatic, unspecific (W) | 1.68E−09 kg | n. a. |
| Hydrocarbons, unspecific (W) | 1.73E−04 kg | n. a. |
| Oil (W) | 1.47E−03 kg | n. a. |
| Phenol (W) | 1.50E−05 kg | n. a. |
| Tributyl phosphate (W) | 1.75E−09 kg | n. a. |

*(continued overleaf)*

Table 3.18 (continued)

| Emissions (water) | Mass/fU Energy/fU | Impact category |
|---|---|---|
| Compounds organic, dissolved (W) | 1.00E−04 kg | n. a. |
| *PAH (W)* | | |
| Benzo (a) pyrene (W) | 2.61E−13 kg | n. a. |
| PAH without B (a) P (W) | 1.37E−11 kg | n. a. |
| PAH, unspecific (W) | 3.19E−06 kg | n. a. |
| *Compounds organic, halogen (W)* | | |
| Dichloroethene (W) | 4.39E−12 kg | n. a. |
| PCB (W) | 9.17E−12 kg | n. a. |
| Compounds organic, chlorine, unspecific (W) | 5.34E−07 kg | n. a. |
| Compounds organic, halogen, unspec. (W) | 2.12E−10 kg | n. a. |
| Compounds organic, unspecific (W) | 6.39E−08 kg | n. a. |
| Vinyl chloride (W) | 7.44E−11 kg | n. a. |
| *Indicator parameter* | | |
| AOX (W) | 1.28E−03 kg | n. a. |
| BOD-5 (W) | 5.86E−02 kg | n. a. |
| COD (W) | 3.31E−01 kg | Eutrophication water |
| TOC (W) | 1.13E−01 kg | n. a. |
| Leachate water, diffuse (W) | 5.41E−02 kg | n. a. |
| Leachate water coll. (W) | 1.40E−05 kg | n. a. |
| Organo tin compound as Sn (W) | 1.97E−10 kg | n. a. |
| Organo silicone (W) | 3.85E−19 kg | n. a. |

n. a.: not assigned.

Table 3.19 Inventory data of radionuclides.

| Radionuclides (W) and (A) | Bq/fU | Impact category |
|---|---|---|
| Americium 241 (A) | 1.20E−03 Bq | n. a. |
| Americium 241 (W) | 1.29E−02 Bq | n. a. |
| Antimony 124 (A) | 1.43E−05 Bq | n. a. |
| Antimony 124 (W) | 1.85E−02 Bq | n. a. |
| Antimony 125 (W) | 2.38E−03 Bq | n. a. |
| Barium 140 (A) | 8.81E−03 Bq | n. a. |
| Cesium 134 (A) | 8.16E−02 Bq | n. a. |
| Cesium 134 (W) | 1.73E+04 Bq | n. a. |
| Cesium 137 (A) | 1.67E−01 Bq | n. a. |
| Cesium 137 (W) | 1.02E+05 Bq | n. a. |
| Cerium 141 (A) | 2.43E−03 Bq | n. a. |
| Cerium 144 (A) | 2.47E−02 Bq | n. a. |
| Cerium 144 (W) | 6.07E+03 Bq | n. a. |

Table 3.19  (continued)

| Radionuclides (W) and (A) | Bq/fU | Impact category |
|---|---|---|
| Chrome 51 (A) | 2.99E−03 Bq | n. a. |
| Chrome 51 (W) | 8.71E−03 Bq | n. a. |
| Curium alpha (A) | 1.90E−04 Bq | n. a. |
| Curium alpha (W) | 2.47E+02 Bq | n. a. I |
| Iron 59 (A) | 3.93E−04 Bq | n. a. |
| Iodine 129 (A) | 3.98E−01 Bq | n. a. |
| Iodine 129 (W) | 2.28E+04 Bq | n. a. |
| Iodine 131 (A) | 1.04E+00 Bq | n. a. |
| Iodine 131 (W) | 1.82E−02 Bq | n. a. |
| Cobalt 58 (A) | 1.03E−03 Bq | n. a. |
| Cobalt 58 (W) | 1.95E−03 Bq | n. a. |
| Cobalt 60 (A) | 3.67E−02 Bq | n. a. |
| Cobalt 60 (W) | 3.60E+04 Bq | n. a. |
| Carbon 14 (A) | 9.51E+02 Bq | n. a. |
| Carbon 14 (W) | 7.78E+03 Bq | n. a. |
| Condensate | 3.28E−01 kg | n. a. |
| Krypton 85 (A) | 5.88E+06 Bq | n. a. |
| Lanthanum 140 | 2.44E−02 Bq | n. a. |
| Manganese 54 (A) | 5.05E−03 Bq | n. a. |
| Manganese 54 (W) | 1.76E−03 Bq | n. a. |
| Manganese 55 (W) | 5.31E+03 Bq | n. a. |
| Neptunium 237 (A) | 3.04E−07 Bq | n. a. |
| Neptunium 237 (W) | 8.54E+01 Bq | n. a. |
| Niobium 95 (A) | 3.05E−03 Bq | n. a. |
| Niobium 95 (W) | 2.00E−03 Bq | n. a. |
| Nuclide mixture (W) | 1.26E−01 Bq | n. a. |
| Palladium 234 m (A) | 1.06E−01 Bq | n. a. |
| Palladium 234 m (W) | 1.97E+00 Bq | n. a. |
| Plutonium 241 beta (A) | 1.06E−01 Bq | n. a. |
| Plutonium 241 beta (W) | 6.26E+04 Bq | n. a. |
| Plutonium alpha (A) | 4.36E−03 Bq | n. a. |
| Plutonium alpha (W) | 2.09E+03 Bq | n. a. |
| Praseodymium 147 (A) | 3.04E−09 Bq | n. a. |
| radioactive metal nuclides, total (A) | 8.15E−06 Bq | n. a. |
| Radionuclides, total (A) | 1.70E+05 Bq | n. a. |
| Radionuclides, total (W) | 1.01E+03 kBq | n. a. |
| Radionuclides, unspecific (A) | 4.31E+02 Bq | n. a. |
| Radionuclides, unspecific (W) | 2.56E+00 kBq | n. a. |
| Radium 226 (A) | 2.31E+00 Bq | n. a. |
| Radium 226 (W) | 1.39E+03 Bq | n. a. |
| Radon 220 (A) | 2.38E+01 Bq | n. a. |
| Radon 222 (A) | 2.16E+07 Bq | n. a. |
| Ruthenium 103 (A) | 2.43E−03 Bq | n. a. |
| Ruthenium 103 (W) | 4.76E−04 Bq | n. a. |
| Ruthenium 106 (A) | 3.79E+00 Bq | n. a. |

(continued overleaf)

**Table 3.19** (continued)

| Radionuclides (W) and (A) | Bq/fU | Impact category |
|---|---|---|
| Ruthenium 106 (W) | 2.28E+05 Bq | n. a. |
| Strontium 90 (A) | 1.20E−01 Bq | n. a. |
| Strontium 90 (W) | 4.55E+04 Bq | n. a. |
| Technetium 99 (W) | 3.98E+03 Bq | n. a. |
| Thorium 230 (A) | 1.19E+00 Bq | n. a. |
| Thorium 230 (W) | 7.89E+02 Bq | n. a. |
| Thorium 234 (A) | 1.06E−01 Bq | n. a. |
| Thorium 234 (W) | 1.97E+00 Bq | n. a. |
| Tritium (A) | 1.02E+03 Bq | n. a. |
| Tritium (W) | 1.63E+08 Bq | n. a. |
| Uranium 234 (A) | 1.34E−01 Bq | n. a. |
| Uranium 234 (W) | 1.40E−03 Bq | n. a. |
| Uranium 235 (A) | 6.65E−03 Bq | n. a. |
| Uranium 235 (W) | 6.27E−05 Bq | n. a. |
| Uranium 238 (A) | 4.48E+00 Bq | n. a. |
| Uranium 238 (W) | 1.45E+02 Bq | n. a. |
| Uranium 2381 (W) | 8.50E+00 Bq | n. a. |
| Uranium alpha (A) | 9.37E−03 Bq | n. a. |
| Uranium alpha (W) | 1.31E+02 Bq | n. a. |
| Uranium alpha, total (W) | 5.20E+02 Bq | n. a. |
| Technetium 99 (A) | 1.83E−13 Bq | n. a. |
| Xenon 133 (Eq) (A) | 5.59E+03 Bq | n. a. |
| Xenon-133 (Eq) | 3.33E+02 Bq | n. a. |
| Zinc 65 (A) | 4.67E−03 Bq | n. a. |
| Zinc 65 (W) | 1.94E−01 Bq | n. a. |
| Zirconium 95 (A) | 2.24E−03 Bq | n. a. |
| Zr95 and Nb95 (W) | 3.79E+02 Bq | n. a. |
| Nuclides, total | 3.27E−03 Bq | n. a. |
| Uran1 (W) | 2.00E−01 Bq | n. a. |

n. a.: not assigned.

## References

Arvidsson, P., Carlson, R., and Pålsson, A.-C. (1999) An Interpretation of the CPM Use of SPINE in Terms of ISO 14041 Standard. Gothenburg, Sweden, CPM Update Report.

Association Française de Normalisation (AFNOR) (1994) Analyse de cycle de vie. Norme NF X 30–300. 3/1994.

Atherton, J. (2007) Declaration by the metals industry on recycling principles. *Int. J. Life Cycle Assess.*, **12** (1), 59–60.

Ayres, R.U. and Ayres, L.W. (1996) *Industrial Ecology. Towards Closing the Materials Cycle.* With contributions by Frankl, P., Lee, H., Weaver, P.M. and Wolfgang, N., Edward Elgar Publishing, Cheltenham.

Baccini, P. and Bader, H.-P. (1996) *Regionaler Stoffhaushalt. Erfassung, Bewertung und Steuerung*, Spektrum Akademischer Verlag, Heidelberg.

Baccini, P. and Brunner, P.H. (1991) *Metabolism of the Anthroposphere*, Springer-Verlag, Berlin.

Bauer, C., Buchgeister, J., and Schebek, L. (2004) German network on life cycle inventory data. *Int. J. Life Cycle Assess.*, **9** (6), 360–364.

de Beaufort-Langeveld, A.S.H., Bretz, R., van Hoof, G., Hischier, R., Jean, P., Tanner, T., and Huijbregts, M. (eds) (2003) *Code of Life-Cycle Inventory Practice*, SETAC Press, Pensacola, FL ISBN: 1-880611-58-9.

Berna, J.L., Cavalli, L., and Renta, C. (1995) A lifecycle inventory for the production of linear akylbenzene sulphonates in Europe. *Tenside Surfactant Deterg.*, **32**, 122–127.

Boguski, T.K., Hunt, R.G., Cholaski, J.M., and Franklin, W.E. (1996) LCA methodology, in *Environmental Life-Cycle Assessment* (ed. M.A. Curran), McGraw-Hill, New York, pp. 2.1–2.37. ISBN: 0-07-015063-X.

Boustead, I. (1992) *Eco-Balance Methodology for Commodity Thermoplastics*. Report to, The European Centre for Plastics in the Environment (PWMI), Brussels, December 1992. * later: Association of Plastic Manufacturers in Europe (APME).

Boustead, I. (1993a) Eco-Profiles of the European Plastics Industry. Report 2: Olefin Feedstock Sources. Report to The European Centre for Plastics in the Environment (PWMI), Brussels, May 1993.

Boustead, I. (1993b) Eco-Profiles of the European Plastics Industry. Report 3: Polyethylene and Polypropylene. Report to The European Centre for Plastics in the Environment (PWMI), Brussels, May 1993.

Boustead, I. (1993c) Eco-Profiles of the European Plastics Industry. Report 4: Polystyrene, Report to The European Centre for Plastics in the Environment (PWMI), Brussels, May 1993; 2nd edn (to APME) Brussels, April 1997.

Boustead, I. (1994a) Eco-Profiles of the European Polymer Industry. Report 6: Polyvinyl Chloride, Report for APME's Technical and Environmental Centre, Brussels, April 1994.

Boustead, I. (1994b) *Eco-Profiles of the European Polymer Industry: Co-Product Allocation in Chlorine Plants*, Report for, APME's Technical and Environmental Centre, Brussels, April 1994.

Boustead, I. (1995a) Eco-Profiles of the European Polymer Industry. Report 8: Polyethylene Terephthalate (PET), Report for APME's Technical and Environmental Centre, Brussels, July 1995.

Boustead, I. (1995b) in *Enzyklopädie der Kreislaufwirtschaft, Management der Kreislaufwirtschaft* (Hrsg. K.J. Thomé-Kozmiensky), EF-Verlag für Energie- und Umwelttechnik, Berlin, pp. 320–327.

Boustead, I. (1996a) Eco-Profiles of the European Polymer Industry. Report 9: Polyurethane Precursors (TDI, MDI, Polyols), A Report for ISOPA, The European Isocyanate Producers Association. Brussels, June 1996.

Boustead, I. (1996b) LCA – How it came about. The beginning in UK. *Int. J. Life Cycle Assess.*, **1** (3), 147–150.

Boustead, I. (1997a) Eco-Profiles of the European Polymer Industry. Report 13: Polycarbonate, Report for APME's Technical and Environmental Centre, Brussels, September 1997.

Boustead, I. (1997b) Eco-Profiles of the European Polymer Industry. Report 14: Polymethyl methacrylate, A Report for the Methacrylates Technical Committee. Methacrylates Sector Group, CEFIC; APME, Brussels, September 1997.

Boustead, I. (1997c) Eco-Profiles of the European Polymer Industry. Report 10: Polymer Conversion, Report for APME's Technical and Environmental Centre, Brussels, in Collaboration with EuPC and supported by EUROMAP, Brussels, May 1997.

Boustead, I. (2003) *Eco-Profiles of the European Plastics Industry: Methodology*. A report for, APME, Brussels.

Boustead, I. and Fawer, M. (1994) Eco-Profiles of the European Polymer Industry. Report 7: Polyvinylidene Chloride, Report for APME's Technical and Environmental Centre, Brussels, December 1994.

Boustead, I. and Hancock, G.F. (1979) *Handbook of Industrial Energy Analysis*, Ellis Horwood Ltd., Chichester.

Braunschweig, A. and Müller-Wenk, R. (1993) *Ökobilanzen für Unternehmungen. Eine Wegleitung für die Praxis*, Verlag Haupt, Bern.

Bretz, R. (1998) SPOLD (Society for the Promotion of LCA Development). *Int. J. Life Cycle Assess.*, **3** (3), 119–120.

Bretz, R. and Frankhauser, P. (1996) Screening LCA for large numbers of products:

estimation tools to fill data gaps. *Int. J. Life Cycle Assess.*, **1** (3), 139–146.

Brunner, P.H. and Rechberger, H. (2004) *Practical Handbook of Material Flow Analysis*, Lewis Publishing: CRC Press, Boca Raton, FL. ISBN: 1566 706 041.

BUWAL (1996, 1998) Habersatter, K., Fecker, I., Dall'Aqua, S., Fawer, M., Fallscher, F., Förster, R., Maillefer, C., Ménard, M., Reusser, L., Som, C., Stahel, U., and Zimmermann, P. in *Ökoinventare für Verpackungen*, Schriftenreihe Umwelt, Nr. 250/Bd. I und II. 2. erweiterte und aktualisierte Auflage, Bern 1998, 1. Auflage (Hrsg. Bundesamt für Umwelt, Wald und Landschaft), ETH Zürich und EMPA, St. Gallen für BUWAL und SVI, Bern.

BUWAL (1991) Habersatter, K. and Widmer, F.) *Oekobilanzen von Packstoffen. Stand 1990*, Schriftenreihe Umwelt Nr. 132, Bundesamt für Umwelt, Wald und Landschaft (BUWAL) (Hrsg.), Bern.

Canadian Standards Association (CSA) (1992) CAN/CSA-Z760. *Environmental Life Cycle Assessment*. 5th Draft Edition, Canadian Standards Association, May 1992.

Carlson, R., Löfgren, G. and Steen, B. (1995) SPINE – A Relation Database Structure for Life Cycle Assessment. Report B 1227, Swedish Environmental Research Institute, Gothenburg.

Chevalier, J.-L. and Le Téno, J.-F. (1996) Life cycle analysis with ill-defined data and its application to building products. *Int. J. Life Cycle Assess.*, **1** (2), 90–96.

Christiansen, K. (1997) Simplifying LCA: Just a Cut?. Final Report of the SETAC-Europe LCA Screening and Streamlining Working Group, SETAC-Europe, Brussels, May 1997.

Ciroth, A. (2007) ICT for environment in life cycle applications. Open LCA – a new open source software for life cycle assessment. *Int. J. Life Cycle Assess.*, **12** (4), 209–210.

Ciroth, A., Fleischer, G., and Steinbach, J. (2004) Uncertainty calculation in life cycle assessments. A combined model of simulation and approximation. *Int. J. Life Cycle Assess.*, **9** (4), 216–226.

Coulon, R., Camobreco, V., Teulon, H., and Besnaimou, J. (1997) Data quality and uncertainty in LCI. *Int. J. Life Cycle Assess.*, **2** (3), 178–182.

Curran, M.A. (ed.) (1996) *Environmental Life-Cycle Assessment*, McGraw-Hill, New York. ISBN: 0-07-015063-X.

Curran, M.A. (2007) Co-product and input allocation approaches for creating life cycle inventory data a literature review. *Int. J. Life Cycle Assess.*, **12** (Special Issue 1), 65–78.

Curran, M.A. (2008) Development of life cycle assessment methodology a focus on co-product allocation. PhD thesis. Erasmus University Rotterdam, 26 June 2008.

De Smet, B. and Stalmans, M. (1996) LCI data and data quality. Thoughts and considerations. *Int. J. Life Cycle Assess.*, **1** (2), 96–104.

Deutsche Normen (1978a) *Einheiten – Einheitennamen, Einheitenzeichen.* DIN 1301 Teil 1, Oktober 1978.

Deutsche Normen (1978b) *Einheiten – Allgemein Angewendete Teile und Vielfache.* DIN 1301 Teil 2, Februar 1978.

Ecoinvent Centre 2003 Ecoinvent Data v1.01, Swiss Centre for Life Cycle Inventories, Dübendorf, www.ecoinvent.ch (accessed 28 Ocotber 2013).

Ekvall, T. (1999) System Expansion and Allocation in Life Cycle Assessment. ARF Report 245, Chalmers University of Technology, Göteborg.

Ekvall, T. and Finnveden, G. (2001) Allocation in ISO 14041 – a critical review. *J. Cleaner Prod.*, **9**, 197–208.

Ekvall, T. and Tillman, A.-M. (1997) Open-loop recycling: criteria for allocation procedures. *Int. J. Life Cycle Assess.*, **2** (3), 155–162.

EPA (1993) Vigon, B.W., Tolle, D.A., Cornaby, B.W., Latham, H.C., Harrison, C.L., Boguski, T.L., Hunt, R.G., and Sellers, J.D. *Life Cycle Assessment: Inventory Guidelines and Principles.* EPA/600/R-92/245, Office of Research and Development, Cincinnati, OH.

EPA (2006) Scientific Applications International Corporation (SAIC) *Life Cycle Assessment: Principles and Practice.* U.S. EPA, Systems Analysis Branch, co-ordinated by Curran, M.A., National Risk Management Research Laboratory, Cincinnati, OH.

European Aluminium Association (2000) *Ecological Profile Report for the European Aluminium Industry*, EAA, Brussels.

European Aluminium Association (2006) *Update of Ecological Profile Report for the European Aluminium Industry*. EAA, Brussels www.eaa-online.org (accessed 17 December 2013).

European Commission (2010) *European Commission – Joint Research Centre – Institute for Environment and Sustainability: International Reference Life Cycle Data System (ILCD) Handbook – General Guide for Life Cycle Assessment – Detailed Guidance*, 1st edn, ISPRA.

ecoinvent 1 (2004) Frischknecht, R., Jungbluth, N., Althaus, H.-J., Doka, G., Dones, R., Heck, T., Hellweg, S., Hischier, R., Nemecek, T., Rebitzer, G., and Spielmann, M. Overview and Methodology. Eecoinvent Report No. 1, Swiss Centre for Life Cycle Inventories, Dübendorf.

ecoinvent 2 (2004) Frischknecht, R., Jungbluth, N., Althaus, H.-J., Doka, G., Dones, R., Hellweg, S., Hischier, R., Nemecek, T., Rebitzer, G., and Spielmann and M. Code of Practice (Data v1.1). Ecoinvent Report No. 2, Swiss Centre for Life Cycle Inventories, Dübendorf.

ecoinvent 3 (2004) Frischknecht, R., Jungbluth, N., Althaus, H.-J., Doka, G., Dones, R., Hellweg, S., Hischier, R., Humbert, S., Margni, M., Nemecek, T., and Spielmann, M. Implementation of Life Cycle Impact Assessment Methods. Ecoinvent Report No. 3, Swiss Centre for Life Cycle Inventories, Dübendorf.

Eyrer, P. (Hrsg.) (1996) *Ganzheitliche Bilanzierung. Werkzeug zum Planen und Wirtschaften in Kreisläufen*, Springer, Berlin. ISBN: 3-540-59356-X.

Falbe, J. and Reglitz, M. (Hrsg.) (1995) *Römpp Chemie Lexikon*, 9. Auflage, Paperback-Ausgabe, Georg Thieme Verlag, Stuttgart. ISBN: 3-13-102759-2.

Fava, J., Jensen, A.A., Lindfors, L., Pomper, S., De Smet, B., Warren, J., and Vigon, B. (eds) (1994) *Conceptual Framework for Life-Cycle Data Quality*. Workshop Report. SETAC and SETAC Foundation for Environmental Education, Inc., Wintergreen, VA, October 1992, SETAC.

Fawer, M. (1996) Life Cycle Inventory for the Production of Zeolite A for Detergents. EMPA-Bericht Nr. 234, EMPA, Ecology Section, Commissioned by ZEODET a Sector Group of CEFIC, Swiss Federal Labo-ratories for Materials Testing and Research (EMPA), St. Gallen.

Fawer, M. (1997) *Life Cycle Inventories for the Production of Sodium Silicates* Report by, Eidgenössische Materialprüfungs- und Forschungsanstalt, St. Gallen, to Centre Européen d'Etude des Silicates (CEES) a Sector Group of CEFIC, Bruxelles.

FEFCO (2003) European Database for Corrugated Board Live Cycle Studies, Brüssel.

Fernandez, I. and Le Téno, J.-F. (1997) A generic, object-oriented data model for life cycle analysis applications. *Int. J. Life Cycle Assess.*, **2** (2), 81–89.

Finkbeiner, M., Wiedemann, M., and Saur, K. (1998) A comprehensive approach toward product and organisation related environmental management tools – life cycle assessment (ISO 14040) and environmental management systems (ISO 14001). *Int. J. Life Cycle Assess.*, **3** (3), 169–178.

Fleischer, G. (1993) Der "ökologische breakeven- point" für das Recycling, in *Modelle für Eine Zukünftige Siedlungsabfallwirtschaft* (ed. K.J. Thomé-Kozmiensky), EF-Verlag für Energie- und Umwelttechnik, Berlin.

Fleischer, G. (1995) in *Enzyklopädie der Kreislaufwirtschaft, Management der Kreislaufwirtschaft* (Hrsg. K.J. Thomé-Kozmiensky), EF-Verlag für Energie- und Umwelttechnik, Berlin, S. 360–369.

Fleischer, G. and Hake, J.-F. (2002) *Aufwands- und Ergebnisrelevante Probleme der Sachbilanzierung*, Reihe Umwelt/Environment, Bd. **30**, Schriften des Forschungszentrums Jülich. ISBN: 3-89336-293-2.

Fleischer, G. and Schmidt, W.-P. (1996) Functional unit for systems using natural raw materials. *Int. J. Life Cycle Assess.*, **1** (1), 23–27.

Frankl, P. and Rubik, F. (2000) *Life Cycle Assessment in Industry and Business. Adoption Patterns, Applications and Implications*, Springer, Berlin. ISBN: 3-540-66469-6.

Frischknecht, R. (1997) The seductive effect of identical units. *Int. J. Life Cycle Assess.*, **2** (3), 125–126 (Comment to the editorial, loc. cit. Klöpffer, 1997).

Frischknecht, R. (2000) Allocation in life cycle inventory analysis for joint production. *Int. J. Life Cycle Assess.*, **5** (2), 85–95.

Frischknecht, R., Althaus, H.-J., Bauer, C., Doka, G., Heck, T., Jungbluth, N., Kellenberger, D., and Nemecek, T. (2007a) The environmental relevance of capital goods in life cycle assessments of products and services. *Int. J. Life Cycle Assess.*, **12** (Special Issue 1), 7–17.

Frischknecht, R., Althaus, H.J., Dones, R., Hischier, R., Jungbluth, N., Nemecek, T., Primas, A., and Wernet, G. (2007b) Renewable Energy Assessment within the Cumulative Energy Demand Concept: Challenges and Solutions, http://www.esu-services.ch/fileadmin/download/frischknecht-2007-CED-SETAC-Gothenburg.pdf (accessed 26 July 2011).

Frischknecht, R., Jungbluth, N., Althaus, H.-J., Doka, G., Dones, R., Heck, T., Hellweg, S., Hischier, R., Nemecek, T., Rebitzer, G., and Spielmann, M. (2005) The ecoinvent database: overview and methodological framework. *Int. J. Life Cycle Assess.*, **10**, 3–9.

Fritsche, U. R., Buchert, M., Hochfeld, C., Jenseits, W., Matthes, F. C., Rausch, L., Stahl, H. and Witt, J. (2011) Gesamt-Emissions- Modell integrierter Systeme (GEMIS), Version 3.0, Öko-Institut e. V. Darmstadt, Freiburg, Berlin im Auftrag des Hessischen Ministeriums für Umwelt, Jugend, Familie und Gesundheit, 1997. Most recent version: 4.7, 2011.

GEMIS (2001) Fritsche, U. et al. Gesamt-Emissions- Modell Integrierter Systeme, Darmstadt/ Kassel, Version 4.1, http://www.oeko.de/service/gemis/deutsch/index.htm.

Geisler, G., Hofstetter, T.B., and Hungerbühler, K. (2004) Production of fine chemicals: procedure for estimation of LCIs. *Int. J. Life Cycle Assess.*, **9** (2), 101–113.

Giegrich, J., Fehrenbach, H., Orlik, W., and Schwarz, M. (1999) *Ökologische Bilanzen in der Abfallwirtschaft*, UBA Texte 10/99, Umweltbundesamt, Berlin.

Goedkoop, M. (1995) The Eco-indicator 95. Final Report (No. 9523) to National Reuse of Waste Research Programme (NOH), National Institute of Public Health and Environment Protection (RIVM) and Netherlands Agency for Energy and the Environment, Amersfoort, November 1995.

Goedkoop, M., Hofstetter, P., Müller-Wenk, R., and Spriensma, R. (1998) The eco-indicator 98 explained. *Int. J. Life Cycle Assess.*, **3** (6), 352–360.

Grahl, B. and Schmincke, E. (1996) Evaluation and decision-making processes in life cycle assessment. *Int. J. Life Cycle Assess.*, **1** (1), 32–35.

Grießhammer, R., Bunke, D., and Gensch, C.-O. (1997) *Produktlinienanalyse Waschen und Waschmittel*, Forschungsbericht 102 07 202, UBA-FB 97-009. UBA-Texte 1/97, Öko-Institut e. V., Freiburg, Berlin.

Guinée, J. B., (final editor) Gorée, M., Heijungs, R.; Huppes, G.; Kleijn, R., Koning, A. de; Oers, L. van; Wegener Sleeswijk, A.; Suh, S.; Udo de Haes, H. A.; Bruijn, H. de; Duin, R. van; Huijbregts, M. A. J.: *Handbook on Life Cycle Assessment – Operational Guide to the ISO Standards*. Kluwer Academic Publishers, Dordrecht. 2002. ISBN: 1-4020-0228-9

Guinée, J., Heijungs, R., and Huppes, G. (2004) Economic allocation: examples and derived decision tree. *Int. J. Life Cycle Assess.*, **9** (1), 23–33.

Günther, A. and Holley, W. (1995) Aggregierte Sachökobilanz-Ergebnisse für Frischmilch und Bierverpackungen. *Verpackungs-Rundschau*, **46** (3), 53–58.

Hau, J.L., Yi, H.-S., and Bakshi, B.R. (2007) Enhancing life-cycle inventories via reconciliation with the laws of thermodynamics. *J. Ind. Ecol.*, **11** (4), 5–25.

Hauschild, M. and Wenzel, H. (1998) *Environmental Assessment of Products*, Scientific Background, vol. 2, Chapman & Hall, London. ISBN: 0-412-80810-2.

Heijungs, R. (1997) Economic drama and the environmental stage. Formal derivation of algorithmic tools for environmental analysis and decision-support from a unified epistemological principle. Proefschrift (PhD Dissertation). Institute of Environmental Sciences, Leiden. ISBN: 90-9010784-3.

Heijungs, R. (2001) *A Theory of the Environmental and Economic Systems – a Unified*

Framework for Ecological Economic Analysis and Decision Support, Edward Elgar, Cheltenham ISBN: 1-84064-643-8.

Heijungs, R. (2005) On the use of units in LCA. Int. J. Life Cycle Assess., 10 (3), 173–176.

Heijungs, R. and Frischknecht, R. (1998) On the nature of the allocation problem. A special view on the nature of the allocation problem. Int. J. Life Cycle Assess., 3 (6), 321–332.

Heijungs, R. and Suh, S. (2002) The Computational Structure of Life Cycle Assessment, Kluwer Academic Publishers, Dordrecht. ISBN: 1-4020-0672-1.

Heijungs, R. and Suh, S. (2006) Reformulation of matrix-based LCI: from product balance to process balance. J. Cleaner Prod., 14 (1), 47–51.

Heintz, B. and Baisnée, P.-F. (1992) Life-Cycle Assessment. Workshop Report, 2–3 December 1991, Leiden Society of Environmental Toxicology and Chemistry – Europe (ed.), SETAC-Europe, Brussels, pp. 35–52.

Hemming, C. (1995) SPOLD-Directory of Life Cycle Inventory Data Sources. Report by Chrisalis Environmental Consulting (UK) to the, Society for the Promotion of LCA Development (SPOLD), Brussels.

Heyde, M. and Kremer, M. (1999) Recycling and recovery of plastics from packaging, in Domestic Waste. LCA-Type Analysis of Different Strategies, vol. 5 (eds W. Klöpffer and O. Hutzinger), LCA Documents, Ecoinforma Press, Bayreuth. ISBN: 3-928379-57-7.

Hindle, P. and de Oude, N.T. (1996) SPOLD Society for the promotion of life cycle development. Int. J. Life Cycle Assess., 1 (1), 55–56.

Huijbregts, M.A.J., Norris, G., Bretz, R., Ciroth, A., Maurice, B., von Bahr, B., Weidema, B., and de Beaufort, A.S.H. (2001) Framework for modelling data uncertainty in life cycle inventories. Int. J. Life Cycle Assess., 6 (3), 127–132.

Hulpke, H., Koch, H.A. and Wagner, R. (Hrsg.) (1993) Römpp Lexikon Umwelt, Georg Thieme Verlag, Stuttgart. ISBN: 3-13-736501-5.

Hulpke, H. and Marsmann, M. (1994) Ökobilanzen und Ökovergleiche. Nachr. Chem. Tech. Lab., 42 (1), 11–27, 8 Seiten (nicht paginierter Sonderdruck).

Hunt, R.G., Boguski, T.K., Weitz, K., and Sharma, A. (1998) Case studies examining streamlining techniques. Int. J. Life Cycle Assess., 3, 36–42.

Hunt, R.G., Sellers, J.D., and Franklin, W.E. (1992) Resource and environmental profile analysis a life cycle environmental assessment for products and procedures. Environ. Impact Assess. Rev., 12, 245–269.

Huppes, G. and Schneider, F. (eds) (1994) Proceedings of the European Workshop on Allocation in LCA. Leiden, February 1994, SETAC-Europe, Brussels.

IFEU (2004) Detzel, A., Ostermayer, A., Böß, A. and Gromke, U. Ökobilanz Getränkekarton für Saft – Bezugsjahr 2002, Instituts für Energie- und Umweltforschung (IFEU). Im Auftrag des Fachverband Kartonverpackungen, Wiesbaden 2004 (unpublished).

IFEU (2004a) Detzel, A., Giegrich, J., Kiuger, M.) Ökobilanz PETEinwegverpackungen und sekundäre Verwertungsprodukte, IFEU, Heidelberg, August 2004 Im Auftrag von PETCORE, Brüssel.

INFRAS (2004b) Handbuch für Emissionsfaktoren des Straßenverkehrs Version 2.1, im Auftrag des, Umweltbundesamts (UBA), Berlin und Bundesamts für Umwelt, Wald und Landwirtschaft (BUWAL), Bern.

IFEU (2006) Detzel, A. and Böß, A. Ökobilanzieller Vergleich von Getränkekartons und PETEinwegflaschen. Endbericht, Institut für Energie und Umweltforschung (IFEU) Heidelberg an den Fachverband Kartonverpackungen (FKN), Wiesbaden, August 2006.

Institut für Energie- und Umweltforschung Heidelberg GmbH (IFEU) (2000) Ökologischer Vergleich graphischer Papiere Eine Ökobilanz zur Entsorgung graphischer Altpapiere sowie zu den Produktgruppen Zeitungen, Zeitschriften und Kopien, UBA Texte 22/00, Projekt FKZ 103 501 20, UBA, Berlin.

Institut für Energie- und Umweltforschung Heidelberg (2006) TREMOD: Transport Emission Model. Energy Consumption and Emissions of Transport in Germany 1960–2030. UFOPLAN 204 45 139, Final Report to UBA, Dessau, March 2006.

International Organization for Standardization (ISO) (2000): *Life Cycle Assessment – Examples of the Application of Goal and Scope Definition and Inventory Analysis*. Technical Report ISO TR 14049, International Organization for Standardization (ISO), Geneva.

ISO (International Standard Organization) (1998a) ISO EN 14041 *Norme Européenne (CEN): Environmental Management – Life Cycle Assessment: Goal and Scope Definition and Inventory Analysis (Festlegung des Ziels und des Untersuchungsrahmens sowie Sachbilanz)*, International Standard (ISO), Geneva.

ISO (International Standard Organization) (1998b) ISO 1000 *SI Units and Recommendations for the Use of their Multiples and of Certain other Units*, International Standard, Geneva.

ISO (International Standard Organization) (2002) *Environmental Management – Life Cycle Assessment. Data Documentation Format*. Technical Specification ISO/TS 14048, ISO/TC 207/SC 5, ISO, Geneva.

ISO (International Standard Organization) (2004) EN ISO 14001 *Umweltmanagement Systeme – Anforderungen mit Anleitung zur Anwendung*, ISO, Geneva.

ISO (International Standard Organization) (2006a) ISO 14040 *Environmental Management – Life Cycle Assessment – Principles and Framework*, ISO, Geneva, October 2006

ISO (International Standard Organization) (2006b) ISO 14044 *Environmental Management – Life Cycle Assessment – Requirements and Guidelines*, ISO, Geneva, October 2006.

IVV (2001) Bez, J., Goldhahn, G. and Buttker, B.) Methanol aus Abfall. Ökobilanz bescheinigt gute Noten. Fraunhofer IVV. Freising. *Müll Abfall*, **33** (3), 158–162.

Janzen, D.C. (1995) Methodology of the European surfactant life-cycle inventory for detergent surfactants production. *Tenside Surfactants Deterg.*, **32**, 110–121.

Kennedy, D.J., Montgomery, D.C., and Quay, B.H. (1996) Data quality. Stochastic environmental life cycle assessment modeling – a probabilistic approach to incorporating variable input data quality. *Int. J. Life Cycle Assess.*, **1** (4), 199–207.

Kennedy, D.J., Montgomery, D.C., Rollier, D.A., and Keats, J.B. (1997) Data quality. Assessing input data uncertainty in life cycle assessment inventory models. *Int. J. Life Cycle Assess.*, **2** (4), 229–239.

Kim, S. and Overcash, M. (2000) Allocation procedure in multi-output processes: an illustration of ISO 14041. *Int. J. Life Cycle Assess.*, **5** (4), 221–228.

Kindler, H. and Nikles, A. (1980) Energieaufwand zur Herstellung von Werkstoffen – Berechnungsgrundsätze und Energieäquivalenzwerte von Kunststoffen. *Kunststoffe*, **70**, 802–807.

Klöpffer, W. (1994) in *Integrating Impact Assessment into LCA* (eds H.A. Udo de Haes, A.A. Jensen, W. Klöpffer, and L.-G. Lindfors). Proceeding of the LCA Symposium held at the Fourth SETAC-Europe Annual Meeting, 11–14 April 1994, Brussels, Society of Environmental Toxicology and Chemistry – Europe, Brussels, pp. 11–15.

Klöpffer, W. (1995) Exposure and hazard assessment within life cycle impact assessment. *Environ. Sci. Pollut. Res.*, **2** (1), 38–40.

Klöpffer, W. (1996a) Allocation rules for openloop recycling in life cycle assessment – A review. *Int. J. Life Cycle Assess.*, **1**, 27–31.

Klöpffer, W. (1996b) Environmental hazard assessment of chemicals and products. Part V. Anthropogenic chemicals in sewage sludge. *Chemosphere*, **33**, 1067–1081.

Klöpffer, W. (1997) In defense of the cumulative energy demand (Editorial). *Int. J. Life Cycle Assess.*, **2** (2), 61.

Klöpffer, W. (2002) The second Dutch LCAguide, published as book (Guinée et al. 2002). Book review. *Int. J. Life Cycle Assess.*, **7**, 311–313.

Klöpffer, W. and Grahl, B. (2009) *Ökobilanz (LCA) – Ein Leitfaden für Ausbildung und Beruf*, Wiley-VCH Verlag GmbH, Weinheim. ISBN: 978-3-527-32043-1.

Klöpffer, W., Grießhammer, R., and Sundström, G. (1995) Overview of the scientific peer review of the European life cycle inventory for surfactant production. *Tenside Surfactants Deterg.*, **32**, 378–383.

Klöpffer, W. and Renner, I. (1995) *Methodik der Wirkungsbilanz im Rahmen von ProduktÖkobilanzen unter Berücksichtigung nicht oder nur schwer quantifizierbarer Umwelt-Kategorien*, UBA-Texte 23/95, Bericht der C.A.U. GmbH, Dreieich, an das Umweltbundesamt (UBA), Berlin. ISSN: 0722-186X.

Klöpffer, W., Sundström, G., and Grießhammer, R. (1996) The peer reviewing process – a case study: European life cycle inventory for surfactant production. *Int. J. Life Cycle Assess.*, **1** (2), 113–115.

Klöpffer, W. and Volkwein, S. (1995) in *Enzyklopädie der Kreis laufwirtschaft, Management der Kreislaufwirtschaft* (Hrsg., K.J. Thomé-Kozmiensky), EF-Verlag für Energie- und Umwelt technik, Berlin, pp. 336–340.

Kougoulis, J.S. (2007) Symmetric functional modeling in life cycle assessment multiple-use energy modules. PhD thesis. TU Berlin; publiziert im Shaker Verlag, Aachen 2008.

Lichtenvort, K. (2004) Systemgrenzenrelevante Änderungen von Flussmengen in der Öko bilanzierung. PhD thesis. TU Berlin.

Lindfors, L.-G., Christiansen, K., Hoffmann, L., Virtanen, Y., Juntilla, V., Leskinen, A., Hansen, O.-J., Rønning, A., Ekvall, T., and Finnveden, G. (1994a) LCA-Nordic, Technical Reports No. 1–9. Tema Nord 1995:502. Nordic Council of Ministers, Copenhagen.

Lindfors, L.-G., Christiansen, K., Hoffmann, L., Virtanen, Y., Juntilla, V, Leskinen, A., Hansen, O.-J., Rønning, A., Ekvall, T., Finnveden, G., Weidema, B.P., Ersbøll, A.K., Bomann, B. and Ek, M. (1994b) LCA-Nordic. Technical Reports No. 10 and Special Reports No. 1–2. Tema Nord 1995:503, Nordic Council of Ministers. Copenhagen. ISBN: 92-9120-609-1.

Lindfors, L.-G., Christiansen, K., Hoffmann, L., Virtanen, Y., Juntilla, V., Hanssen, O.-J., Rønning, A., Ekvall, T., and Finnveden, G. (1995) *Nordic Guidelines on Life-Cycle Assessment*. Nord 1995: 20, Nordic Council of Ministers, Copenhagen.

Mauch, W. and Schaefer, H. (1996) in *Ganzheitliche Bilanzierung. Werkzeug zum Planen und Wirtschaften in Kreisläufen* (Hrsg. P. Eyrer), Springer, Berlin, S. 152–180. ISBN: 3-540-59356-X.

Meadows, D.L., Meadows, D.H., Zahn, E., and Milling, P. (1973) *Die Grenzen des Wachstums. Bericht des Club of Rome zur Lage der Menschheit* Ts. rororo Taschenbuch, Hamburg 1973; neue Auflage im, dtv Taschenbuchverlag, pp. 101–200. ISBN: 3-499-16825-1.

Mill, J.S. (1848) *Principles of Political Economy, with some of their Applications to Social Philosophy*. First published by, J. W. Parker, London.

Mills, I., Cvitaš, T., Homann, K., Kallay, N., and Kuchitsu, K. (eds) (1988) *IUPAC. Quantities, Units and Symbols in Physical Chemistry*, Blackwell Scientific Publications, Oxford. ISBN: 0-632-01773-2.

Österreichisches Statistisches Zentralamt (1995) *Statistisches Jahrbuch für die Republik Österreich*. 46. Jhg., Neue Folge, Österreichische Staatsdruckerei, Wien. ISBN: 3-901400-04-4.

PlasticsEurope (2005a) Boustead, I. Eco-Profiles of the European Plastics Industry – Low Density Polyethylene (LDPE), data last calculated March 2005, report prepared for PlasticsEurope, Brussels, 2005 (download August 2005 von: http://www.lca.plasticseurope.org/ index.htm (accessed 28 October 2013)).

PlasticsEurope (2005b) Boustead, I. Eco-Profiles of the European Plastics Industry – High Density Polyethylene (HDPE), data last calculated March 2005, report prepared for PlasticsEurope, Brussels, 2005 (download August 2005 von: http://www.lca.plasticseurope.org/ index.htm (accessed 28 October 2013)).

PlasticsEurope (2005c) Boustead, I. Eco-Profiles of the European Plastics Industry – Polypropylene (PP), data last calculated March 2005, report prepared for PlasticsEurope, Brussels 2005 (download August 2005 von: http://www.lca.plasticseurope.org/ index.htm (accessed 28 October 2013).

PlasticsEurope (2005d) Boustead, I. Eco-Profiles of the European Plastics Industry – Nylon 66, data last calculated March 2005, report prepared for PlasticsEurope, Brussels 2005 (download August 2005 von: http://www.lca.plasticseurope.org/index.htm (accessed 28 October 2013)).

Pohl, W. (1992) *Lagerstättenlehre*, 4 Auflage, E. Schweizerbart'sche Verlagsbuchhandlung, Stuttgart.

Rebitzer, G. (2005) Enhancing the Application Efficiency of Life Cycle Assessment for Industrial Uses. Thèse No. 3307. École Polytechnique Féderale de Lausanne.

Rice, G., Clift, R., and Burns, R. (1997) LCA software review. Comparison of currently available European LCA software. *Int. J. Life Cycle Assess.*, **2** (1), 53–59.

Riebel, P. (1955) *Die Kuppelproduktion. Betriebs- und Marktprobleme*, Westdeutscher Verlag, Köln.

Ross, S., Evans, D., and Webber, M. (2002) How LCA studies deal with uncertainty. *Int. J. Life Cycle Assess.*, **7** (1), 47–52.

SETAC (1991) Fava, J.A., Denison, R., Jones, B., Curran, M.A., Vigon, B., Selke, S. and Barnum, J. (eds) *SETAC Workshop Report: A Technical Framework for Life Cycle Assessments, August 18–23, 1990. Smugglers Notch, Vermont*, SETAC, Washington, DC.

SETAC-Europe (1996) *Inventory Analysis* (ed. R. Clift) Report of the, SETAC-EuropeWorking Group, Brussels (unpublished).

Schaltegger, S. (ed.) (1996) *Life Cycle Assessment (LCA) – Quo Vadis?*, Birkhäuser Verlag, Basel. ISBN 3-7643-5341-4 (Basel), ISBN 0-8176-5341-4 (Boston).

Schmidt, W.-P. (1997) Ökologische Grenzkosten der Kreislaufwirtschaft. Dissertation. Technische Universität Berlin, September 1997.

Schmidt, M. and Schorb, A. (1995) *Stoffstromanalysen in Ökobilanzen und Öko-Audits*, Springer-Verlag, Berlin. ISBN: 3-540-59336-5.

Schmidt-Bleek, F. (1993) MIPS re-visited. *Fresenius Environ. Bull.*, **2**, 407–412.

Schmidt-Bleek, F. (1994) *Wie viel Umwelt braucht der Mensch? MIPS – Das Maß für ökologisches Wirtschaften*, Birkhäuser Verlag, Berlin.

Schmitz, S., Oels, H.-J., and Tiedemann, A. (1995) Ökobilanz für Getränkeverpackungen. Teil A: Methode zur Berechnung und Bewertung von Ökobilanzen für Verpackungen. Teil B: Vergleichende Untersuchung der durch Verpackungssysteme für Frischmilch und Bier hervorgerufenen Umweltbeeinflussungen, UBA Texte 52/95, UBA, Berlin.

SIA (2010) Swiss Standard on Grey Energy of Buildings: SIA 2032.

Siegenthaler, C.P., Linder, S., and Pagliari, F. (1997) *LCA Software Guide 1997. Market Over View – Software Portraits*, ÖBU Schriften reihe 13, ÖBU, Adliswil (Schweiz). ISBN: 3-908233-14-3..

Singhofen, A., Hemming, C.R., Weidema, B.P., Grisel, L., Bretz, R., De Smet, B., and Russel, D. (1996) Life cycle inventory data; development of a common format. *Int. J. Life Cycle Assess.*, **1** (3), 171–178.

Society of Environmental Toxicology and Chemistry – Europe (ed.) (1992) *Life-Cycle Assessment*. Workshop Report, 2–3 December 1991, Leiden, SETAC-Europe, Brussels.

Society of Environmental Toxicology and Chemistry (SETAC) (1993) *Guidelines for Life- Cycle Assessment: A "Code of Practice"*, 1st edn, From the SETAC Workshop held at Sesimbra, Portugal, 31 March–3 April 1993, Society of Environmental Toxicology and Chemistry (SETAC), Brussels and Pensacola, FL, August 1993.

UNEP/SETAC (2011) Sonnemann, G. and Vigon, B. (eds) *Global Guidance Principles for Life Cycle Assessment Databases – "Shonan Guiding Principles"*, UNEP/SETAC Life Cycle Initiative, Paris www.unep.org/pdf/Global Guidance -principles for LCA.pdf (accessed 17 December 2013).

Spatari, S., Betz, M., Florin, H., Baitz, M., and Faltenbacher, M. (2001) Using GaBi 3 to perform life cycle assessment and life cycle engineering. *Int. J. Life Cycle Assess.*, **6** (2), 81–84.

Stalmans, M., Berenbold, H., Berna, J.L., Cavalli, L., Dillarstone, A., Franke, M., Hirsinger, F., Janzen, D., Kosswig, K., Postlethwaite, D., Rappert, T., Renta, C., Scharer, D., Schick, K.-P., Schul, W., Thomas, H., and Van Sloten, R. (1995) European lifecycle inventory for detergent surfactants production. *Tenside Surfactant Deterg.*, **32**, 84–109.

Tiedemann, A. (Hrsg.)2000) *Ökobilanzen für graphische Papiere*, UBA Texte 22/2000, Umweltbundesamt, Berlin.

Tukker, A. (1998) *Frames in the Toxicity Controversy. Risk Assessment and*

Policy Analysis Related to the Dutch Chlorine Debate and the Swedish PVC Debate, Kluwer Academic Publishers, Dordrecht. ISBN: 0-7923-5554-7.

Tukker, A., Kleijn, R. and van Oers, L. (1996) A PVC Substance Flow Analysis for Sweden. TNO-report STB/96/48-III, Report by TNO Centre for Technology and Policy Studies and Centre of Environmental Science (CML) Leiden to Norsk Hydro, Apeldoorn, November 1996.

UBA (2000a) Plinke, E., Schonert, M., Meckel, H., Detzel, A., Giegrich, J., Fehrenbach, H., Ostermayer, A., Schorb, A., Heinisch, J., Luxenhofer, K. and Schmitz, S. *Ökobilanz für Getränkeverpackungen II*, UBA-Texte 37/00, Zwischenbericht (Phase 1) zum Forschungsvorhaben FKZ 296 92 504 des, Umweltbundesamt, Berlin – Hauptteil, September 2000. ISSN: 0722-186X.

UBA (2000b) Umweltbundesamt (Hrsg.) *Ökobilanzen für graphische Papiere*, UBA-Texte 22/00, Umweltbundesamtes, Berlin.

UBA (2000c) Umweltbundesamt (Hrsg.) *Ökologische Bilanzierung von Altölverwertungswegen*, UBA-Texte 20/00, Umweltbundesamtes, Berlin.

UBA (2001) Umweltbundesamt (Hrsg.) *Grundlagen für eine ökologisch und ökonomisch sinnvolle Verwertung von Verkaufsverpackungen*, Umweltbundesamt, Berlin.

UBA (2002) Schonert, M., Metz, G., Detzel, A., Giegrich, J., Ostermayer, A., Schorb, A., and Schmitz, S. *Ökobilanz für Getränkeverpackungen II, Phase 2*, UBA-Texte 51/02. Forschungsbericht 103 50 504 UBA-FB 000363 des, Umweltbundesamtes, Berlin, Oktober 2002. ISSN 0722-186X.

Udo de Haes, H.A., and de Snoo, G.R. (1996) Environmental certification. Companies and products: two vehicles for a life cycle approach? *Int. J. Life Cycle Assess.*, **1** (3), 168–170.

VDI (1997) *VDI-Richtlinie 4600: Kumulierter Energieaufwand (Cumulative Energy Demand). Begriffe, Definitionen, Berechnungs methoden.* Deutsch and Englisch, Verein Deutscher Ingenieure, VDI-Gesellschaft Energietechnik Richtlinienausschuss Kumulierter Energieaufwand, Düsseldorf.

Vigon, B.W. (1996) in *Environmental Life-Cycle Assessment* (ed. M.A. Curran), McGraw-Hill, New York, pp. 3.1–3.25. ISBN: 0-07-015063-X.

Weidema, B. (1998) Remark: standards for data exchange or for databases? *SETAC-Europe News*, **9** (3), 14.

Weidema, B. (2001) Avoiding co-product allocation in life-cycle assessment. *J. Ind. Ecol.*, **4** (3), 11–33.

Weidema, B.P., Frees, N., and Nielsen, A.-M. (1999) Marginal production technologies for life cycle inventories. *Int. J. Life Cycle Assess.*, **4** (1), 48–56.

Weidema, B. (2000) Avoiding co-product allocation in life-cycle assessment. *J. Indust. Ecology*, **4** (3), 11–33.

Wenzel, H., Hauschild, M., and Alting, L. (1997) *Environmental Assessment of Products*, Methodology, Tools and Case Studies in Product Development, vol. 1, Chapman & Hall, London. ISBN: 0-412-80800-5.

Werner, F. and Richter, K. (2000) Economic allocation in LCA a case study about aluminium window frames. *Int. J. Life Cycle Assess.*, **5** (2), 79–83.

White, P., Franke, M., and Hindle, P. (1995) *Integrated Solid Waste Management: A Life Cycle Inventory*, Blackie Academic & Professional, London.

Wrisberg, N., Udo de Haes, H.A., Triebswetter, U., Eder, P., and Clift, R. (eds) (2002) *Analytical Tools for Environmental Design and Management in a Systems Perspective*, Kluwer Academics Publishers, Dordrecht.

# 4
# Life Cycle Impact Assessment

*It must be emphasized that these methods of analysis do not indicate that actual impacts will be observed in the environment because of the life cycle of the product or process under study, but only that there is a potential linkage between the product or process life cycle and the impacts*

*Reinout Heijungs and Jeroen B. Guinée*[1]

## 4.1
### Basic Principle of Life Cycle Impact Assessment

The life cycle impact assessment (LCIA) is the second predominantly scientific phase of life cycle assessment (LCA), together with the life cycle inventory (LCI) analysis inserted between the two scientifically 'softer' components 'Definition of goal and scope' before the LCI and 'Interpretation' after LCIA.[2]

ISO 14044 refers to two types of studies: LCA studies and LCI studies. Inventory studies do not contain an impact assessment, they do however contain the phases 'definition of goal and scope' as well as 'interpretation'. An inventory study is therefore not to be confused with the LCA phase 'life cycle inventory analysis' (see Chapter 3).

Why is an impact assessment necessary for a full LCA?

1. An LCA or *eco*balance[3] requires considering and quantifying substantial environmental aspects, which refers to inputs and outputs that can interact with the environment, and the consequential potential environmental impacts, related to an examined product system. The inventory supplies environmental aspects of the defined product system as inputs and outputs per functional unit (fU). In order to derive potential environmental impacts from these data

---

1) Heijungs and Guinée (1993).
2) Klöpffer (1994a, 1997a, 1998a).
3) The term 'ecobalance' was frequently used in the time of the 'proto-LCAs', it is still used in Japan to name the yearly LCA conference and it is the English equivalent to the official German name of LCA: Ökobilanz; ISO 14040+44 (official German translation); Klöpffer and Grahl (2009).

*Life Cycle Assessment (LCA): A Guide to Best Practice*, First Edition.
Walter Klöpffer and Birgit Grahl.
© 2014 Wiley-VCH Verlag GmbH & Co. KGaA. Published 2014 by Wiley-VCH Verlag GmbH & Co. KGaA.

further work is necessary, the impact assessment. A summary of different definitions of the impact assessment can be found at.[4]

2. With a complete inventory, numerous data on mass flows, emissions, resource consumption and energy demand are present, which are difficult to handle and therefore, make aggregations desirable (see Section 3.7).
3. An inventory supplies more information than can be expected at first sight of a listing of raw input and output data.
4. An ecological product comparison must not imply that, for example, a product system A using less energy in its life cycle than product system B, but with emissions of (environmental) toxic substances with a small mass flow but substantial impact, performs better than product system B.

For these reasons there has been a continuing effort to develop a type of impact related aggregation of the inventory results which goes beyond a cumulative energy demand (CED) (see Section 3.2.2). Also the sum of solid wastes can be viewed as an aggregation and as sum parameter at inventory level. It may be used as a measure for the material throughput. Besides, this value gives a reference to the primarily technical field of waste disposal, but this issue with its negative side effects is traditionally regarded as an environmental problem area. The CED and sum of solid waste were typical aggregations during the time of the 'proto-LCA'.

The best-known earliest proposal of an aggregated impact estimation is the Swiss method 'of critical volumes' (c.V.), which is discussed in Section 4.2. Started around 1992 it was increasingly replaced by the method of environmental problem fields or impact categories, developed by CML (Centrum voor Milieukunde Leiden), Leiden[5] (see Section 4.4). Environmental problem fields or impact categories are, for example, 'acidification' or 'climate change'.

The standardisation of the LCA in the ISO 14040 series of standards corresponds by structure and content to a large extent to the ideas developed by CML, although only with a general definition of impact assessment[6] and no concrete recommendations for impact categories:

*Life cycle impact assessment LCIA:*
*Phase of life cycle assessment aimed at understanding and evaluating the magnitude and significance of the potential environmental impacts for a product system throughout the life cycle of the product*

In ISO 14044, the formulation *potential environmental impacts* emphasises that the LCIA is *not* to be confused with an environmental risk assessment; in this case substance-immanent properties would have to be correlated with the concentration of these substances at the site of impact.[7]

---

4) Owens (1998).
5) Gabathuler (1998).
6) ISO (2006a, Section 3.4).
7) Such a risk assessment is, for example, requested by the European chemicals law REACH, EC (2006).

> **Example**
>
> **Potential Environmental Impact**
>
> An impact is always by definition related and unambiguously assignable to a cause. The environmental impacts of a product system in its life cycle have their cause in consumption (inputs) and releases (outputs), which are determined in the inventory. If, for example, within various processes of a life cycle of a product acids (substance-immanent property: release of $H_3O^+$-ions in aqueous solution and thus decrease of pH value) are released into the air, which thenes reach the soil and rivers, then acids are the cause for acid rain as well as soil and water acidification. The decreased pH value can have a set of impacts, like skin damage, fish mortality, remobilisation of heavy metals and much more. Insofar cause–effect relationships exist.
>
> As consumptions and releases of the product system which are listed in the inventory can rarely be assigned to a single definable location, the extent of damage at a certain place cannot or can only rarely be quantified: Concerning an environmental impact to be expected, it is a substantial difference whether 1 kg hydrogen chloride (HCl) eludes within a short time from only one chimney into the neighbourhood or whether during the entire life cycle of the product small quantities are released from many plants distributed over a large geographical area resulting in a 1 kg release, calculated on the overall system and applied to the fU. As the fU is chosen by convenience, results of the inventory can amount to a multiple or a fraction of 1 kg. The results of an inventory can therefore not be correlated to existing concentrations. Two product systems with correctly defined fUs can however be compared to one another concerning the output 'HCl into air'.
>
> To adequately account for the uncertainty of the exposition, we speak of 'potential environmental impacts' in LCIA. If a differentiated exposition analysis is accomplished and thus a risk assessment is feasible, this has to be explicitly described in the context of the impact assessment (see also Section 4.5.3).

## 4.2
## Method of Critical Volumes

The method of critical volumes, although outdated, is shortly appreciated for its impact on the methodological development, as also on the CML method. It was suggested for the first time in the famous Swiss 'BUS report' in 1984[8] and contains an aggregation of the emissions into air and into water in the case of existing regulations indicating threshold values. The method can also be applied to the soil compartment but was only rarely used for lack of threshold values. The method can, in principle, be applied to ground waters also.

---

[8] BUS (1984), BUWAL (1991) and Klöpffer and Renner (1995).

## 4 Life Cycle Impact Assessment

An aggregation to critical volumes starts with the 'critical volume' (c.V.) of an emission $i$ per functional unit (fU) into an environmental medium $j$ according to Equation 4.1:

$$c.V._{ij} = \frac{\text{Emission}_i/\text{functional unit emitted into medium } j \text{ (mass)}}{\text{limit value for } i \text{ in medium } j \text{ (mass/volume)}} \text{ (volume)} \quad (4.1)$$

$j$: air, water and soil (and/or groundwater).

The emission per fU refers to the *entire life cycle*, therefore corresponds to the mass aggregation of all unit processes corresponding to the individually released substances $i$ in the inventory. As such, only direct releases into the same medium $j$ are aggregated. A further distribution in the sense of multimedia models (see Section 4.5.3.2.4) is not made. As most limit values have the dimension mass per volume, the aggregation has the dimension volume and is called *critical volume*. A definition of the limit values as mass per mass, for instance in the soil compartment, would lead to a 'critical mass', which however is not common.[9]

The $c.V._i$ of one substance $i$ has a clear meaning: it represents the volume of pure air, pure water or pure soil, which is needed to dilute a released pollutant quantity $c.V._i/fU$ in order to just obtain the threshold value concentration. Here, particularly very high values can occur for the medium air: A released mass of $1 \text{ kg fU}^{-1}$ of a substance $i$ with a threshold value of $1 \text{ µg m}^{-3}$ results in:

$$c.V._i = 10^9 \text{m}^3 \equiv 1 \text{ km}^3$$

For an aggregation of several emitted substances this visual evidence is lost, since all compounds occupy the same virtual volume together. With an aggregation for each compartment $j$ the sum of c.V. of all individual emissions $i$ is formed for which both the data of the inventory and limit values are present (Equation 4.2):

$$c.V._{\cdot j} = \sum_i c.V._{\cdot i,j} \quad (4.2)$$

As an example air pollutants of a non-specified packaging are considered here (Table 4.1). The fU is the filling, packing and the transport of 1000 l of fruit juice.

### 4.2.1
### Interpretation

Since existing limit values determine the size of the c.V., those air pollutants with the lowest limit values specified in Table 4.1 dominate the sum value. In regulations the lowest limits can be found for substances damaging human health in low concentrations. Limits that encompass the damage to ecological systems as a whole do not exist. Thus in the above example the aggregated value is dominated by the c.V. for $SO_2$ and $NO_x$. The Hydrocarbons (CH) are ranked as less toxic (higher limit value) and are almost negligible for the final result. As no limit value is specified for $CO_2$, it is not considered at all.

---

9) And give an association to the atomic bomb.

**Table 4.1** Critical volume air (example).

| Pollutant | Load (mg fU$^{-1}$) | Limit value$^a$ (mg m$^{-3}$) | c.V. (m$^3$ fU$^{-1}$) |
|---|---|---|---|
| Sulphur dioxide SO$_2$ | 467 000 | 0.03 | 15.6 × 10$^6$ |
| Nitrogen oxides NO$_x$ | 199 000 | 0.03 | 6.63 × 10$^6$ |
| Hydrocarbons HC | 323 000 | 15 | 0.02 × 10$^6$ |
| Carbon dioxide CO$_2$$^b$ | 77 775 000 | None ('∞') | 0 |
| Sum | — | — | 22.25 × 10$^6$ |

$^a$Limit values according to BUWAL (1990): Swiss MIC values are the basis for sulphur dioxide and nitrogen oxides. The value for HC was approximated from MAK-values. (MIC: maximum immission concentration; MAK[10] (OEL).
$^b$CO$_2$ is not considered as a toxic gas; in the BUWAL method, climate change was not yet recognised as an impact.

The 22 million cubic metres of 'c.V. air' do not have a descriptive meaning because the individual pollutants occupy the same volume, an addition of the volumes physically, thus, does not make sense. For weighting however this is meaningless, it is only important that substances with low limit values are weighted stronger than those with higher limit values. The dimension is for formal reasons a volume, the unit either (m$^3$) (with air) or (l) (with water), which however has no practical significance. The way the c.V. are calculated permits an aggregation and a relative grading within a comparison of different product systems.

The representation of the results according to the BUWAL method is usually done numerically or by bar charts in an 'eco-profile' (Figure 4.1).

The eco-profile has the advantage of a simple representation and a direct, also visual, comparability. In a highly aggregated form advantages and weaknesses of compared systems are now comparable with regard to emissions with existing limit values. In view of a successive system optimisation the detailed data should not get lost during the aggregation. Otherwise it is not possible to conclude during which part of the life cycle the load has occurred and where to start to make improvements. Representation in bar charts with differently specified life cycle stages is common practice to date even if other units are used (see Figure 4.4).

### 4.2.2
### Criticism

This old method, ideal in view of feasibility, simplicity and reproducibility, which could even be easily expanded in case of existing limit values, has however been critically rated since about 1993 (CML method of impact categories)[11] and is

---

10) MAK (Maximale Arbeitsplatzkonzentration): The highest legal permissible concentration of a substance in the air at the workplace; corresponds to OEL (Occupational Exposure Limit).
11) Heijungs et al. (1992).

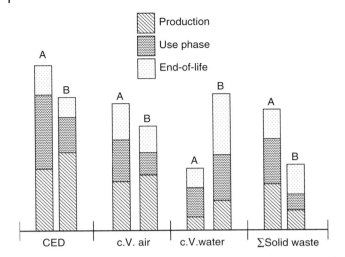

**Figure 4.1** Schematic representation of two fictitious eco-profiles (two product systems related to the functional unit): Cumulative energy demand (CED); c.V. air; c.V. water; Sum of solid wastes (example of an early 'impact assessment').

thus but rarely being applied outside of Switzerland[12] (in parts by BASF's[13] eco-efficiency method).[14] The most important reasons for this declining attitude despite operational advantages are:

1. Purely scientifically deduced limit values occur in the rarest cases; they contain elements of feasibility, analytic detectability, cognitive boundaries of science, of social or economic desirability, and so on. Even if some limits may be close to scientifically acknowledged toxicologically deduced effect thresholds, for others political objectives prevail. An evaluating element is particularly integrated into impact assessment by limit values that are partly defined by political reasoning. This results in a blend of impact assessment and valuation, for example, included in the so-called ecopoint methods.[15] Within these a border line between impact assessment in the strict sense and valuation (weighting) is no longer perceptible.
2. The limits vary by country. If no limits are available, values are designed using various auxiliary assumptions (see hydrocarbons/HC according to BUWAL in Table 4.1). This arbitrariness in use of self-made 'limits' has played a key role in discrediting the method.
3. For many substances, especially in case of impacts without effect threshold, no legal limits exist or, if they do, they exist only in the form of technical indicative values, as with carcinogenic working substances. The same is true for many

---

12) Klöpffer (1994a) and Klöpffer and Renner (1995).
13) Saling et al. (2002) and Landsiedel and Saling (2002).
14) German: Ökoeffizienzmethode.
15) BUWAL (1998, 1990), Steen and Ryding (1992), Goedkoop (1995) and Goedkoop et al. (1998).

substances that are either only produced in very small quantities or considered as harmless because they have so far never been correlated with poisoning or environmental problems.
4. Most limit values only consider human health and thus do not cover ecological toxicity. Limit values at the ecological system level do not exist at all. An extreme example is carbon dioxide, which for humans is only poisonous in very high concentrations. The gas to date is regarded as the most important greenhouse gas (GHG, see Section 4.5.2.2). It was, however, not included in BUWAL 1991 for a computation of the c.V. of air and thus not evaluated[16], for lack of limit values.

## 4.3 Structure of Impact Assessment according to ISO 14040 and 14044

### 4.3.1 Mandatory and Optional Elements

The LCIA phase according to ISO 14040 and 14044[17] has got a structure, which is composed of mandatory and optional elements.

*Mandatory elements*:
- Selection of impact categories, category indicators and characterisation models;
- Assignment of LCI results (classification);
- Calculation of category indicator results (characterisation).

The impact category indicator results are the results of the mandatory elements of the impact assessment. They are generated according to scientific rules.

*Optional elements*:
- Calculation of the magnitude of impact category indicator results relative to reference information (normalisation);
- Grouping;
- Weighting.

An application of optional elements to the impact category indicator results leads to weighted data. The priority criteria of the optional elements can be scientifically justified only in parts (see Section 4.3.3).

### 4.3.2 Mandatory Elements

#### 4.3.2.1 Selection of Impact Categories – Indicators and Characterisation Factors

The terms *impact category*, *category indicator* and *characterisation factor* are defined as follows in ISO 14044:

---

16) Neither in US-American legislation $CO_2$ has been considered as pollutant until recently.
17) ISO (2006a, Section 5.4) and ISO (2006b, Section 4.4).

> *impact category*: Class representing environmental issues of concern to which life cycle inventory analysis results may be assigned.*impact category indicator*: Quantifiable representation of an impact category.*characterisation factor*: Factor derived from a characterization model which is applied to convert an assigned inventory analysis result to the common unit of the category indicator.

This terminology in ISO 14040 and 14044 is somewhat bulky and can best be illustrated by an example. This is done for the scientifically best substantiated impact category 'climate change'[18] (see also Section 4.5.2.2):

1. *Impact category*: Climate change.
2. *Inventory results*: Amount of a GHG per fU.
3. *Characterisation model*: Baseline model of 100 a of the International Panel of Climate Change.[19]
4. *Category indicator*: Infrared *radiative forcing* (W m$^{-2}$).
5. *Characterisation factor*: global warming potential ($GWP_{100}$)[20] for each GHG (kg $CO_2$-equivalents/kg gas).
6. *Category indicator result* (unit): Kilograms of $CO_2$-equivalents per fU.
7. *Category endpoints*: for example, Coral reefs, forests, crops harvests. (It should be noted here, that many, probably most, endpoints are not yet known. They concern other obvious geological formations (glacier, arctic ice). Here changes or disappearances will be of major importance to the living world including humans.)
8. *Environmental relevance*: Infrared radiative forcing is a proxy for potential effects on the climate, depending on the integrated atmospheric heat absorption caused by emissions, and the heat absorption over time.

As ISO 14044 does not provide a list of impact categories, does not even recommend one, the selection of the categories depends on the authors of the LCA. Table 4.2 shows two sample lists for a selection of impact categories. On the right side of the table impact categories that can be assigned to the results of inventories (*Mid-point Categories*), which can be further bundled (*Damage Categories*) are defined.

As the selection of impact categories must correspond to the goal and scope of the study, the selection of categories should be made in the first phase of the LCA. This is particularly important because the data to be procured in the inventory must comply with the demands of the impact assessment. On the other hand, an LCA is basically an iterative process for good reasons. Therefore the following approach is recommended:

- Selection of impact categories plus category indicators as well as assignable inventory parameters as far as possible in the first phase of LCA (definition of goal and scope).
- Data collection in view of selected impact categories in the phase LCI analysis.

---

[18] ISO (2006b, Table 1).
[19] Intergovernmental Panel on Climate Change (IPCC).
[20] GWP is not an impact category but a category indicator result.

## 4.3 Structure of Impact Assessment according to ISO 14040 and 14044

**Table 4.2** Two sample lists for a selection of impact categories.

| Impact category[a] | Impact category[b] | |
|---|---|---|
| | Mid-point categories | Damage categories |
| Human toxicity | Human toxicity | Human health |
| Ecotoxicity | Impact on respiration | |
| Eutrophication (aquatic) | Ionising radiation | |
| Eutrophication (terrestrial) | Ozone layer destruction[c] | |
| Land use | Photochemical oxidation | |
| Ozone formation (near-surface) | Aquatic ecotoxity | Quality of ecosystems |
| Resources demand | Terrestrial ecotoxity | |
| Ozone depletion (stratospheric) | Aquatic acidification | |
| Greenhouse effect | Aquatic eutrophication | |
| Acidification | Terrestrial acidification and eutrophication | |
| | Land use | |
| | Global warming | Climate change |
| | Non-renewable energy | Resources |
| | Mining of minerals | |

MEMO Verlag: Gestaltung wie Table 4.2 deutsch.
[a]Würdinger et al. (2002).
[b]Jolliet et al. (2003); Damage Category is frequently called *Area of Protection* or *Safeguard Subject*; expression 'Climate Change' is used by ISO 14040/14044 for (Mid-point) impact category.
[c]The impact category 'ozone layer destruction' was erroneously only assigned to the *Damage Category* 'human health' because of a proven correlation between skin cancer illnesses and short-wave UV radiation; however there are also correlations to 'quality of ecosystems' and particularly to 'climate change' which is classified here as *Damage Category*, which deviates from ISO 14044 definition.

- Fine selection of category indicators and characterisation models in the first part of the phase LCIA, reasoning for selection, references to literature.
- Complement the definition of goal and scope if necessary.
- Completion of inventory data if necessary.

The standard 14044 emphasises an obligation to supply comprehensive information regarding the selection of impact categories, category indicators, indicator models and characterisation factors, which seems to be exaggerated for all standard categories usually applied in LCA like climate change and acidification, but is more than justified for those rarely used. Hence, the use on an equal footing of home built methods besides those that are internationally accepted, without referring to different levels of development, is avoided.

Although ISO does not prescribe an impact category list, ISO 14044 refers to sample categories and indicator models of the technical guideline ISO 14047[21] (no standard!). It can offer assistance for a selection of impact categories and indicator

---

21) ISO (2002).

models, it should not, however, replace a thorough study of literature on the current state of the art.

ISO 14044 further recommends that impact categories, category indicators and characterisation models are to be internationally accepted, based on international agreement or are recognised by an authorised international board. Possible candidates at present are the SETAC (Society of Environmental Toxicology and Chemistry) and the UNEP/SETAC Life Cycle Initiative. These are mere recommendations. If taken literally, new categories, indicators, and so on, under development cannot be tested in practice. Actually, this excerpt of the standard often serves as an excuse to exclude certain impact categories. An analysis of the LCIA methods (per impact category) and methodologies (sets of impact methods) is provided in the handbook published electronically by the European Commission (EC).[22] It is planned to publish recommended methods that may be binding for official LCAs ordered by the Commission. Such a binding list would, of course, violate the international standard (see above) stating that the impact categories, and so on, are selected in the phase Goal and Scope. Similar considerations may be appropriate for the recently proposed 'Product Environmental Footprint' (PEF)[23] to be tested in 2014–2017.

It is further noted that categories and indicators should be based on as little value choices and assumptions as possible (that means scientifically objective), double counting should be avoided, environmental relevance exists and so on. These partially redundant enumerations of ISO 14044, Section 4.4.2.2.3 can be explained by the fear of manipulation of the method, an underlying issue of all LCA standards.[24]

### 4.3.2.2 Classification

Classification is a correlation of inventory items to impact categories, for example, GHGs to the impact category climate change or acid-forming gases to the impact category acidification. Besides output-relevant releases from the technosphere into the environment, inputs from the environment into the technosphere have to be assigned to the extent of their procurement in the inventory. An example is the assignment of fossil raw materials to the impact category resource demand. The most important impact categories are discussed in detail in Section 4.5 with regard to their indicators and characterisation models.

Figure 4.2 shows the principle of classification and the subsequent phase of characterisation.

Classification according to ISO 14044 includes as a mandatory component[25] a differentiation between inventory results that can be assigned to only one impact category, and an identification and assignment of results that refer to more than one impact category. Within the latter it should be possible to distinguish between parallel impact mechanisms (e.g. $SO_2$ as a toxic substance and as acid-forming gas,

---

22) European Commission (2010).
23) European Commission (2013).
24) Klöpffer (2005, 2012a).
25) ISO (2006b, Section 4.4.2.3).

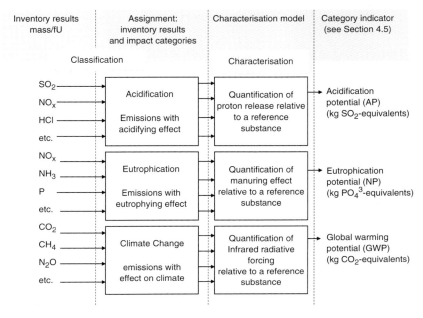

**Figure 4.2** Principle of classification and characterisation in the phase life cycle impact assessment.

or $NO_x$ as acid-forming gas and as gas with fertilisation effect) and serial impact mechanisms (e.g. $NO_x$ as acid-forming gas *after* the formation of photo oxidants in the summer smog).[26] Because this distinction is not always straightforward, this requirement of the standard is rarely fulfilled. We nevertheless recommend exactly reflecting possible impact mechanisms because this is the only way to attain a deeper understanding of the environmental impact of the studied product systems.

### 4.3.2.3 Characterisation

Characterisation is the core item of LCIA. Its somewhat bulky definition, according to ISO 14044,[27] reads:

> *the calculation of indicator results (characterization) involves the conversion of LCI results to common units and the aggregation of the converted results within the same impact category. This conversion uses characterization factors. The outcome of the calculation is a numerical indicator result.*

The characterisation factor according to ISO 14040[28] is a

> *factor derived from a characterization model which is applied to convert an assigned life cycle inventory analysis result to the common unit of the category*

---

26) In the process of the photochemical smog formation, little or non-toxic NO is converted into $NO_2$ forming nitric acid ($HNO_3$) and nitrous acid ($HNO_2$) with water droplets in the atmosphere.
27) ISO (2006b, Section 4.4.2.4).
28) ISO (2006a, Section 3.37).

indicator. Note: *The common unit allows calculation of the (impact) category indicator result.*

In the example of the impact category climate change (see Section 4.3.2.1) this translates as follows: The masses per fU assigned to this impact category from the inventory are multiplied with a specific characterisation factor ($GWP_{100}$, e.g. equals 1 for $CO_2$ and 25 for $CH_4$) and converted into (kg $CO_2$-equivalents). Thus a common unit is obtained. With it, different GHGs can be added into an overall (impact) category indicator value of the respective impact category (see Figure 4.2). Characterisation models and characterisation factors are developed by specialised sciences. The scientific basis of the more important impact categories in LCAs at present are introduced in Section 4.5.

The calculation of category indicator results based on inventory data is automatically accomplished by relevant software. This explains a widespread thoughtless way of conducting the LCIA. ISO 14044 therefore requires that procedures used for the calculations must be documented, including applied value choices and assumptions. This demand is not trivial, because the basics are often not kept in mind. It is also pointed out that the *complexity of environmental impact mechanisms* (often not fully investigated) also cover spatial and temporal characteristics, for example the persistence of a substance in the environment[29] and dose-effect characteristics. It will usually be impossible to include all these factors into the impact assessment; only if the complexity is adequately addressed can the results be relevant and meaningful and overinterpretation can be avoided.

### 4.3.3
### Optional Elements of LCIA

#### 4.3.3.1 **Normalisation**
Normalisation is defined according to ISO 14044[30] as the

> *... calculation of the magnitude of the category indicator results relative to some reference information. The aim of the normalization is to understand better the relative magnitude for each indicator result of the product system under study ...*

Normalisation means that category indicator results – thus the numerical results of the characterisation – are divided by selected reference values.

As reference values national, regional (e.g. European Union and North America) and international values (e.g. Organisation for Economic Co-operation and Development, OECD) are used with respect to an approximate accordance with geographical system boundaries. The principle and benefits of normalisation are illustrated in the following three examples:

---

29) Klöpffer and Wagner (2007a,b).
30) ISO 14044 206: 4.4.3.2.1

## Example 1

Impact categories 'climatic change' and 'stratospheric ozone depletion' with reference to an annually released mass of $CO_2$-equivalents in Germany (GWP – see Section 4.5.2.2) respectively R11[31]-equivalents (ozone depletion potential (ODP)[32] – see Section 4.5.2.3): specific contribution[33]

The following results are obtained following a transformation of inventory data of a fictitious product system: for the impact category 'climatic change' a category indicator result of 500 kg $CO_2$-equivalents/fU (GWP = 500 kg) and for impact category 'stratospheric ozone depletion' a category indicator result of 0.0000022 kg R11 equivalents/fU (ODP = $2.2 \times 10^{-6}$ kg). The geographical system boundary is Germany.

*Normalisation of the category indicator result '$CO_2$-equivalents':*
- The category indicator result for an annual release of $CO_2$-equivalents in Germany is 1 017 916 500 t (year of reference 2003[34]). The selected reference year should correspond to the reference period of the study.
- The normalisation consists in dividing the category indicator result of the fU of the product system by the category indicator result of the overall annual release:

| Category indicator results $CO_2$-equivalent | | Normalised value |
|---|---|---|
| Releases caused by the product system per fU | Annual release in Germany | Specific contribution |
| 500 kg | 1.02E+12 kg | 4.91E−10 |

The result of the standardisation is the specific contribution of the fU of the product system to the total load of the selected geographical reference area, here Germany. The specific contribution is dimensionless according to the definition of inputs.[35] As the fU with respect to its reference flow can be selected arbitrarily within wide limits, the absolute figure (e.g. $4.91 \times 10^{-10}$) signifies little as such, but should be compared to the appropriate numerical values of other impact categories: if a specific contribution of another impact category amounts to, for example $10^{-15}$, the examined product system contributes relatively less to this impact in the reference area (here Germany). The normalisation thereby permits a relative structuring of impacts. A category indicator result is therefore categorised as more important if it is larger when compared to an annual measured total load in the reference area.

---

31) Refrigerant 11 (CFC, chlorotrifluoromethane).
32) *Ozone Depletion Potential* (ODP), see above, Section 4.5.2.3.
33) Schmitz and Paulini (1999).
34) IFEU (2006).
35) Considering one year reference (the usual timeframe in statistics); generally the quotient would have the dimension time with common unit (a).

The usefulness of such relative structuring is illustrated by a comparison of the standardised value for $CO_2$-equivalents with those of R11-equivalents as category indicator for stratospheric ozone depletion.

Because of international conventions, substances known by their stratospheric reactions for the ozone depletion (persistent halogenated gases; see Section 4.5.2.3) were only used in larger quantities in product systems with freons as propellant or coolant or with methyl bromide as pesticide. The produced quantities have, since the 1990s, substantially diminished (significance of $N_2O$ see Section 4.5.2.3). These substances will therefore rarely appear in recent primary data. However, in generic data sets stored in data bases, these substances are often present in small quantities, as a multiplicity of processes with all side processes were aggregated into one data set.

*Normalisation of the category indicator result 'R11-equivalents':*

- Contrary to $CO_2$-equivalent, release data to R11-equivalents as reference for standardisation are not available in Germany. If as approximation a total amount of ozone layer-damaging substances produced in Germany in 2004 is used, which corresponds to 9364 t,[36] R11-equivalents, the following normalisation result follows:

| Category indicator results R11-equivalent | | Normalised value |
| --- | --- | --- |
| Emissions caused by the product system per functional unit | Annual release in Germany | Specific contribution |
| 2.20E−06 kg | 9.36E+06 kg | 2.35E−13 |

The relative significance of the product system concerning the category indicator '$CO_2$-equivalent' is thus about three orders of magnitude higher than that of category indicator 'R11-equivalent'. Because the result of the normalisation is determined by the reference quantity, the relative significance of $2.2 \times 10^{-6}$ kg R11-equivalents of the product system rises if the emitted amount as reference quantity decreases. These influences must be critically discussed in the interpretation (see Chapter 5).

Because of its global and regional significance, the impact category 'stratospheric ozone depletion' is classified as very important, and hence, the estimated specific contributions within comparative LCAs could strongly modify the final result. The result could, for example, be: Product A has a 10 times larger specific contribution concerning the impact category 'stratospheric ozone depletion' than product B. Normalisation in these cases can show that the relative significance of this impact

---

36) Data to the environment at *http://www.uba.de*; in 1990s the corresponding value amounted to 10E+04 tonne R11-equivalents within Netherlands, which is substantially smaller Breedveld, Lafleur and Blonk (1999).

category in the examined product system is orders of magnitude lower than that of other impact categories.

With normalisation groups of related impact categories can also be analysed, for example indicator values for GWP, CED, fossil resource consumption and acidification which often, but not always, correlate by consumption of fossil fuels.

## Example 2

**Impact category 'climate change', with reference to an annually released mass of $CO_2$-equivalents per inhabitant in Germany: resident equivalents (REQs)**

A better illustration of normalised category indicator results can be obtained by a 'per capita of population or a comparable measure'. This possibility is quoted in ISO 14044.

In this case, the category indicator results caused by an inhabitant in Germany or another geographical region are used as reference data. These reference values are called *resident equivalents (REQs)*.

*Normalisation of the category indicator result of $CO_2$-equivalents:*

- The category indicator value for the annual release of $CO_2$-equivalents 2003 in Germany amounted to 1 017 916 500 t, and the number of inhabitants of Germany of the year amounted to 82 532 000.[37]
- The annual release of the $CO_2$-equivalents is divided by the number of inhabitants of the year. The result is the resident equivalent (REQ), with unit kilogram per inhabitant. This REQ is the reference value. In the example this results in an annual $1.23 \times 10^4$ kg of $CO_2$ equivalents caused per inhabitant in Germany.
- The normalisation is performed by dividing the category indicator result of the product system by the resident equivalent. The normalised value means: Per fU of the examined product system, as many $CO_2$-equivalents are released as are caused by an average of $4.05 \times 10^{-2}$ inhabitants in 1 a, or differently expressed, 500 kg fU$^{-1}$ correspond to $4.05 \times 10^{-2}$ REQs. The descriptiveness in this form, however, is not substantially larger than for normalisation by the specific contribution as in the Example 1.

| Category indicator results ($CO_2$-equivalent) | | | Reference value | Normalised value |
|---|---|---|---|---|
| Emissions caused by the product system per fU | Annual emission in Germany | Number of inhabitants in Germany | Emissions per inhabitant: 1 IAV | REQ |
| 500 kg | 1.02E+12 kg | 8.25E+07 Inhabitants | 1.23E+04 kg/ inhabitant | 4.05E−02 Inhabitants |

---

37) IFEU (2006).

## Example 3

**Specific contribution or resident equivalents for the annual production of the examined product system**

In the sense of the standards relatively descriptive results can easily be obtained by a deviation from the functional unit considering an annual production of the examined product system instead. This is easily done with simple fUs and available statistical data by multiplication. Here, it must be remebered that the Dutch LCA-guidelines[38] recommend the annual production as the fU for detailed LCAs; in this case the conversion would be unnecessary.

*Normalisation of the annual production by the specific contribution of the category indicator result of $CO_2$-equivalents:*

| Category indicator result $CO_2$-equivalent | | | | Normalised value |
|---|---|---|---|---|
| Emission caused by the product system per fU | Number of fU produced per annum | Emissions caused by annual production | Reference value annual emissions in Germany | Specific contribution of annual production |
| 500 kg | 1.00E+07 | 5.00E+09 kg | 1.02E+12 kg | 4.91E−03 equal to 0.49% |

*Normalisation of the annual production per resident equivalents of the category indicator result of $CO_2$-equivalents:*

| Category indicator result $CO_2$-equivalent | | | | Normalised value |
|---|---|---|---|---|
| Emission caused by the product system per fU | Number of fU produced per annum | Emissions caused by annual production | Reference value annual releases per average inhabitant in Germany | Resident equivalents (REQ) per annual production |
| 500 kg | 1.00E+07 | 5.00E+09 kg | 1.23E+04 kg/ inhabitant | 4.05E+05 inhabitants |

With normalisation of the annual production by resident equivalents the descriptiveness increases considerably: The fictitious product system serving as example causes as many emissions of GHG per annual production, measured as category indicator $CO_2$-equivalent as 405 000 average inhabitants. Interpretation should, however, be done cautiously since the REQ is a calculated figure whereby the entire industrial production of the product investigated is assigned to inhabitants. If a reference region is thickly populated, but exhibits lower industrialisation than,

---

38) Guinée et al. (2002).

for example Germany, the releases per average inhabitant are smaller and the numerical value 'REQ per annual production' is higher. Before numerical values of different LCAs are compared with one another, it must, therefore, be examined whether a common basis of underlying assumptions exists at all.

As a further possible example for a choice of reference values, ISO 14044 proposes a correlation with inputs and outputs of a reference scenario.

It has been shown in several publications[39] that a correct application of normalisation is by no means trivial.

### 4.3.3.2 Grouping

Grouping as an optional element of LCIA provides an option to summarize the results of the preceding elements. Contrary to 'weighting' (Section 4.3.3.3) no value choices should be included. ISO 14044 definition of grouping is little descriptive:

> *Assignment of impact categories into one or more sets as predefined in the goal and scope definition, and it may involve sorting and/or ranking.*

This suggests a formation of classes, which may include a ranking. Two possibilities are indicated:

- to *sort* impact categories on a nominal basis (e.g. by characteristics such as inputs and outputs or global, regional and local spatial scales) or
- to *rank* the impact categories in a given hierarchy (e.g. high, medium and low priority).

It is expressly pointed out in the standard that ranking does after all depend on value choices and therefore different persons, organisations and social groups may come to different conclusions. These considerations clearly imply that this component and (even more so) the following 'weighting' would be better integrated in the component 'Interpretation'. Unfortunately the revision of the ISO standards, which transferred standards ISO 14040–43: 1997/2000 to ISO 14040–44:2006, neglected these corrections.

Surprisingly the element *Grouping* is allowed for studies with 'comparative assertions' intended to be made available to the public, whereas the next element *Weighting* is not. The same ambiguity applies to the former standard 14042.[40] This encouraged the German Federal Environmental Agency (UBA) Berlin to develop a valuation method in accordance with the ISO standard.[41] It was ensured that ranks for the individual impact categories were characterised verbally, not by figures. These verbal ranks were derived from an analysis of environmental endangering and of distance-to-target[42] between the status quo and political and legal objectives concerning the environment. As an example this distance is very low with regard to the impact category 'stratospheric ozone depletion' (successful implementation of the protocol of Montreal of 1986 and subsequent amendments);

---
39) Seppälä and Hämäläinen (2001), Erlandsson and Lindfors (2003) and Heijungs *et al.* (2007).
40) ISO (2000a)
41) Schmitz and Paulini (1999).
42) BUWAL (1998), Schmitz and Paulini (1999) and Seppälä and Hämäläinen (2001).

whereas the distance with regard to other impact categories is, however, still very high (implementation deficits). These rankings are not completely free of arbitrariness, but can be applied by authorities in a responsible way for one nation (here: Germany).

Several software tools have already integrated the grouping criteria without users being aware of how these criteria were developed and on what value choices they are the based. The following provides a short sample illustration of this approach in the element 'grouping' by the method of the Federal Environmental Agency (UBA).[43]

An overview of todays LCIA methods can be found in Section 4.5. The ranking of impact categories in the method applied[44] by the Federal Environmental Agency (UBA) refers to three criteria: ecological endangering, distance-to-target and specific contribution.

1. **Ecological Endangering** According to this criterion impact categories are put into an order according to the seriousness of the potential damage.
   The following premises are defined for a characterisation of seriousness:
   - Profound impacts at the level of the ecological system are more serious than those at the level of organisms.
   - Irreversible impacts are more serious than reversible ones.
   - Ubiquitous impacts are more serious than spatially limited ones.
   - Large uncertainty for the prognosis of an environmental impact because of unsatisfactory scientific knowledge is serious.
   
   Specialised departments of UBA provided expertise for ranking based on these premises on impact categories. The advisory board graded impact categories on a five-stage scale (A: highest priority to E: lowest priority, Table 4.3). It is expressly stressed that this ranking is based on value choices of the UBA and needs to be examined thoroughly for a scientific upgrade.

2. **Distance-to-target** According to this criterion impact categories are assigned ranks by the seriousness of the distance between the status quo and political and legal objectives concerning the environment. The following premises are defined for a characterisation of seriousness:
   - The larger the distance between the status quo and a quantified quality goal of the environment, the more serious the deviation.
   - A high diminution demand provided by an environmental action goal is serious.
   - Rising loads (e.g. emissions) are regarded as more serious than stagnating or diminishing ones.
   - Small enforceability and technical accessibility of a goal are regarded as serious.

---

43) Schmitz and Paulini (1999).
44) Schmitz and Paulini (1999).

Table 4.3  Ranking of the impact categories according to UBA.

| Impact category | Ranking by the German Federal Environmental Agency | |
|---|---|---|
| | It is stressed that another advisory board may provide another ranking. | |
| | Ecological endangering | Distance-to-target |
| Eutrophication (aquatic) | B | C |
| Eutrophication (terrestrial) | B | B |
| Land use | A | A |
| Photochemical ozone formation | D | B |
| Scarceness of fossil energy sources | C | B |
| Stratospheric ozone depletion | A | D |
| Greenhouse effect | A | A |
| Acidification | B | B |
| Human toxity[a] | | |
| Ecotoxity[a] | | |

[a]The toxicity categories are individually discussed.

The ranking of impact categories according to the criterion 'distance-to-target' is similar to the approach described for the criterion 'ecological endangerment': An interdisciplinary team classes the impact categories on a five-stage scale (A–E). A highest, E lowest (Table 4.3).

3. **Specific Contribution** Specific contributions are used as the third criterion for the ranking of impact categories (see Section 4.3.3.1, Example 1). They are categorised by five classes, where the highest specific contribution serves as the base factor:
A: highest priority 80–100% of the maximum value to
E: lowest priority 0–20% of the maximum value.

4. **Unification of Results** For a final ranking of the impact categories the results are integrated according to the three grouping criteria with a fixed even-weighted pattern to an 'ecological priority'[45]: If, for example, one impact category concerning all three grouping criteria is assigned to group of A (highest priority), these single results are subsumed as of 'very large ecological priority'. This signifies that the environmental loads of the examined product system concerning this impact category are regarded as highly relevant.

The above detailed presentation of the example of the grouping method according to UBA serves to clarify the following:
- The ranking in the element 'grouping' in the context of the impact assessment is not trivial.
- The ranking of the impact categories include value choices. This cannot be avoided. Different committees can at different times present different rankings.

---

45) Schmitz and Paulini (1999).

- Because some subjectivity is inevitable in the course of the ranking it is to be made certain that all phases in an LCA are presented with transparency and comprehensibly.
- If in two LCAs different grouping methods are used, the results of the element 'grouping' are not directly comparable.

### 4.3.3.3 Weighting

The designation weighting can be regarded as replacement for 'valuation', which according to ISO standards is to be strictly avoided and thereby performs the function of an euphemism. In contrast to the element 'grouping', numerical factors are admitted which are based on value choices[46]:

> Weighting is the process of converting indicator results of different impact categories by using numerical factors based on value-choices. It may include aggregation of the weighted indicator results.

In the last sentence of the quotation the possibility of ecopoints and similar aggregations is implicitly suggested. These methods are also called *single point methods* because in the context of weighting all considered impact categories are quantitatively taken into account but only a highly aggregated result is documented. This synopsis of category indicator results cannot be justified scientifically: Rather, value-based decisions must be made as to which weighting factor is applied and to which impact category. Also an equal weighting of all impact categories is a value-based decision. It can thus be concluded that weighting is inappropriate for the phase LCIA and should be integrated into the phase interpretation (our 'ceterum censeo' for many years, see also[47]). In LCAs intended for comparative assertions that are intended to be made accessible to the public, the optional element 'weighting' must not be used (ISO 14044).[48] It can be concluded that 'single point methods' if used for comparative assertions *are only permissible for internal use*, not for marketing or statements in the press or other media.

The much used single point (often called *ecopoint*) methods are the Swiss 'ecofactors'[49], the Swedish EPS (enviro-accounting) method[50] as well as the Dutch eco-indicator.[51] The latter is integrated into the widespread LCA software 'SimaPro'.

Common to all single point methods is the loss of information due to a simplified representation of the final result. For an aggregation, valuation or weighting factors must inevitably be introduced that, even though described by the authors, are often not present to the user. The methods are against the spirit as well against the wording of the standards ISO 14040 and 14044. These disadvantages have, particularly in Germany, led to a widespread refusal of these procedures and initiated

---

46) ISO (2006b, Section 4.4.3.4).
47) Reap *et al.* (2008a,b).
48) ISO (2006b, Section 4.4.5).
49) BUWAL (1990, 1998) and Frischknecht, Steiner and Jungbluth (2009).
50) Steen and Ryding (1992).
51) Goedkoop (1995) and Goedkoop *et al.* (1998); http://www.simapro.de; http://www.pre.nl.

research work to overcome the disadvantages of 'automated' valuation processes. IFEU has proposed 'a verbal-argumentative' valuation procedure,[52] which literally proposes a true alternative for mechanical computation of purely numerical values. Klöpffer and co-workers investigated the origins of the values necessary for a valuation and on how to provide mathematical tools (Hasse-diagrams) for an improved structuring of verbal-argumentative results.[53] An extensive discussion coordinated by the German Federal Ministry of the Environment and UBA Berlin, which was joined by the Federal Association of the German Industry (BDI), did not provide any results.[54] Some aspects capable of consensus, according to ISO 14040, were integrated into the UBA methodology 'Bewertung99'.[55] Further information about this crucial topic is provided in Chapter 5.

#### 4.3.3.4 Additional Analysis of Data Quality

This component of the standard is also completely misplaced, because similar requirements are a basic element of the LCA phase interpretation. The demand for an additional analysis of data quality at this stage of the impact assessment can only be interpreted to mean that the impact assessment with immature methods can lead to misjudgement. This is supported by the fact that this component of the impact assessment is not an option but *mandatory* for use in comparative assertions intended for publication.

Three specific methods are suggested:

- Centre of Gravity analysis,
- Uncertainty analysis,
- Sensitivity analysis.

The results of the analysis can suggest the need for an improvement of the inventory analysis, for example, if the data procured in the inventory are not sufficient for a correct conduct of an impact assessment. The methods for an analysis of data quality are discussed in Chapter 5.

## 4.4
## Method of Impact Categories (Environmental Problem Fields)

### 4.4.1
### Introduction

As the framework, according to international standards of the LCIA, has already been discussed in Section 4.3, the scientific part and the historical development, to some extent, is now elaborated. This latter aspect would be obsolete if all the methods have been fully developed. This being impossible, some 'historical' aspects

---

52) Giegrich *et al.* (1995).
53) Klöpffer and Volkwein (1995), Volkwein and Klöpffer (1996) and Volkwein, Gihr and Klöpffer (1996).
54) BDI (1999) and UBA (1999).
55) Schmitz and Paulini (1999).

are included (see also Section 4.2) so that something can be learned to aid future developments of impact assessment.

The method of the impact categories, mainly developed at the Institute of Environmental Science Leiden (CML)[56] is based on the idea of the 'environmental problem fields'. It has also been proposed to the German Federal Environmental Agency Berlin including some modifications,[57] and has been generally accepted by both the Code of Practice of SETAC[58] and by the international standardisation process of ISO 14042.[59]

The basic idea of the method is to coordinate between existing inventory data and a list of environmental problem fields or impact categories as *quantitatively* as possible by means of classification (Section 4.3.2.2) and characterisation (Section 4.3.2.3). Potential damages or harmful effects of the examined product system are thus to be perceived and to be approximately quantified. A list of environmental problem fields is always incomplete because it can only correspond to the current level of knowledge and the public reception of environmental problems.

## 4.4.2
### First ('Historical') Lists of the Environmental Problem Fields

Experts of a working group of the SETAC Europe LCA symposium in Leiden (December 1991[60]) suggested the following list, which was supposed to cover with the smallest possible overlap the most important acknowledged environmental problems. To date it still represents the basis of most category lists.

- scarce, renewable resources;
- non-renewable resources (raw material);
- global warming;
- ozone depletion;
- human toxicity;
- environmental toxicity;
- acidification;
- eutrophication;
- COD discharge;
- photo oxidant formation;
- space requirement;
- nuisance (smell, noise);
- occupational safety;
- final solid waste (hazardous);
- final solid waste.

The problem field fresh and drinking water was neither listed by the experts nor addressed during subsequent discussions. The same applies for the upstream

---

56) Heijungs *et al.* (1992), Udo de Haes (1996) and Guinée *et al.* (2002).
57) Klöpffer and Renner (1995).
58) SETAC (1993).
59) ISO (2000a)
60) SETAC Europe (1992a,b)

problem field of groundwater pollution. The drinking water problem, like the general exposition via food, is implicitly included in the category human toxicity. On the other hand water also represents an important resource and therefore cannot be only valuated by toxicological criteria. Recently close attention has been paid to this issue, both in research and standardisation (see Section 4.5.1.5).

To this day *occupational safety* is a controversial issue in the LCA community. On the one hand, endangerment in the workplace is part of the technosphere and strictly regulated in many highly developed industrial nations; however, *only* there! On the other hand it is argued that a risky production procedure should also obtain a malus in LCIA. An inclusion of this (environmental) problem field into the impact assessment is demanded particularly by Scandinavian states.[61] The problem will be solved if in the comprehensive sustainability analysis of the product[62] the workplace will be a general issue within the scope of product-related social assessment (see Chapter 6).

The protection of landscape or the demand for natural space has never been explicitly addressed by SETAC experts. It can however be regarded as part of *space requirement*. A severe deficit in this first list is the absence of exposure to hard radiation as an individual impact category. Implicitly is it included in the categories human toxicity and environmental toxicity.

Altogether it cannot be denied that the list was the result of brain storming at the time of the conference. Thus, it has, for example, been pointed out that the chemical oxygen demand (COD) is no environmental problem field as such, but rather an indicator for those of eutrophication and ecotoxicity. Also, the example of *final solid waste* shows that not every indicator is an impact category: quantities of non-usable waste is a result of the LCI and can affect different environmental problem fields depending on the type of waste treatment and landfill.

Table 4.4 shows a further elaboration of the list by a team of experts of the SETAC Europe chaired by Helias Udo de Haes with the prospect of integration into ISO regulations of LCIA. As it turned out, however, for good reasons ISO did not accept the integration of a specific list of categories into the standard 14042.

*Casualties* is a new entry of the list, occupational safety is cancelled (partly overlaps with casualties). A study on behalf the UBA Berlin[63] showed that accidents in power plants can basically be included into LCAs if this is part of the goal definition. A general method however does not exist. 'Solid waste' from the original Leiden list was referred into the LCI.

Numerous lists compiled by different authors and committees differ usually only slightly by designation or structuring. They adequately reproduce environmental problems discussed at the turn of the century but can neither be complete nor free of overlap. A breakdown of complex categories such as human toxicity implies an integration of many '(toxicological) end points', which finally would induce confusing results even in the case of adequate and sufficient data. Consequently there have been attempts to reduce numerous category end points to a few broader

---

61) Lindfors *et al.* (1995) and Udo de Haes and Wrisberg (1997).
62) Klöpffer and Renner (2007).
63) Kurth *et al.* (2004).

**Table 4.4** List of impact categories.

---

Designation according to SETAC Europe[64]
(A) Input-related categories (resource depletion or competition)
    Abiotic resources (deposits, funds and flow)
    Biotic resources (funds)
    Land
(B) Output-related categories (pollution)
    Global warming[a]
    Depletion of stratospheric ozone
    Human toxicological impacts
    Ecotoxicological impacts
    Photo-oxidant formation
    Acidification
    Eutrophication (including BOD and heat)
    Odour[b]
    Noise
    Radiation
    Casualties[b]

---

[a] Later renamed as 'climate change'.
[b] Proposed without operationalisation method.

*safeguard subjects*[65] or *areas of protection (AoP)*[66] such as human health, integrity of ecological systems and resources. However, confusion remains since these are also often called *end points (or damage categories)*.

From the very beginning, the significance of 'feasibility' in the impact assessment has been addressed.[67] The definition of the number of impact categories and the level of detail of the linkage with potential impacts is a tight-rope walk between the desired scientific accuracy on the one hand and feasibility with limited data and system information on the other hand.

*From the aspect of precaution it has to be pointed out that such lists always and only correspond to present knowledge and reception (they differ!). In this context it would be useful to consider such a list that was drawn up 20, 40 or 60 a ago.*

When were individual environmental problem fields acknowledged as such, (i) scientifically and (ii) social-politically? Answers are summarised in Table 4.5. It should be noted that the exact time cannot always be as easily determined as for stratospheric ozone depletion (1974). In this case scientific knowledge and public awareness nearly coincided – a rare case.

With the data from Table 4.5 it would be possible to compile a 'list' for each decade in the recent past. An average 'incubation period' from scientific discovery to public attention of approximately 10 a is discernible. A compilation of avoidable

---

64) Udo de Haes (1996).
65) Beltrani (1997).
66) Udo de Haes et al. (2002).
67) Klöpffer (1994b).

**Table 4.5** Temporal occurrence of the environmental problems.

| Environmental problem field (impact category) | Scientific discovery | Entering public consciousness |
|---|---|---|
| Resource depletion | Approximately 1965–1970 Limits to growth[a] | First oil crisis of 1973 |
| Greenhouse effect (Climate change)[b] | Approximately 1975–1980 | UN world conference in Rio de Janeiro 1992; 'Agenda 21' |
| Stratospheric ozone depletion | Rowland and Molina[c] 1974; Discovery of the Antarctic 'ozone hole'[d] around 1985 | Prohibition of CFC ('freons') in sprays (USA, approximately 1978)[e]; Convention of Vienna (1985) and Protocol of Montreal (1987)[f] |
| Toxic endangering of humans (human toxicity) | Knowledge of poisons is age-old | Chemical laws in US (TSCA)[g] and France around 1975, EEC 1977, FRG 1981; EU: REACH 2006 |
| Toxic damage of organisms, ecotoxicity | Rachel Carson 'Silent Spring' 1962[h] | For example, DDT law, FRG 1972 |
| Summer smog | Approximately 1950, Los Angeles[i] | Catalyst regulation in California, approximately 1975 |
| Acidification | First phase (direct harm of acid gases): second half of the nineteenth century (Stöckert, Stoklasa)[j] Second phase (indirect impacts, esp. in forests) approximately 1970 | International forest damage conferences around 1900 (Stoklasa, 1923, loc. cit.) Article in the 'Spiegel' covering 'The Dying Forest' 1980; Acidification of remote Swedish lakes |
| Eutrophication | Massive growth of algae in lakes, 1960s; oxygen loss by BOD | Measures of restoration starting about 1970; Detergents without Phosphate starting 1990 |
| Annoyances (smell, noise) | Everyday life experience, dating is difficult | Noise perceived as environmental problem No. 1 (especially traffic noise[k]) |
| Hard radiation[l] | Since the use of nuclear energy, military (1945) and civilian (1950s); Disaster of Chernobyl 1986 | Consciousness of endangering since the beginning of military use;[m]civil nuclear power esp. since 1986 |
| Waste | Starting approximately 1960 | Early LCAs around 1970[n] |

[a]Meadows et al. (1972).
[b]Recently 'climate disruption' has been proposed.
[c]Molina and Rowland (1974); Rowland and Molina (1975).
[d]McIntyre (1989).
[e]UBA (1979).
[f]Deutscher Bundestag (1988).
[g]USA (1976).
[h]USA (1976).
[i]McCabe (1952).
[j]Stoklasa (1923).
[k]Including noise by aeroplanes near airports.
[l]High energy per particle or photon ($\alpha$, $\beta$, $\gamma$ and $h\nu$).
[m]Not however in the research phase: Marie Curie died from leukaemia from impacts of her experiments.
[n]See Chapter 1.

undesirable trends on the basis of case studies was published by the European Environmental Agency.[68] Most of these negative environmental trends could have been avoided by an improved attention to the precautionary principle. Some false alarms in science, for example, the aluminium/Alzheimer discussion in the late 1980s[69], however, cannot be ignored.

The speed of development of new impacts is illustrated by the fact that in earlier LCAs (around 1990) the greenhouse effect (or global warming, or climate change) was not even listed. Since the inaugural, seemingly 'immortal' list of Leiden, at least four impacts have emerged, which are strongly discussed today:

- hormone disrupters: Substances, which either imitate or displace natural hormones (blocking at site of impact); this mechanism is part of ecotoxicity, possibly also human toxicity, reproduction damaging impacts;
- possible harmful impacts of genetically modified organisms (GMOs) (microorganisms, crops) on the environment;
- invasive species (a subset of the neophytes and neozoa characterised by abundant proliferation, superseding indigenous species and changing ecosystems);
- freshwater as a regionally scarce resource.

These issues have been discussed in scientific literature for the last 10–20 a; a (sometime heated) public discussion has been started, but the topics are still insufficiently represented in LCIA.

Above all it should be learned from these considerations that it makes no sense to presume that today all impacts of human activity on complex systems like ecosystems are known. The list of the impact categories must therefore be amended or updated from time to time, furthermore the precautionary principle should, by suitable indicators,[70] be represented in LCA. LCIA will therefore remain a permanent 'building site'.

### 4.4.3
**Stressor-Effect Relationships and Indicators**

The following discussion relates more to group B of the newer SETAC Europe list (Table 4.4) than to scarcity of resources (group A). The problems are, however, similar in both groups. In both groups the question is on how to coordinate between the results of the inventory and the impact categories. Thereby two issues need to be clarified at the beginning:

- hierarchy of impacts (which approach to be chosen for characterisation, which indicator to be used for quantification) and
- potential versus actual impacts.

---

68) EEA (2001).
69) Krishnan et al. (1988).
70) Schmidt-Bleek (1993, 1994), Klöpffer and Renner (1995) and Klöpffer and Volkwein (1995).

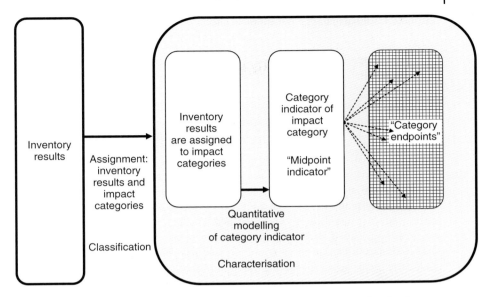

**Figure 4.3** Impact category, classification/characterisation and assignment of end points (schematic): following the characterisation framework according to ISO CD 14042.3.[69]

#### 4.4.3.1 Hierarchy of Impacts

The stressor concept and the impact hierarchy were introduced during the SETAC Workshop in Sandestin, Florida.[72] The term *stressor* is disputed and has not been generally accepted. A stressor within the scope of an LCA was defined as a chemical or physical factor from the inventory that can interact with the animate and inanimate environment by diverse impacts at multiple system levels like single organisms, species, communities and ecosystems (mostly but not always above a single species effect threshold). As an overall expression for all types of influences on the environment the word *intervention* is used in the literature. In the context of literature on operational environmental management (also in ISO 14040/44) the term *environmental aspects* is used to describe influences on the environment with potential impacts. The basic idea behind these expressions will be discussed by the following examples.

An interrelation of these impacts and end points differing for each category is schematically illustrated in Figure 4.3. An indicator can be defined as 'closer to the inventory' or 'closer to individual endpoints' (see also Figure 4.2).

Impacts can be arranged according to the SETAC Sandestin Workshop (Fava et al., 1993, loc. cit.) within a hierarchy of primary, secondary and tertiary impacts as illustrated by the following examples:

---

71) A corresponding scheme was adapted, though not by form, to the standard 14042:2000; see ISO 14044 (Figure 3).
72) Fava et al. (1993).

## Example 1: Climate change (greenhouse effect)[73]

*Primary impact*: Increased radiation absorption by molecules of the atmosphere within the infrared (IR) window of approximately 10–15 μm (radiative forcing).
*Secondary impact*: Increase in the average temperature of the troposphere.
*Tertiary impact*: Melting of glaciers and arctic ice, climate instabilities, shift of climate zones, rise of the Sea level, spreading of diseases, changes in ecosystems, and so on.

It is important to note that while the primary effect is a well measurable phenomenon, the secondary impact can be more or less adequately described by the atmospheric life time of the GHGs including scenario-like assumptions, but tertiary effects can only be poorly quantified. The lesson to be learned is that GHGs that were quantified in the inventory have to be as closely correlated to primary and secondary effects as possible, and not to the far more uncertain tertiary effects. These are the basics of the so-called midpoint-method.

## Example 2: Acidification

*Primary impact*: Deposition of airborne acids on lakes, (bare) soil, trees (leaves, roots, etc.) and other vegetation.
*Secondary impact*: Change of pH value in case of insufficient buffering.
*Tertiary impacts*: Fish mortality by acid or by the release of $Al^{3+}$ ions; contribution to the 'new type of forest damage', damage to the vegetation by mineral depletion of the soils (e.g. $Na^+$, $K^+$ and $Mg^{2+}$), contamination of groundwater by (re)mobilised heavy metals, and so on. Change of aquatic and terrestrial ecosystems.

The position of the (mid-point) indicator is chosen as 'closest possible to releases'. Subsequent quantification is done by stoichiometric conversion of inventory items to the mass of protons set free ($H^+$) or the equivalent mass of $SO_2$ as anhydride of sulphurous acid ($H_2SO_3$) and precursor to sulphuric acid ($H_2SO_4$) in the atmosphere. Equivalence factors can clearly be determined from the chemical formulas unambiguously. The different quantifications proposed ($H^+$ or $SO_2$-equivalents) have identical validity, they merely differ by numerical values. Other possibilities of quantification and regionalisation are introduced in Section 4.5.2.5.

---

73) IPCC (1990, 1992, 1995a,b,c, 1996a,b, 2001, 2007).

## Example 3: Ecotoxicity

*Primary impact*: Impact of pollutants on organisms, following the intake of the pollutant or by a change of abiotic living conditions induced by the pollutant.

*Secondary impact*: Harmful impacts on single organisms, populations, species, biocoenoses or ecosystems, impacts of transformation products and metabolites of the pollutant.

*Tertiary impacts*: Harmful impacts on the level of ecosystems; drastic changes to ecosystems, for example, by starvation of single species due to organ lesions, but also by subtler impacts like, for example, the perturbation of chemical communication systems or the hormone system; changes in biodiversity, that is type and variety of species, changes in nutrition, nutrient cycles, energy flows of ecosystems, and so on.

The primary impact of a pollutant following its intake by an organism is linked to exposition. The bioconcentration factor (BCF) ($C_{\text{pollutant X in the organism}} / C_{\text{pollutant X in it the surrounding medium}}$) is a useful indicator directly connected to exposition but does not allow a differentiation of the toxicological potentials of the pollutants. Since the level of genuine importance for impact assessments – the tertiary impacts – is however generally not retraceable to quantifiable single events, substances are mostly assessed on the level of secondary impacts (usually level of organisms, e.g. daphnia, fish and alga tests).

In a similar way all impact categories can be analysed. Noise can, for example, be designated as an annoyance, but a chronic impact by noise may result in psychological damage (continuous stress by traffic noise) or hearing defect. For LCA it is important to realize that a mid-point quantification (closer to the inventory) can be accomplished more easily, and the number of potential categories can be reduced to a manageable quantity. Besides, indicators closer to the releases better correspond to the precautionary principle since many of the possible subsequent impacts are included in the assessment without detailed knowledge, which is often not available, of causal chains. (see Figure 4.3).

### 4.4.3.2 Potential versus Actual Impacts

The approximately quantified effects in LCIA are usually considered as 'potential' impacts.[74] ISO 14040 (2006a), for example, says:

*LCA addresses the environmental aspects and potential environmental impacts\* ( … ) thoughout a product's life cycle from raw material acquisition through production, use, end-of-life treatment, recycling and final disposal (i.e. cradle to grave).*

---

74) Udo de Haes (1996), Heijungs and Guinée (1993), ISO (2006a) and Finnveden et al. (2009).

The same standard specifies in a footnote (*):

> *The potential environmental impacts' are relative expressions as they are related to the fU of a product system.*

The most important reasons for this judgement and this designation are:

1. LCA usually has a low temporal and spatial resolution. Only the framework is designated by system boundaries.
2. The numerical values of the LCI refer to a fU which is freely scalable within large margins; it is, for example, completely unimportant whether we refer in a practical example to 1 l, 1 hl or 1000 l (most frequently applied for LCAs of beverage packaging) or to 1 million litres! The pollutant load per fU and concentrations derived from the emission data differ proportional to the fU by many orders of magnitude.
3. For materials bought on the free market (i.e. without supply contract with a specific supplier), for example *commodities*, their origin is rarely known (the same applies to sources of energy), which *excludes in principle* a tight geographical link to the product tree 'further above' – or according to Clift[75] to the 'background'. Alternative material procurement by long-term supply contracts increases the probability of an assignment to the region of origin. Specific instead of generic data can be used in such cases, the suppliers are part of the 'foreground'.

An assignment of the impacts to space and time for an entire life cycle is therefore usually not possible. Expositions caused by a specific analysed product system are rarely determinable (see also Section 4.1 – Example 4.1). Nevertheless the determination of concentration-dependent 'risks' is repeatedly demanded in the impact assessment, which has led to detailed and astute analysis on the limits of the impact assessment as part of LCA.[76] It is occasionally possible to actually identify a part of the life cycle (in the 'foreground') with sufficient accuracy so that for this section a risk analysis can be accomplished. In this case, however, assessments are not based on an arbitrary fU but on real material and substance quantities, for example, on the basis of the annual production of the examined product. For example, given the efficiency of the waste water sewage plants, the known annual consumption of a surfactant in a country can be converted into an average concentration in surface waters (*Predicted Environmental Concentration* – PEC). This value can be compared with impact thresholds or *Predicted No Effect Concentrations* (PNECs) to (PEC/PNEC) that corresponds to the common procedure in risk assessment of chemicals.[77]

Actual impacts are also determinable if the location of the manufacturing plant of the examined product is known and as well the specific contamination caused by the specific production of the examined product. In this case a site-specific risk

---

75) SETAC (1996)
76) White *et al.* (1995), Owens (1996) and Potting and Hauschild (1997a,b).
77) TGD (1996, 2003); see also ECHA (European Chemical Agency).

analysis can be integrated into the impact assessment. It is, however, to be made certain within comparative studies that the symmetry principle is not violated.

To justify the demand for site-specific risk analysis in case of short-ranged or local impacts (usually not to be accomplished), it is frequently referred to effect thresholds as follows: it should be of no toxicological and ecological relevance if pollutants were released in quantities not exceeding PNEC or limits values.[78] This perception is countered by the *less-is-better* concept[79]: In compliance with the precautionary principle, minimisation of all harmful releases is aimed at, even if laws are not violated and no harmful impacts should occur according to present knowledge.[80] Particularly persistent and accumulating substances can later have harmful impacts, even if emitted in small quantities only.[81] This LCA-type of thinking is also called *beyond compliance*.

With the greenhouse effect or with the stratospheric ozone depletion the problem does not occur because it is unimportant where a molecule $CO_2$ or R11 is released.

In summary, regarding the impact assessment, two schools of thought can be distinguished:

1. The impact assessment refers to potential impacts and adheres to the precautionary principle.[82]
2. The impact assessment, as far as possible, should depend on actual impacts (polluter pays principle) and its evaluation should base on scientifically derived or legally specified thresholds and limit values (no endangerment below these values).[83]

The first school is emphasised in Europe, the second in the USA. A certain freedom of choice remains and cannot be solved scientifically. The issue of subjectivity in LCA,[84] in particular in the impact assessment phase is discussed in Chapter 5. Here, it should be remembered that the problem of subjectivity is relevant even for the scientifically 'hard' phase LCI in the form of the allocation problem.

For a discussion of the impact levels (mid-point vs endpoint) the 'less-is-better' debate has the following consequences:

In the context of the first school of thought quantification should be done 'closest possible to the inventory interface' because at that point an exact knowledge of secondary and tertiary impacts is not necessary (*mid-point*).

However, in the context of the second school of thought, if an assessment of causal chains to single impacts deem to be necessary, quantification occurs close to end points. This will necessarily imply a larger number of subcategories and, as mentioned above, a more detailed knowledge of space and time dependencies.

---

78) Hogan, Beal and Hunt (1996).
79) White et al. (1995).
80) Hertwich (1996).
81) Klöpffer (1989, 1994c); this aspect of the precautionary principle has for the first time also been incorporated into the European chemical regulation act (REACH).
82) Klöpffer and Renner (1995), White et al. (1995), Udo de Haes (1996) and Hertwich (1996).
83) Hogan et al. (1996), Owens (1996) and Barnthouse et al. (1998).
84) Klöpffer (1998a) and Owens (1998).

In view of the state of the art of LCA, the second approach is not applicable with consistency in an entire life cycle. In what follows the impact assessment will therefore be mainly discussed starting from the CML approach[85], also applied by the Danish EDIP method[86] which has subsequently been modified and rearranged for the UBA, Berlin.[87] Further approaches at the stage of research and testing are discussed in the context. The development of impact assessment in SETAC Europe working group Impact Assessment 2 (1998–2001),[88] in the context of the UNEP/SETAC Life Cycle Initiative,[89] did not provide fundamentally new categories. It did provide a thorough valuation of the status quo and a partial harmonisation of the existing methods and a careful opening to endpoint methods. The most recent evaluation of LCIA categories has been performed by the EC.[90]

## 4.5
## Impact Categories, Impact Indicators and Characterisation Factors

In numerous LCA software tools the mandatory phases of the impact assessment 'selection of the impact categories, classification and characterisation' are integrated to a point where their background is not apparent. However, already within the first phase of an LCA 'definition of goal and scope' it has to be indicated how these mandatory elements are to be elaborated. These definitions have an influence on the data to be procured in the inventory. An adequate selection of impact categories, classification and characterisation regarding the goal of the study requires an exact knowledge of the background of the selected impact indicators and characterisation models.

In this section the impact categories used at present in LCAs are presented and the scientific background for the selection of impact indicators and characterisation models is described. There are impact categories with an existing broad consensus with regard to useful indicators and models and others with a multiplicity of competitive approaches. Besides, the two schools of thoughts often choose different impact indicators and characterisation factors with respect to the goals of the impact assessment (precautionary principle versus polluter pays principle).

### 4.5.1
### Input-Related Impact Categories

#### 4.5.1.1 Overview
This group of impact categories aims at the preservation and sparing use of natural resources. Not all human activities that contribute to resource consumption actually imply an irreversible destruction of appropriate resources (as with incineration of

---

85) Heijungs *et al.* (1992), Udo de Haes (1996) and Guinée *et al.* (2002).
86) Hauschild and Wenzel (1998).
87) Klöpffer and Renner (1995).
88) Udo de Haes *et al.* (1999a,b, 2002).
89) Töpfer (2002) and Jolliet *et al.* (2003, 2004), *http://lcinitiaitve.unep.fr*
90) EC (2010)

**Table 4.6** Systematics of Resources[91]

| Type of resources | Examples |
| --- | --- |
| Abiotic finite | Minerals, fossil raw materials |
| Abiotic regenerative | Groundwater, surface (fresh) water; oxygen[a]; not however: fossil groundwater |
| Biotic finite | Tropical wood from primary forests, species threatened by extinction |
| Biotic regenerative | Wild plants, wild animals (e.g. Sea fish); not however: agricultural and forestry products and fish from fish farms, since these products are generated within the technosphere. |

[a] As far as not irreversibly chemically bound.

fossil fuels), but 'only' imply contamination or dispersion (a kind of entropy increase). It is not always easy to distinguish between these two usage types of resources and at the same time to prevent the method of impact assessment getting intolerably complicated. We will try to follow a path in between.

According to SETAC Europe[92] (see Table 4.4) this group includes:

- abiotic resources,
- biotic resources and
- land use.

The first two impact categories can be divided into finite and regenerative resources, depending on their regenerative capacity (Table 4.6).

As indicated by their name and common to all categories of this group, inventory data to be classified and characterised occur at the input side. They thus concern raw materials and similar factors in the ecosphere, which are used, dispersed, stained or converted for the production of chemicals, materials and goods, fuels, and so on. As the overarching notion for these different uses, the expression 'consumption' is employed, that often does not strictly apply physically or chemically. However, it does apply if the appropriate raw materials are regarded as economic goods (see e.g. the commonly used expression 'energy consumption' which physically makes no sense).[93] The characterisation of resource consumption is one in view of the *scarceness, regeneration ability and significance for the ecosystems.*

With regard to abiotic resources with the exception of water, a primarily anthropocentric view cannot be overlooked[94]: The exhaustion of oil, coal and ore mines would have a greater impact on humankind than on the earth's ecosystem and its subsystems. As for the use of biotic resources, water and land, humans and nature

---

91) Klöpffer and Renner (1995).
92) Udo de Haes (1996) and Udo de Haes *et al.* (1999a,b, 2002).
93) (Apart from Einstein's $E = mc^2$.).
94) Udo de Haes (1996).

are about evenly affected. It is to be noted that side-effects with a superficial concern for nature 'only' can indirectly react on humankind and challenge its survival.[95]

A comprehensive representation of resources within LCIA is yet to be done. An overview of the suggested characterisation factors as well as an attempt at a uniform treatment of resources in view of functionality and quality (with respect to quality reduction) has been published by Steward and Weidema.[96] However, in addition to a uniform representation, fundamental differences between plantation forests and jungles, breeding and wild animals, and so on, were not considered. There are also no *back-up technologies*,[97] for example, for extinct species, as requested by this method. With regard to the metallic resources a concern for quality (of the deposits) is adequate as once they are extracted they are not really consumed, but rather dispersed (if not recycled).

#### 4.5.1.2 Consumption of Abiotic Resources

Abiotic resources considered in LCIA are:

- Fossil fuels (oil, natural gas, hard coal and lignite in deposits); spatial reference: predominantly global (by worldwide trade), for lignite predominantly regional.
- Uranium ores (in deposits); spatial reference: global.
- Mineral raw materials (ores, sand, clay, gravel, limestone, rock salt, phosphates, etc.); spatial reference: global (ores) to local (sand, gravel,etc.).
- (fresh) water; spatial reference: local to regional (is not traded on worldwide basis).
- Air and its components; Spatial reference: global (by the nature of the atmosphere).

**4.5.1.2.1 Impact Indicators** The list above implies that a simple addition of inventory data, of the consumption of resources per fU, would be completely inadequate because practically inexhaustible resources like sand, salt, air, and so on, would be equally weighted as resources that will be exhausted in foreseeable time like crude oil. In addition, oil, coal and natural gas are irreversibly lost during the dominant use by incineration whereas other raw materials, water above all, are polluted but in the long run are preserved. In principle, fossil raw materials can also be regarded as renewable materials, albeit with a slow regeneration rate. Since the associated time periods are extremely long, such a designation would be purely theoretical.

The concept of impact indicators and characterisation models was developed for output-based impact categories and is supposed to provide a quantitative relation between the data procured in the inventory and potential negative impacts. A transfer of these models to the assessment resources (here: abiotic resources) first requires an answer to the question: which is the common impact which then requires a quantitative model, as simple as possible, to be found. This common impact is the scarcity of the resource that will first affect the technosphere by an increase in the prices of raw materials. In the ecosphere, the lowering of

---

95) Klöpffer (1993) and Beltrani (1997).
96) Steward and Weidema (2004); see also Müller-Wenk (1999a); *http://www.iwoe.unisg.ch/servic*
97) Techniques, which serve the re-establishment of a condition before human interference.

the groundwater table because of excessive water use can induce a change in biodiversity, in produced biomass and in extreme cases, by steppe formation or desertification, a fundamental change of the original ecological system. Because endpoint modelling is always problematic as future events and conditions must be considered, shortage or scarcity as such is taken as the mid-point indicator. The weighting factors discussed below attempt a quantitative description for both cases – finite or non-regenerative and regenerative, respectively. The weighting factors are used as characterisation factors.

**4.5.1.2.2 Indicator Model and Characterisation Factors** The simplest model for the exhaustion of non-renewable abiotic resources relates consumption per fU to the total reserves[98]:

$$\text{Exhaustion of abiotic resources} = \frac{\sum_i \text{Consumption}_i (\text{kg fE}^{-1})}{\text{Reserves}_i (\text{kg})} \qquad (4.3)$$

This formula however does not consider that even small reserves can practically be inexhaustible if the total consumption (not only for the examined product system) is accordingly small. And enormous reserves will be quickly exhausted with a very large total consumption.

A more adequate model should quantify the scarceness as follows: The consumption of resources which will, for example, be exhausted in 100 a' time assuming present-day overall consumption should be classified twice as scarce compared to resources with a 200-a period exhaustion considering a constant overall consumption. The available time ('static range') for a certain not renewable resource is the quotient of reserves and consumption per time unit (Equation 4.4) if despite the increasing shortage a constant annual consumption is assumed:

$$\text{Static range (a)} = \frac{\text{World reserves (kg or J)}}{\text{World annual consumption (kg a}^{-1}\text{ or J a}^{-1}\text{)}} \qquad (4.4)$$

Energetically usable resources can be indicated in energy units (J, MJ, …) or mass units (kg, t, …); with natural gas also, a volume unit ($Nm^3$) is used. Here unfortunately the barrel must be mentioned: the quantity of crude oil is often reported in the outdated (US) volume unit (Barrel, 159 l).

As both natural gas as well as oil are chemical mixtures, they do not have a uniform density and the indication of the volume (in the case of gas at standard pressure and temperature) as a primarily measured figure does make sense. Average (empirical) values of density are the starting point for a conversion to mass. Something similar is valid for conversion to energy which is again based on the average of the lower heating value (LHV) or higher heating value (HHV) (see Section 3.2.3.2). For all other abiotic resources only mass units make sense.

---

98) Heijungs *et al.* (1992).

In order to accomplish meaningful aggregations, resource consumption per fU (out of LCI) is divided by the range of the resource.

$$\text{Scarcity abiotic resource}_i = \frac{\sum_i \text{Consumption}_i(m_i) \,(\text{kg fU}^{-1})}{\text{Range}_i(a)} \quad (4.5)$$

In Equation 4.5 the (static) range is the time for the complete exhaustion of resource $i$ according to Equation 4.4. The resulting dimension for the shortage is (mass/time), related to the functional unit.

As characterisation factor, the reciprocal static range (a), typical for a special resource ($i$) is defined as *resource scarcity factor* $R_i$ [$a^{-1}$] (see Table 4.7). It is multiplied with the consumption of resources $i$ per fU ($m_i$) determined in the inventory. The result is the *resource consumption factor R*. The resource consumption factor for the entire product system considers all abiotic finite resources. It is the sum of all resource consumption factors of all abiotic resources procured in the inventory, which were classified in the impact category 'consumption of abiotic resources':

$$R(\text{abiotic finite}) = \sum_i (m_i \times R_i)(\text{kg a}^{-1}) \quad (4.6)$$

---

**Example**

In the inventory of a product system the following consumptions of abiotic resources were determined per fU ($R_i$ in accordance with Table 4.7):

Crude Oil 6 kg
Hard coal 4 kg
Resource consumption factor (oil) = 6 kg × 0.023 ($a^{-1}$) = 0.138 kg $a^{-1}$
Resource consumption factor (hard coal) = 4 kg × 0.0125 ($a^{-1}$) = 0.05 kg $a^{-1}$
Resource consumption factor ($\Sigma$ abiotic
final) = 0.138 kg $a^{-1}$ + 0.05 kg $a^{-1}$ = 0.188 kg $a^{-1}$.

The unit (kg $a^{-1}$) for the resource consumption factor $R$ can easily be misinterpreted as real mass flow. As the resource scarcity factor $R_i$ necessary for the computation of $R$ is a figure weighted by static range, this is however not true, rather $R$ is an impact indicator. The considered impact is scarcity.

---

*Water* as a regenerative abiotic resource is usually separately recorded and assessed. Formally it can be treated like a biotic regenerative resource (see Section 4.5.1.4). The inclusion of water into the impact assessment is promoted by the UNEP/SETAC Life Cycle Initiative.[99]

A combined formula for finite + regenerative resources, which is also to be used as first approximation for the resource 'water' is presented in the next section.

---

99) Koellner and Scholz (2008), Milá i Canals et al. (2009), Berger and Finkbeiner (2010), Pfister, Koehler and Hellweg (2009) and Koellner and Geyer (2013).

**Table 4.7** Resources shortage factors of important abiotic resources.

| Resources | Known reserves | World annual production (output) | Static range$^a$ (a) | Resource scarcity factor $R_i$ (a$^{-1}$) |
|---|---|---|---|---|
| **Source of Energy** | | | | |
| Oil | $1.968 \times 10^{11}$ m$^3$ | $3.906 \times 10^9$ t a$^{-1}$ (2007)$^{bc}$ | 43 | 0.023 |
| | $1.687 \times 10^{11}$ t (at the end of 2007)$^b$ | | | |
| Natural gas | $1.774 \times 10^{14}$ m$^3$ (at the end of 2007)$^b$ | $2.94 \times 10^{12}$ m$^3$ a$^{-1}$ (2007)$^b$ | 60 | 0.017 |
| Hard coal$^d$ | $4.309 \times 10^{11}$ t (at the end of 2007)$^b$ | $5.37 \times 10^9$ t a$^{-1}$ (2006)$^e$ | 80 | 0.0125 |
| Brown coal$^f$ | $4.166 \times 10^{11}$ t (at the end of 2007)$^b$ | $9.14 \times 10^8$ t a$^{-1}$ (2006)$^e$ | 456 | 0.0022 |
| Uranium (U) | $2 \times 10^6$ t (2008)$^g$ | $6.22 \times 10^4$ t a$^{-1}$ (2007)$^h$ | 32 | 0.031 |
| **Metals/Ores** | | | | |
| Aluminium (bauxite) | $3.2 \times 10^{10}$ t (2006)$^i$ | $1.77 \times 10^8$ t (2007)$^i$ | 181$^j$ | 0.0055 |
| Chrome (Cr$_2$O$_3$) | $3.7 \times 10^9$ t (2006)$^k$ | $1.9 \times 10^7$ t a$^{-1}$ (2006)$^k$ | 195 | 0.0051 |
| Iron ore (raw) | $3.3 \times 10^{11}$ t (2004)$^l$ | $1.238 \times 10^9$ t a$^{-1}$ (2003)$^l$ | 267 | 0.0037 |
| Manganese (Mn) | $6.7 \times 10^8$ t (2004)$^b$ | $6.22 \times 10^6$ t a$^{-1}$ (newer estimation) | 108 | 0.0093 |
| Nickel (Ni) | $1.44 \times 10^8$ t (2006)$^m$ | $10^6$ t a$^{-1}$ (1990 till now)$^n$ | 144 | 0.0069 |
| Tin (Sn) | $1.1 \times 10^7$ t (2004)$^o$ | $2.53 \times 10^4$ t a$^{-1}$ (2003)$^p$ | 435 | 0.0023 |
| Copper (Cu) | $4.8 \times 10^8$ t (2002)$^q$ | $1.34 \times 10^7$ t a$^{-1}$ (2002)$^q$ | 36 | 0.028 |
| Lead (Pb) | $8.5 \times 10^7$ (approximately 2007)$^r$ | $6 \times 10^6$ t a$^{-1}$ (approximately 2007)$^r$ | 14 | 0.071 |
| Zinc (Zn) | $4.6 \times 10^8$ t (2004)$^o$ | $9.2 \times 10^6$ t a$^{-1}$ (2003)$^p$ | 50 | 0.02 |
| Indium (In) | 6000 t (approximately 2007)$^s$ | 476 t a$^{-1}$ (approximately 2007)$^s$ | 13 | 0.079 |
| Mercury (Hg) | $1.2 \times 10^5$ t (2000)$^q$ | 1800 t a$^{-1}$ (2000)$^q$ | 67 | 0.015 |
| Silver (Ag) | $5.5 \times 10^5$ t (2006)$^s$ | $1.95 \times 10^4$ t a$^{-1}$ (2006)$^s$ | 28 | 0.035 |
| Cadmium (Cd) | 6.105 t (2002)$^s$ | $1.87 \times 10^4$ t a$^{-1}$ (2002)$^s$ | 32 | 0.031 |
| Platinum Group (PGM) | $8 \times 10^4$ t (2006)$^t$ | Pt: 223 t a$^{-1}$ Pd: 222 t a$^{-1}$ Σ (2006) 445 t a$^{-1}$ | 180 | 0.006 |

*(continued overleaf)*

**Table 4.7** (Continued)

| Resources | Known reserves | World annual production (output) | Static range[a] (a) | Resource scarcity factor $R_i$ (a$^{-1}$) |
|---|---|---|---|---|
| **Other raw materials** | | | | |
| Antimony (Sb) | >5 × 10$^6$ t (approximately 2007)[u] | 5 × 10$^4$ t a$^{-1}$ (approximately 2007)[u] | >100 | 0.01 |
| Arsenic (As) | 1.2–2.0 × 10$^6$ t (2007)[s] | 5.9 × 10$^4$ t a$^{-1}$ (2007)[s] | 25 | 0.040 |
| Phosphate (Rock) | 5 × 10$^7$ t (2004)[o] | 1.47 × 10$^5$ (2003)[o] | 340 | 0.003 |
| Bismuth (Bi$_2$O$_3$) | 1.1 × 10$^5$ t (not dated)[s] | 5700 (2006)[v] | 19 | 0.052 |

[a] Internationally also designated as *reserves/production (R/P) ratio*
[b] BP 2008; Statistical Review of World Energy 2008: *http://www.bp.com/statisticalreview*
[c] The world oil consumption according to the same source (b) 2007 was about 3.953 10$^9$ t, that is, consumption and delivery are even-balanced.
[d] Hard coal = anthracite + bituminous (BP, 2008).
[e] World Coal of Institutes: Coal Facts 2007, *http://www.worldcoal.org*
[f] Sub-bituminous = lignites + brown coal (BP, 2008).
[g] With dismantling costs upto $80 kg$^{-1}$; 5 Mt with dismantling costs up to $130 kg$^{-1}$; *http://www.euronuclear.org*
[h] Correspond to 622 Mt of oil equivalents according to (b), 1 Mt $U \approx 14 \times 10^9$ t Stein–Kohle–Einheit (SKE) $\approx$ 1010 t of oil equivalents.
[i] Reuters Alertnet *http://www.alertnet.org/thenews/newsdesk/L15774125.htm*
[j] The value is almost constant since the 1950s, shortage is however predicted (Meyer, 2004) due to an exponential growth of consumption (5% per year!).
[k] *http://www.icdachromium.com*
[l] *http://www.mapsofworld.com/minerals*; source for reserves data: Mineral Commodities Summaries 2004.
[m] Mineral Commodities Summaries 2007.
[n] Constant within ±7% since 1990, *http://www.em.csiro.au/news/facts/nickel*
[o] *http://www.mapsofworld.com/minerals*, source: Mineral Commodities Summaries 2004.
[p] Web page such as (o); source: World Mineral Production 1999–2003.
[q] *http://www.usgs.gov/minerals*
[r] Via Internet: Santos Ribeiro, J.A. – DNPM/BA; jose.rebeiro@dnpm.gov.br
[s] *http://minerals.usgs.gov/minerals*
[t] Via Internet: Ricciardi, O. de P. – DNPM/BA; osmar.ricciardi@dnpm.gov.br
[u] *http://www.lenntech.com/Periodic-chart-elements*
[v] Wikipedia; The indicated number refers only to the mining industry – more is exploited by refining of ores with bi-minerals as admixtures (as Cd with Zn).
[w] Encyclopaedia of the elements, Wiley VCH 2004; published online 23 January, 2008.
[x] *http://environmentalchemistry.com/yogi/periodic/Mn.html*

## 4.5 Impact Categories, Impact Indicators and Characterisation Factors

For a derivation of Equations 4.4–4.6 a continuous annual consumption is presumed that may be valid as a rough approximation. For all resources it is difficult to determine the supplies or world reserves. Known supplies depend on the respective status of exploration which again depends on economic factors. Beyond that there are estimations of world reserves justifying exploitation and estimations of entire occurrences (those justifying exploitation at present + those not (yet) justifying exploitation). The problem remains that a justification of exploitation is a function of the price of the respective raw material. If the demand rises, for example, due to a real or politically caused shortage, the reserves increase because less productive resources or those difficult to obtain will also be exploited. However, even a free-market economy needs some time to transact the investments necessary for the discovery of new resources and an operation of mines, and hence, reliability of data cannot be assumed. Because many metals, for example, relatively noble ones like copper are not really consumed but are accumulated in the technosphere[100], it is unclear whether these stocks are to be included as reserves or not. Another question arises: Where is the starting point for mine exploitation of, for example, landfills ('waste mining')?

Furthermore, the employment of weighting factors requires tables, which lists data on reserves and consumption for as many raw materials as possible. For this reason, mostly *explored* (safe) reserves are used for the determination of the 'static range'. For oil this 'time' amounts to approximately 40–45 a, that is, it is as much newly explored as consumed. By comparison with comparable figures for coal it is nevertheless possible to state that the static ranges for coal are twice (hard coal) or 10 times (lignite) as large as those of oil. For an aggregation, *relative* data are completely sufficient as long as they were procured by uniform methods. In Table 4.7 the static ranges and resource scarcity factors for the most important abiotic resources are provided.

For the following elements 'high to extremely high' static ranges are indicated by Crowson[101]: beryllium (Be), gallium (Ga), kaolin, lithium (Li), magnesium (Mg), phosphate, rare earths and silicon (Si); 'large' for germanium (Ge). These elements and minerals, therefore, do not have to be included in the weighting of resources. The inclusion of the rare earths in this list has recently been challenged because a scarcity of several of these elements is beginning to cause serious concerns. This scarcity seems to be economically/politically caused and not due to a real physical shortage. A shortage of lithium seems to be possible, too, if production and use of electric cars increase.

A large static range can also occur because of very small annual consumptions that may increase with new technologies, for example, indium in the cell phone technology or if gallium were largely used as semiconductor material in the future.

On the basis of resources consumption in agriculture (e.g. phosphates) Brentrup and co-workers do not recommend an untimely aggregation of abiotic resources

---

100) Brunner and Rechberger (2004).
101) Crowson (1992).

because of their different operational areas. They do nonetheless recommend a resource consumption index.[102]

The recent developments of natural gas and oil exploration using the 'fracking' technology are not discussed in this section. This technology will cause new environmental problems, but may increase the static range of the two most valuable fossil resources.

**4.5.1.2.3 Further Abiotic Resources** *Abiotic flow resources*[103] also belong to abiotic resources in a wider sense: Solar radiation, wind, tide and currents, rain and river waters. These resources are at present not yet recorded and assessed routinely; however, they can be of crucial importance with regard to individual objectives, particularly with LCA studies on renewable energies.

### 4.5.1.3 Cumulative Energy and Exergy Demand

The *cumulative energy demand*[104] (CED) is often integrated into the impact assessment as means of a measure for a primary energy demand per fU. This was common practice in the time of the proto-LCAs (Section 4.1). Because any kind of energy demand, according to ISO criteria, does not correspond to an impact category, a CED cannot, strictly speaking, be an indicator. In the first round of international standardisation, CED was neither considered (not even mentioned) in the inventory (LCI) phase (ISO 14041) nor in the impact assessment (LCIA) phase (ISO 14042). If the metaphor is allowed, the CED was somehow trapped between the two standards and, inspite of an integration of 14041–43 into one new standard (14044: 2006), has since remained there.

The CED is however approved by the Dutch guidelines with an explicit reference to ISO standards.[105] It is a very useful characterising figure[106] that can be determined with relatively small uncertainty and it designates the overall energy demand, including renewable forms of energy. It is therefore an ideal supplement to the information provided by impact categories like resource consumption and climate change concerning the fossil and nuclear energy carriers. It mainly serves to support and assess energy saving measures. For an ecological product comparison a product consuming less energy should obtain a bonus even if renewable energy sources contribute to that. It is however not suitable as the sole criterion.

If CED is not integrated into the inventory where it does not, strictly speaking, belong because of issues related to system boundaries,[107] it should be integrated with input-oriented impact categories as an assessment value as well as into the interpretation.

---

102) Brentrup et al. (2002a).
103) Udo de Haes (1996). Fossil raw materials and minerals serving as depot resources are characterised by finiteness. Flow resources can however only change permanently with drastic changes of the environment.
104) VDI (1997).
105) Guinée et al. (2002).
106) Klöpffer (1997b), Finnveden and Lindfors (1998) and Huijbregts et al. (2010).
107) Frischknecht reply to Klöpffer and Poster Göteborg.

The quantification of CED has been thoroughly discussed in the chapter on inventory (Section 3.2.2).

There is a relation between CED and the *cumulative exergy demand* (CExD).[108] Whereas CED designates the overall primary energy per functional unit of a product system, exergy[109] designates the available amount of energy and is thus related to the 'free energy' or 'free enthalpy' of physical chemistry. The laws of thermodynamics imply that even though the total energy of a system cannot be lost (first principle), heat may be produced or lost during the transformation of one form of energy into the other (e.g. frictional heat) that can no longer be employed for work in a physical sense. *Exergy quantifies that part of the total energy that is available for work*. As such it is the opposite of entropy, which quantifies a tendency of the system to be transformed into a non-ordered type and is not usable for work, (second principle, see Equation 4.7). A small variation of enthalpy $dH$ (energy at constant volume), for example, within a chemical reaction is composed of a variation of free enthalpy $(dG)_T$ at temperature $T$ plus a further amount of energy $TdS$ where $dS$ signifies the variation of entropy:

$$dH = (dG)_T + TdS \tag{4.7}$$

$H$ (J) enthalpy
$G$ (J) free enthalpy
$S$ (J K$^{-1}$) entropy
$T$ (K) temperature

With each energy conversion, the free energy, respectively, the free enthalpy constitutes the maximum that can be converted into work. In technology this thermodynamic figure is called *exergy* and can also be applied to non-energetic resources, above all, minerals and ores.[110] Thus, a loss of resources by dispersion without actual consumption can be integrated into a uniform figure. Exergy can be assigned to all raw materials that can prove their applicability in LCA and databases.[111] Like most thermodynamic figures, exergy cannot be computed absolutely but needs a reference compound, which usually corresponds to the lowest state of energy of the element, for each material. The exergy then corresponds to the work necessary for the formation of the desired substance – mineral, fresh water, and so on, – or the maximum work generated in the case of the reverse reaction. Reference compounds and – energies obtain an exergy value of zero. Such assignments cannot be made without certain arbitrariness, and therefore, require some convention to be followed. It can be '*de facto*' provided in the form of a large table, which is inserted into a database.[112] Furthermore, assumptions concerning their composition must be made for ores and something similar is also valid for chemical substances, which do not represent pure compounds, but mixtures.

---

108) Finnveden and Östlund (1997), Dewulf and Van Langenhove (2002), De Meester et al. (2006), Bösch et al. (2007), Koroneos, Rovas and Dompros (2011) and Koroneos et al. (2011).
109) Szargut, Morris and Steward (1988) and Szargut (2005).
110) Finnveden and Östlund (1997), Szargut (2005) and Bösch et al. (2007).
111) Bösch et al. (2007).
112) Szargut (2005), Bösch et al. (2007) and Koroneos et al. (2011).

With energetic raw materials reference energies must be specified. Different resources store exergy in different forms of energy, such as chemical, thermal, kinetic, potential and nuclear energy. Which form is to be assigned to a specific material, and so on, depends on the use of the resource[113]:

- chemical exergy for all material resources, biomass, water and fossil fuels (all materials with exception of the reference compounds in a reference state – these obtain the value zero);
- thermal exergy for geothermal energy (no material transfer);
- kinetic exergy for wind energy (wind generator);
- potential exergy for water in hydro-electric power plants;
- nuclear exergy for nuclear fission in nuclear power stations;
- radiating exergy for solar radiation (solar panel).

According to this listing, exergy is allocated to both scarce resources (some material resources, water in many parts of the world, potential energy for conventional water power) as well as to those with practically unlimited reserves (solar radiation, wind energy). Therefore a CExD as scarceness indicator for energy resources is only of limited use. This valuation might vary for material resources because here a large dilution, which implies a large expenditure for the mining, must be included in the result.

Due to missing experience with CExD in real LCA, this indicator should be regarded as a highly interesting area of research within the impact assessment. Exergy values suitable for the setting up of characterisation factors can be extracted from the quoted work papers and have already been integrated into the ecoinvent data base.

### 4.5.1.4 Consumption of Biotic Resources

Biotic resources are living natural beings and communities, which grow without direct human effort, reproduce and have a specific function within the natural ecological systems.[114] To these belong the fish of the seas, rainforests (more general: natural forests), and their plants and animals, *not* products of agriculture and forestry plus related techniques like commercial aquaculture ('fish farms'), all kinds of plantation economy, keeping of domestic cattle, and so on. The reason for this separation is the general system boundary of LCA: The technosphere is separated from the ecosphere, and all anthropogenic activities are based *within* the technosphere. The environment is by this definition everything that is not part of the technosphere.

Biotic resources are mostly, but not always regenerative. For example, tropical rain forests cannot be sustainably cultivated because it is already heavily damaged by the building of roads necessary for development, and hence, tropical wood from primary forests cannot be regarded as regenerative. Game in the wild is a border line case in cultural forests, for example, in the high mountains. This game, fit

---

113) Bösch et al. (2007).
114) Müller-Wenk (2002a).

for hunting, is mostly 'fostered', that is, only hunted at certain times and fed on in the winter. Such activities belong to the technosphere. Similar considerations are valid for the fishery in rivers and lakes. Too large game populations imply substantial damage to vegetation and these animals do not belong to species threatened with extinction. Non-fostered wild game and those, which inspite of protective regulations are illegally poached can belong to threatened species. In general, animals and plants in nature with a high commercial value are threatened, for example, as hunting trophies, nutrition, source for medical active substances or for certain cultural or superstitious practices. The quoted subchapter of the final report of the second SETAC Europe working group 'Impact Assessment'[115] besides a detailed discussion of the protection goal provides a list of animals and plants, threatened with extinction (extreme case) by overfishing, or by a drastic population decrease, and so on. These species are, if relevant for a specific LCA, to be recorded in the inventory and subsequently be assessed in the impact assessment as 'biotic resources'. Mueller Wenk estimates that of the many millions of species of (wild) animals and plants only some thousands are used by humans as resources and only some hundreds (above all, fish and tropical plants) are threatened by direct use. The threat to a variety of species by destruction of habitats due to anthropogenic land use is not considered. Also Mueller Wenk points to the fact that apart from the shortage caused by land use a contribution to a reduction of the variety of species or biological variety (biodiversity) is to be considered.[116] Its impairment is an important impact category, unfortunately with no clear indicator yet (see Section 4.5.1.6), so that it can only indirectly be covered by other categories.

**Impact Indicators and Characterisation Factors**
If to be recorded separately, non-regenerative (finite) biotic resources, for example, ecological systems like the tropical rain forest with its specific spectra of species can be elaborated with similar impact indicators as discussed for abiotic non-regenerative resources. In addition however tables with static ranges would have to be present.

Shortage or scarceness as impact indicator is also valid for regenerative biotic resources. Shortage occurs, if withdrawal – globally or in a specific region – exceeds generation. For a quantification of regenerative biotic resources their *formation rate* must be known. Contrary to finite resources – which by continuing withdrawal will get exhausted in any case, being only a question of time – *a sustainable use can be achieved for regenerative resources, if the following permanently applies*:

Withdrawal per time unit (world annual consumption) ≤ formation rate

The natural measure for the scarceness of regenerative resources is thus the difference between the world's annual consumption and the formation rate, related to world reserves. In this case the resources scarcity factor $R_i$ can be computed according to Equation 4.8.

As with abiotic finite resources the world reserves of biotic resources can only with difficulty be precisely established. Also the determination of formation rates

---

[115] Mueller Wenk, in: Udo de Haes *et al.* (2002).
[116] Koellner and Geyer (2013) special issue land use.

is complicated. Such procurement is, for example, done in the fishery for the computation of fishing quotas. If biotic resources are important in an LCA, adequate research on those is necessary.

$$R_i(a^{-1}) = \frac{\text{World annual consumption}\,(t\,a^{-1}) - \text{Formation rate}\,(t\,a^{-1})}{\text{World reserves}\,(t)} \qquad (4.8)$$

The resource ($i$) is scarce if $R_i > 0$, if consumption exceeds the formation rate. Only such resources should be evaluated as scarce. The resource scarcity factor is calculated as in Section 4.5.1.2 for abiotic finite resources described in Equation 4.9:

$$R(\text{biotic, scarce}) = \sum_i (m_i \times R_i), \qquad R_i > 0\,(\text{kg}\,a^{-1}) \qquad (4.9)$$

The formulas for worldwide shortage can mutatis mutandis also be applied for smaller, approximately closed regions, for example, for a fresh water lake, a marginal sea with small fluctuations or a closed jungle area if it is required by the objective and scope of the LCA and if the data can be determined.

### 4.5.1.5 Use of (Fresh) Water

Fresh water is a regenerative abiotic resource, only in a few processes is it irreversibly used (cement → concrete, hydrolyses). For some applications water only gets heated (cooling in thermal power stations) or it supplies potential energy (hydro-electric power plants). Evaporation (e.g. during irrigation for agricultural use) withdraws water temporarily from human use, it is however not removed from the geological cycle. It may rain down over land or into the sea, but also evaporate from the surface of the sea and come down over land again. In reality the water cycle is much more complicated than mentioned here (in a nutshell) and so are the impacts of water.[117] In some countries water is abundant to an extent that no consciousness on its (global) scarceness can be observed. Recently however the regional scarcity of fresh water is a highly regarded topic and within the ISO 14000 series the standard ISO 14046 'water footprint' was developed[118] (see below).

A scientific discussion on organisation and quantification of resources distinguishes[119]:

1. *Deposits*, no regeneration during a time span comparable to human lifetimes (e.g. minerals, fossil energy carriers and raw materials).
2. *Funds*, these are resources that regenerate within relatively short times (within the measure of human life times) (e.g. game and wild plants and cultivated forests belong to the technosphere).
3. *Flows*, these are resources which continuously regenerate (e.g. wind and solar radiation).

Water can be part of each category depending on local/regional issues: fossil groundwater belongs to deposits, non-fossil groundwater to funds, surface water,

---
117) Milá i Canals et al. (2009), Pfister et al. (2009) and Berger and Finkbeiner (2010).
118) ISO/DIS 14046: 2013. Environmental management – Water footprint – Principles, requirements and guidelines.
119) Guinée et al. (2002) and Udo de Haes et al. (2002).

especially rivers, to flows. The latter is true only in case of a plentiful supply or of use not followed by consumption (e.g. as cooling water).

As water is already regionally traded, though not yet worldwide, and shows extremely inhomogeneous local distribution, a global reference basis (as in Equation 4.8) does usually not make sense. Already within relatively homogeneous economic areas (EU, USA) extreme differences in the available fresh water supply occur, so that only a regional view is applicable. This requires a sufficiently elaborate LCI and also includes a definition of use or 'consumption' of this resource. Consumption mostly consists in the use of water with contamination, which makes it useless for further use, for example, as drinking water, another loss being evaporation (especially during irrigation). Water purification belongs to the *end-of-life* of many product systems and can be regarded as resources recovery.

The inclusion of water use seems particularly important in view of LCAs where geographical system boundaries of countries, respectively, to regions with scarce clean fresh water supply are included. The resource types (e.g. surface water (river, lake), fossil and non-fossil groundwater, precipitation) considered and the forms of water use (e.g. drinking, cooling, irrigation) must be defined. The volume of water considered depends on whether only the so-called blue water (surface water and groundwater) is included or the so-called green water (precipitate and soil moisture that is evaporated by plants) is also included. The volume of water calculated for the production of, for example, 1 kg wheat will differ significantly. Hence, particular attention must be paid when comparing LCA results and not exactly the same LCI and LCIA methods have been used.

The quantification of water scarcity can principally be done according to Equation 4.8, albeit under consideration of water withdrawal and regional availability of fresh water supply (withdrawal-to-availability ratio – WTA). Regional reserves of the special case as well as formation rates for all used water categories are to be determined. Regarding the availability of fresh water supply a recourse to data banks is possible, for example, WaterGAP2.[120] An approach to include water scarcity is also included in the Swiss ecoscarcity model.[121]

The inclusion of water as a resource is essential whenever consumption exceeds the formation rate ($R_i > 0$ in Equation 4.9), or if different uses compete, for example, irrigation in agriculture, drinking water or water supply to a humid biotope (wetlands).

A comprehensive survey of water as a resource in the context of the impact estimation has been elaborated by a working group of the UNEP/SETAC Life Cycle Initiative.[122] An aspect that has so far been neglected was considered: often (fresh) water is not only a scarce resource for humans but also essential to the life of all organisms. It therefore serves a more substantial function than the resources discussed above, all fossil and most mineral resources being of interest to humans only or at least predominantly. In this function, water belongs to a protected

---

120) Alcamo *et al.* (2003).
121) Frischknecht *et al.* (2009).
122) Koellner and Scholz (2008) and Koellner and Geyer (2013).

target of ecological systems and should be characterised by an additional suitable *Mid-point* indicator.

Recent scientific research and ISO 14046 discuss water as an essential natural resource and consider both the increasing scarcity of fresh water in many regions and the degradation of water quality.

The published characterisation methods show two general lines of development[123),124)]:

- Primary impact (mid-point indicator): Scarcity is considered to be the impact. Examples are the 'Swiss Ecoscarcity Method'[125)] or the method of Ridoutt and Pfister considering a regional 'water stress index' (WSI) caused by the consumptive water use (CWU).[126),127)]
- Secondary (or higher order) impact (damage indicator), for example, damage of human health or of ecosystems:
Examples are the method of Motoshita *et al.*[128),129)] and the consideration of qualitative aspects by including degradative water use (DWU) in the method of Ridoutt and Pfister[130)], following the eco-indicator 99 approach.[131)]

Two continuative methodological approaches are to be mentioned:

Mila i Canals *et al.*[132),133)] propose to consider the 'Ecosystem Water Requirements', that means the amount of water used by the respective ecosystem in the region of interest, because this has to be subtracted from the availability of fresh water supply. The method of Boulay et al. addresses water scarcity caused by pollution.[134),135)] The quality of input and output water as well as the benefit for potential downstream riparians is considered.

All published impact assessment methods demand differentiated inventory data and ISO 14046 includes requirements on the documentation of elementary flows in the LCI: water quantity (inputs and outputs), resource type of used water, displacement of water from one resource type into another (e.g. groundwater to surface water), water quality characteristics, designated use of water, geographical location of water withdrawal and discharge including information of relevant drainage basin, temporal aspects and water quality.

ISO/DIS 14046 states that the communication of a 'water footprint' shall cover both the quantitative aspect (water availability, respectively, scarcity footprint) and the qualitative aspect (water footprint addressing water degradation). A specific characterisation model is not mandatory but the methods used shall be chosen

---

123) Berger and Finkbeiner (2010).
124) Berger and Finkbeiner (2012).
125) Frischknecht *et al.* (2009).
126) Ridoutt and Pfister (2010).
127) Ridoutt and Pfister (2013).
128) Motoshita, Itsubo and Inaba (2008).
129) Motoshita *et al.* (2009).
130) Pfister *et al.* (2009) and Pfister, Saner and Koehler (2011).
131) Goedkoop and Spriensma (2001).
132) Milá i Canals *et al.* (2009).
133) Milá i Canals *et al.* (2010).
134) Boulay *et al.* (2011a).
135) Boulay *et al.* (2011b).

according to the goal and scope of the study and shall be described precisely. If a full LCA is conducted, that is, impact categories are included indicating qualitative water pollution like eutrophication, acidification, ecotoxicity and thermal pollution, the water scarcity footprint can be used as additional impact category.

### 4.5.1.6 Land Use

Land use is discussed in the impact assessment as discrete impact category. This is not to be confused with the importance of land use and land use change for the impact category climate change. In this context land use is analysed on the properties to function as source and sink of GHGs, mainly $CO_2$ and $N_2O$.

Depending on the safeguard subject the focus of the view can be different with the consequence that different impacts are addressed and thus different indicators apply. If the view focuses on the safeguard subject 'protection of natural areas without anthropogenic intervention' this can be seen as an umbrella goal, for example, for biodiversity. If the view focuses on 'preservation of soil fertility' obviously different indicators are meaningful. Both approaches are explained below.

The most recent 'state of the art' with regard to land use and LCIA is presented in a special issue of the International Journal of Life Cycle Assess.[136] It is based on a UNEP/SETAC guideline on land use in LCIA.[137]

**4.5.1.6.1 The Hemerobic Level Approach** The approach to land use assessment discussed here goes back to ecological landscape assessment and may be older than LCIA. The primary subject for safeguard is the nativeness of land, which is seen as another scarce resource particularly in densely populated countries and regions: natural spaces of sufficient size are becoming scarce. For many years now, natural spaces have become smaller worldwide because of land requirement for intensive agriculture and renewable raw materials . In a broader sense any occupation and/or transformation of soil, natural or used by humans in different intensity is assessed. Many animals and plants depend on the presence of larger areas, either natural ones or those which are only extensively used. An ongoing settlement, increasing populations, de-fragmentation of landscapes by roads, intensive agriculture, forestry by means of plantations, and so on imply extinction of species and in the worst case the desertification of landscapes.

The variety of species and biodiversity as an admitted environmental target can be mapped within an LCA only with great difficulty. Therefore an attempt was made to indirectly define a criterion that covers at least some aspects for the protection of species. Besides, a demand for natural space can be regarded as criterion applicable for the protection of nature and landscape, soil and groundwater and similar goods, where natural soil is an indispensable prerequisite. A detailed consideration of possible and hypothetical consequences of land use by means of 'endpoints' where causal relations to the examined product system could hardly be provided would be far out of scope of an LCA. This is rather the task of environmental compatibility assessment and local planning. In LCIA, the category land use is above all regarded

---

136) Koellner and Geyer (2013).
137) Koellner et al. (2013a,b).

**Table 4.8** List of hemerobic levels[a,138]

| Hemerobic level | Level of naturalness | Use/examples |
|---|---|---|
| Ahemerobic | Natural | Uninfluenced ecological system |
| Oligohemerobic | Nearly natural | No or occasional use |
| Mesohemerobic | Semi-natural | Forestry (mixed woodlands), meadows and pastures (extensive) |
| β-Euhemerobic | Partly nature-remote | Forest mono-cultures, natural fruit cultivation and biological agriculture |
| α-Euhemerobic | Nature-remote | Arable land and garden areas (conventional agriculture) and viniculture (intensive) |
| Polyhemerobic | Xenonatural[b] | Sporting areas and landfills |
| Metahemerobic | Artificial | Sealed areas |

[a] By insertion of intermediate levels in between levels 2 and 3, 3 and 4, 4 and 5 as well as 5 and 6 an overall of 11 hemerobic levels (H0–H10) according to Brentrup et al. (loc. cit.) are obtained. In Table 1 Brentrup et al., provide examples for each level, which can also be useful for the seven-level system proposed here.
[b] Analogy to 'xenobiotic' (used for persistant man-made chemicals in the environment).[138]

as an indicator for the protection of nature and (terrestrial) species. Korte et al. (1992) and Klöpffer (2012b) Additionally soil has other basic functions like its mere availability for agricultural and other human activities (land occupation) or functions, regulating water regimes, offering recreation areas, and so on.[139]

There are no unambiguous decisions possible either for an operationalisation of this impact category or with regard to the choice of a simple parameter that quantitatively describes all or at least its most important functions. There are numerous grades between the extremes of completely natural and completely sealed areas ('between jungle and parking lot'), which make an application of a simple pattern difficult or impossible. A demand for natural space related to a fU is to be assigned to the type of use and the duration of use. Two of these figures can be expressed by numerical values: space and time. The third figure of qualitative type is used to characterise the nearness or remoteness towards nature. For this purpose, the well-known hemerobic levels of ecological landscape assessment can be used (Table 4.8).[140]

---

138) Brentrup et al. (2002b), suggest a 11-level scale
139) Klöpffer and Renner (1995), Müller-Wenk (1999a), Koellner (2000), Schenk (2001), Brentrup et al. (2002b), Lindeijer, Müller-Wenk and Steen (2002), Pennington et al. (2004), Milà i Canals et al. (2007a), Koellner and Scholz (2007) and Michelsen (2008).
140) Peper, Rohner and Winkelbrandt (1985), Klöpffer and Renner (1995), Kowarik (1999), Giegrich and Sturm (2000) and Brentrup et al. (2002a).

Certainly there is no clear dividing line between the levels, but nevertheless a transition can be perceived half way (approximately at level 4) between (still) natural and remote, with a near coincidence to the border between extensive and intensive agriculture. The first three stages are those, which are usually perceived as 'nature' (especially from an European perspective), even if the designation 'natural' strictly speaking only applies to 1 (ahemerob; wilderness).[141] Starting from level 5 technical aspects of intensive human influence on the ground by intensive agriculture, building of settlements, traffic routes and industrial surfaces clearly predominate. The highest level of remoteness from nature is found for sealed areas (level 7, metahemerobic or artificial). Even though an 11-level scale according to Brentrup permits a more refined assignment of land use, it is doubtful whether real inventories in LCIs provide such depth of detail necessary for local environmental evaluations. In addition, global LCAs also require comparability of hemerobic levels. A restriction to Europe (as in the 11-level scale) is inadmissible.

An at least theoretically attractive alternative to the concept of hemerobic levels is a quality index for soils/ecological systems that considers the variety of species and the productivity of the respective ecological system.[142] An advantage would be a continuous scale which would allow the development of *one* characteristic figure for the category land use; on the other hand *small productivity and low variety of species do by no means indicate inferior ecological systems* as all extreme ecological systems, for example, in high-alpine or arctic areas, steppes, dunes, meagre meadows, and so on, probably correspond to this description. Strictly speaking this is also valid for deserts but these do not fulfil the protection requirement of shortage (at least not on a global scale) and should be regarded separately.

In the hemerobic level concept scarceness is the common denominator of the impact category land use and other input-related categories. If hemerobic levels 1 and 2 in industrialised countries were not so scarce, beautiful old cultural landscapes (often confounded with nature itself) would be placed in the lead position of the natural spaces worth protecting. Such concepts should be employed for a transfer of the hemerobic level approach to other continents without losing sight of the global context (i.e. tropical rain forests, boreal coniferous forests and many other natural spaces).

**4.5.1.6.2 Characterisation Using the Hemerobic Level Concept** For the quantification of land use at least three factors with respect to the proposed characterisation model have to be provided:

1. Area per fU ($m^2$),
2. Utilisation period (a)
3. Type of use (hemerobic levels 1–7).

---

141) Human influences onthe atmosphere, precipitations and water currents cannot completely be excluded; with respect to this there are no areas on earth which are completely natural Klöpffer (2012b).
142) Lindeijer (1998).

Land uses of high impact and frequently applied in LCAs are cultivated areas particularly of agriculture and forestry, mining areas (especially open cast mining), traffic and dumping areas.

The collection of this information in the inventory can be quite laborious. Land use with relevant environmental impacts is however of great importance for both LCAs on nutrition and on renewable raw materials and should under no circumstances be neglected. Examples of renewable raw materials playing an important role in many product systems are wood (e.g. building products and cardboard, paper), oil seeds, sugar cane, sugar beet (e.g. fuel and lubricant) or corn (input material for agragas plants).[143] In addition, infrastructure areas including traffic routes and flooded areas for hydro-electric power plants are to be considered. In the latter case ongoing releases for many years of $CO_2$ and $CH_4$ by decomposition of flooded biomass must also be considered (see 'Climate change').

As soon as every area is quantified by space ($F_i$) and the utilisation period is determined, the impact indicator is formed by multiplication. Areas of the same hemerobic level are added up (Equation 4.10), the results of different hemerobic levels are however not aggregated.

Area specifications for relevant inventory data ($m^2$ a $fU^{-1}$) if necessary by conversion or estimation must be assigned to the selected hemerobic level of the impact assessment. Without further aggregation this results in

$$\text{Land use} = \sum_i (F_i \times \text{utilisation period}) \ (m^2 a)$$

(for each hemerobic level) $i$ \hfill (4.10)

$F_i$: area of hemerobic level $i$ (1–7) per fU
Utilisation period: time used to produce the quantity of material or energy needed per fU.

The evaluation according to Equation 4.10 thus does not provide a total 'naturalness score'.

Brentrup et al. (2002b, loc. cit.) define a 'naturalness degradation potential (NDP)' linearly increasing with hemerobic level from zero (hemerobic level 1 = H0) to one (hemerobic level 7 = H10) in order to obtain a cardinal scale. The designations H0 to H10 refer to the 11-level scale preferred by the authors. Similar to most scoring systems this weighting is arbitrary and serves only for a better adaption to the usual characterisation applied in other impact categories (inventory result × characterising factor = impact indicator result). Following the arguments of the authors the NDP ($i$ = type of use) can be used as characterisation factor for area × time (Equation 4.11):

$$NDI = \sum_i (F_i \times \text{duration of use } i) \ NDP_i \ (m^2 a) \hfill (4.11)$$

---

143) Faulstich and Greiff (2008).

NDI: naturalness degradation indicator.[144]

A further allocation of the NDI to the most important European types of land as suggested by Brentrup requires a larger spatial resolution than at present is common practice in LCI. The use of geographical navigation systems will make this possible in the future, provided the location is known (problem with generic data). In addition such a eurocentric view excludes non-European land or natural areas from the impact assessment. Because of these difficulties it is recommended to accept the interpretation according to Equation 4.10 and regard an interpretation according to Equation 4.11 only as supplementary.

Because land use can also provide improvement for individual cases considered in an LCA ('renaturation') and as the sealing of areas closer to nature should be rated worse than those, for example, of sporting grounds, the earlier status of the land (before transformation) should ideally be known. In this case the *change* of level would additionally have to be indicated. This was for the first time applied in the eco-inventories of energy systems.[145] The problem of irreversible or reversible transformation in contrast to temporary use or occupation of land is discussed below.

**4.5.1.6.3 Advanced Concepts** Demand on natural space and land use are important areas of research within the applied ecosystem research, landscape ecology and protection of species (variety of species = biodiversity), but also includes practical aspects like productivity of soils, groundwater formation, flood prevention, and so on. Land use has become such an important topic within the LCIA that the question arises whether this impact category should not be integrated in greater detail into LCA than can possibly be achieved by the concept of hemerobic levels. The scientific-academic discussion has been taking place within the expert groups of SETAC and the UNEP/SETAC Life Cycle Initiative.[146] It is accompanied by a great thoroughness of the assessment of the most important anthropogenic impacts on the soils as far as these have not been considered in other impact categories (e.g. ecotoxiticity by employment of pesticides). The land use special issue of the International Journal of Life Cycle Assess[147] discusses the following impacts.

Land use impacts:

- on biodiversity: effects on the safeguard subject 'natural environment';
- on biotic production potential: effects on the safeguard subject 'natural resources';
- on ecological soil quality: effects on the safeguard subject 'natural environment'.

It is differentiated between land occupation and transformation. Occupation corresponds to land use (without long-lasting changes). Transformation designates either a permanent change, or in case of not using a changed land, the slow recovery towards the original statusThis last point, in particular, requires knowledge of

---

144) NDI (Brentrup *et al.*, 2002a): Naturalness degradation indicator; NDP: Naturalness degradation potential
145) Suter and Walder (1995).
146) Lindeijer (1998), Udo de Haes *et al.* (1999a,b), Lindeijer *et al.* (2002) and Milà i Canals *et al.* (2007a).
147) Koellner and Geyer (2013).

processes that can extend over long periods. It can be accordingly difficult to find suitable indicators with the help of which quantification is possible.[148]

The published method of Mila i Canals et al.[149], currently widely discussed, refers to the safeguard subject 'life support function (LSF)' of agricultural and forestry areas. The LSF is calculated by building the difference between the 'soil organic matter (SOM)' – calculated based on the carbon content of soil at the beginning and at the end of specified land use. Thus the LSF mirrors the difference of soil fertility reduced to the indicator 'change in carbon content' between two states. For consequential LCA this may be a useful indicator, for attributional LCA assumptions concerning the reference situation are necessary.

This concept was modified by Brandao and Milà i Canals[150]: Instead of the LSF the indicator 'soil organic carbon' (SOC) is directly used and referring to Koellner et al.[151],[152]; the reference situation is defined as the (quasi-)natural vegetation depending on the geographical location and thus biomes and ecoregions are considered. The SOC of the vegetation under study (forest or crop) and the reference situation is calculated based on IPCC data,[153],[154] which correlate specified vegetation with the carbon content of the soil. However the geographic resolution and the differentiation of agricultural crops and forests are low.

The difference in carbon content addresses, according to the authors, the biological production potential.

An enumeration of possible indicators to the impacts specified above has led to a debate on the correct progress.[155] Udo de Haes criticises the absence of a critical discussion on which aspects, if any, of land use are compatible with substantial elements of an LCA (as for instance *quantitative* analysis 'from cradle to grave', to the comparison on the basis of a $fU$, generic – thus not local – treatment of space, and to *flow equilibria*). Further elements that absolutely do *not* fit into an LCA should be identified, and proposals should be prepared on how these aspects or impacts could be handled outside of LCA. However, Milà i Canals et al. counter that site-dependent characterisation factors are increasingly considered in the impact assessment; a flow equilibrium can be established and kept by appropriate temporal average values; (still) no specific indicators have been suggested as on date; however probably the most important impacts have been named, at least the most important: the variety of species. It is stressed, by the Dutch guideline[156] that the admission of an unweighted inventory parameter surface × time ($m^2$ a) is better than to completely omit land use in the impact assessment. It is only one step from here to land use weighted by means of hemerobic levels.

---

148) Koellner and Scholz (2007).
149) Milá i Canals, Romanya and Cowell (2007).
150) Brandao and Milà i Canals (2013).
151) Koellner et al. (2013a).
152) Koellner et al. (2013b).
153) IPCC (2003).
154) IPCC (2006).
155) Udo de Haes (2006), Guinée et al. (2006) and Milà I Canals et al. (2007b).
156) Guinée et al. (2002).

Other methods address the biodiversity[157] or a set of soil parameters[158],[159] in order to quantify impacts of land use. A proposal for a systematic classification of inventory data useful for all impact assessment methods is published by Koellner et al.[160]

The most important result of this high level debate is probably that more research and experience will be necessary in LCAs for this important category to be comprehensively and practicably integrated into the impact assessment. Theoretical models are inevitable for a clarification of terms but cannot replace applicable indicators in LCA with respect to impact assessment (LCIA).

## 4.5.2
## Output-Based Impact Categories (Global and Regional Impacts)

### 4.5.2.1 Overview

Output-related impact categories are those which do not assess the consumption of natural goods but the loads in the environment by releases from the technosphere in a wider sense[161] (stressors, interventions). This is supplemented by threats to human health and nuisances. Categories which describe global or regional impacts are particularly important. Within these there is a pretty good international agreement on indicators and characterisation models, especially in the case of climate change and stratospheric ozone depletion.[162]

Categories of global–regional, in some cases also of local impact characteristic are (see also Table 4.2):

1. climate change (global)
2. stratospheric ozone depletion (global)
3. formation of photooxidants (continental/regional/local)
4. acidification (continental/regional/local)
5. eutrophication (continental/regional/local).

The reasons for an overall acceptance of indicators and characterisation models of *global* impacts (1 and 2) should be based on the following:

- For global impacts the location of the releases can be neglected.
- Proposals for quantification have been elaborated by recommended scientific committees (IPCC, WMO) for a selection of indicators and characterisation models that can be adopted for use in LCIA.
- For individual partial impacts causal chains have been experimentally detected or at least been made very probable.

---

157) de Baan, Alkemade and Koellner (2013).
158) Baitz (2002).
159) Saad et al. (2011).
160) Koellner et al. (2013a).
161) Releases in a more narrow sense are defined as 'emissions to air and discharges to water and soil'.
162) Klöpffer et al. (2001a), Potting et al. (2001, 2002) and Udo de Haes et al. (2002).

- Quantification has to be done at the starting point of the impact hierarchy for global categories because only based on these, can relatively secure model calculations be accomplished (these are typical *mid-point* categories).

For these reasons global impact categories are considered to be of greater objectivity than others.[163] As for regional impact categories there are requirements for a stronger consideration of geographical release, distribution, and impact modes even if this implies higher requirements in the inventory.[164].

#### 4.5.2.2 Climate Change

*The waves of heat speed from our earth through the atmosphere towards space. These waves dash in their passage against the atoms of oxygen and nitrogen, and against the molecules of aqueous vapour. Thinly scattered as these latter are, we might naturally think meanly of them, as barriers to the waves of heat.*

*(John Tyndall 1863.)*[165]

There is worldwide a broad consensus on the impact category climate change and its quantification. However, characterisation factors slightly change with time as shown in reports published by the IPCC (Intergovernmental Panel on Climate Change)[166], especially owing to varying knowledge and subsequent assessments of indirect impacts. This is a common scientific process: we thereby apprehend that science approaches 'truth' at best, without ever, according to Popper,[167] reaching it.

The following can be learned from the category climate change for development of methods in LCIA:

- Quantification of the selected indicator in a form suited for LCA, here as GWP;
- Deduction of scientific relations including equivalence factors in specialised disciplines;
- Thorough scrutinising of the methods by an international Peer Review;
- Publication of results on behalf of a respected scientific committee (IPCC), which is accountable to the United Nations only.

To the first point: this turned out by chance, because usability of the GWP for LCIA was surely the last thing the IPCC was concerned about. To the second: LCA practitioners are often generalists and should delegate an elaboration of indicators plus quantification to specialists. Unfortunately their proposals are often unfeasible because they are usually not familiar with the LCA methodology.[168] The

---

163) Owens (1996, 1998).
164) Potting and Hauschild (1997a,b), Owens (1997), Bare, Pennington and Udo de Haes (1999) and Potting et al. (2001, 2002).
165) Tyndall (1873). (first edition 1863; the discovery of the natural greenhouse effect by Tyndall dates back to 1859 after predictions by Joseph Fourier 1824 and Claude Pouillet 1827).
166) IPCC (1990, 1992, 1995a, 2001, 2007).
167) Popper (1934).
168) Fava et al. (1993).

most important factor is probably the last: the prestige of the IPCC working on behalf of the UN. The IPCC was awarded the Nobel Peace Prize in 2007.

**4.5.2.2.1 'Greenhouse Effect'** Within the climate discussion and the LCIA, the term *Greenhouse effect* designates an *additional, anthropogenic* greenhouse effect. Life on earth (as we know it and being a part of it)[169] is only made possible by the *natural* greenhouse effect, which is induced by the gases, water vapour and carbon dioxide, in their pre-industrial concentration; without it the average temperature on the surface of the earth would be around $-18\,°C$ instead of $+15\,°C$ as at present. The natural greenhouse effect was already known in the nineteenth century as exemplified by the quotation of John Tyndall (1873, loc. cit.): at least water vapour as natural GHG was already known before 1870.

The additional, anthropogenic greenhouse effect which has already led to an increase of the average surface temperature by about $1\,°C$ is caused by an increased concentration of some trace gases in the troposphere (GHGs) partly identical to 'natural' GHG[170], see also Table 4.10:

- carbon dioxide ($CO_2$)
- water vapour ($H_2O$)
- methane ($CH_4$)
- dinitrogen oxide ($N_2O$)
- ozone ($O_3$, tropospheric)
- synthetic, persistent chemicals (mostly highly halogenated, for example, $CF_4$, $SF_6$, $NF_3$).

Relevant gaseous emissions listed in the inventory as mass per fU have their origin in a multitude of human activities, for example:

- Incineration of fossil fuels or materials produced from fossil raw materials ($CO_2$)
- calcination of minerals ($CO_2$)
- agriculture ($CH_4$, $N_2O$)
- losses during extraction and transport of fossil fuels ($CH_4$)
- industrial processes (halogenated solvents, $CF_4$, $SF_6$, $N_2O$)
- private use (chlorinated solvents, refrigerants: freon substitutes)
- landfilling, waste deposit ($CH_4$, $CO_2$).

Often non-listed in LCI are $CO_2$ emissions due to incineration or aerobic biodegradation of renewable raw materials or fuels that originated only a relatively short time ago (there is controversy about the period of time discussed, mostly between 20 and 100 a) by assimilation of atmospheric $CO_2$. This often so-called $CO_2$ neutrality does not make any sense for the anaerobic degradation under formation of $CH_4$, for example, in waste deposits, even if it originates from renewable sources: this GHG has a much higher GWP than carbon dioxide into which methane finally transforms by incineration, aerobic biochemical oxidation or atmospheric

---

169) It seems that lower forms of life survived periods of low temperatures on earth.
170) Brühl and Crutzen (1988), Deutscher Bundestag (1988, 1992) and IPCC (1990, 1992, 1995a,b,c, 1996a,b, 2001, 2007).

degradation by OH radicals (already to be considered in the inventory!). In order to safeguard a reliable C-balance the $CO_2$ assimilation by photosynthesis and the emissions should be quantified in the inventory.

The main prerequisites for a contribution to the greenhouse effect are the absorption in the atmosphere within the infrared spectral 'window' of about 10–15 µm and a sufficient tropospheric lifetime to allow an even distribution in the atmosphere. Substances with a short lifetime only generate islands of measurable concentrations near the emission sources. These properties imply a contribution to absorption of infrared radiation emitted from the surface of the earth in the direction of space, Tyndall's 'waves of heat'. The GHG have therefore an impact comparable to the walls of glass of a greenhouse, hence the name 'greenhouse effect'. In real greenhouses the solar radiation can pass through windows (with the exception of UV), the infrared heat radiation is, however, only partly emitted from the interior.

**4.5.2.2.2  Impact Indicator and Characterisation Factors**[171]  Impact indicator for the impact category climate change is the enhanced radiative forcing (difference between radiant energy received by the earth and energy radiated back to space) measured or calculated as radiation per area ($W\,m^{-2}$). This is the common and global primary effect which can cause multiple secondary and tertiary effects. The primary effect is related to 'global warming', an increase in the average temperature near the surface of the earth (including the lower troposphere and the surface water of the oceans). Therefore this impact category was formerly (and sometimes even today) called *global warming*, which however neither designates the primary effect correctly nor the multitude of the following effects. Radiative forcing is thus to be used as indicator for the renamed category of climate change. This is a typical *mid-point* indicator that may later be supplemented by *endpoint* indicators if scientific models allow such a correlation. These would for instance include the increase in the sea level and disastrous weather events (additional and strong floods, hurricanes, etc.), changes in the ecosystems and an increase of heat-related illnesses in moderate climate zones. The melting of glaciers and of Arctic ice is already vigorously taking place.

For an impact assessment in an LCA, a measure for a relative scale of the impact is necessary to make it possible for emissions of, for example, methane, nitrous oxide, carbon dioxide and chlorofluorocarbon (CFC) refrigerants to be weighted against each other and aggregated into a weighted sum. This applies to GWPs: they indicate the mass of $CO_2$, which has the same impact as the release of 1 kg of another GHG; for example, 1 kg of methane corresponds to 25 kg carbon dioxide ($GWP_{100}$). However, as the various GHGs have a different tropospheric life time (methane with around 10 a is relatively short-lived) the simulations have to be provided with a *time horizon* indicating the period of validity of the calculation. For LCAs a time horizon of 100 a is usually chosen. It is nevertheless probable that for some objectives (depending on the goal definition) shorter or longer time horizons

---
171) Klöpffer and Meilinger (2001a).

should be used. The respective GWP values are listed in publications of the IPCC and the corresponding updated (i.e. most recent) values should be used.

A prediction of temperature increase does not only depend on scientific principles but also on the development of GHG releases in the future and whether or when suitable measures are taken and whether these measures will not be annihilated by continuing economic growth. These trends can only be simulated with appropriate scenarios whereby – in addition to the diverse atmospheric lifetimes of the GHGs – a further time dependency is introduced.

In Table 4.10 the $GWP_{100}$-values for most common GHGs are listed.[172] GWPs of CFCs and some related chlorinated gases (e.g. HCFC-22, $CCl_4$) are not listed because besides having a usually very high heating effect they also exhibit compensation effects, which cannot be precisely calculated. These can, in rare cases, even result in a calculated cooling effect if aggregated. Should an inclusion of these substances be necessary for a specific LCA these can easily be found in literature.[173] Also GWP values with a shorter or longer time horizon ($GWP_{20}$ and $GWP_{500}$) can be found in the cited papers. In addition to the three most important GHGs ($CO_2$, $CH_4$, $N_2O$), the highly persistent perfluorinated gases and partly hydrogenated fluor-compounds used as freon substitutes must be considered. Compared to $CO_2$ they are only released in small quantities but show a GWP of up to 20 000-fold! This is caused by a high IR absorbance and the persistence of these substances. The perfluorinated substances are extremely hydrophobic and hence, are not washed out by precipitations and do not react with OH radicals for other reasons.[174]

$CO_2$ is the most important contribution to the overall GWP/fU in most LCA studies. For a calculation of the GWP from the inventory only the $CO_2$ that originates from fossils (incineration of coal, oil, etc.) and minerals (calcination of lime, production of cement) are considered, and therefore, have to be separately assigned. $CO_2$ from biological sources should also be assigned separately because this amount is extracted from the atmosphere by photosynthesis relatively short time ago and will again be released into the atmosphere by incineration or aerobic degradation. Such emissions are often called *$CO_2$-neutral* although this is true only over a sufficiently long time scale. The calculation of biogenic C-flow should not be omitted because the investigation of a credible C-balance must be ensured and documented transparently.

The greenhouse effect of the freons CFC-11 and CFC-12, which is not exactly determinable in size due to side effects, is less known and has been addressed by Ramanathan[175] shortly after the discovery of their ozone-depleting impact.

**4.5.2.2.3 Characterisation** The total GWP per fU is the sum of $CO_2$ equivalents which are calculated by a multiplication of GHG loads ($m_i$ per fU) from the

---

172) IPCC/TEAP (2007), Velders and Madronich (2007), Klöpffer and Meilinger (2001b), IPCC (1996, 2001) and WMO (1999).
173) IPCC/TEAP (2007); Table 2.6 in Velders and Madronich (2007).
174) Klöpffer and Wagner (2007a,b).
175) Ramanathan (1975).

inventory, and the respective $GWP_i$:

$$GWP = \sum_i (m_i \times GWP_i) \qquad (4.12)$$

$m_i$ = load of the respective substance $i$ per fU.

The GWP mostly selected for LCAs is $GWP_{100}$ (see Table 4.9). If in an impact assessment, several time horizons are used, it must be ensured that only GWPs of the same time horizon are used in the summing up according to Equation 4.12. GWPs of various time horizons of very long-lived GHGs only differ slightly; therefore, the life times in Table 4.9 are to be considered.

GWP values should be adopted from the latest IPCC or WMO report or from recent secondary literature.

**4.5.2.2.4 'Carbon Footprint'** Carbon footprint (CF) is a popular name for the GWP, especially if global warming is used as the one and only impact category in a LCA-type study. Such studies can be used for estimating the contribution of a product system, process, company or country with regard to $CO_2$-regulations such as the Kyoto-protocol, $CO_2$-certificate trading, and so on. The name is catchy, but clearly a misnomer, because carbon is a chemical element without which life would not be possible (and thus cannot be bad in itself) and there are several strong GHGs without carbon in the molecule ($N_2O$; $SF_6$; $NF_3$). A separate standardisation of CF[176] is nevertheless meaningful, since the international standards ISO 14040 + 14044 do not describe in detail how the GWP = CF has to be determined.

Furthermore it has to be considered that a life cycle study with only one impact category *is not and never will be* an (environmental) LCA and even less so a life cycle sustainability assessment (LCSA).[177] A low CF per fU may be overcompensated by other environmental and social risks and impacts, as clearly shown by nuclear electricity production.

In addition to the cited pre-standard publicly available specification, PAS 2050 and ISO/TS 14067, private initiatives have also been developing CF guidelines in order to enable enterprises to correctly measure, calculate and report the CF of product systems or even companies 'from cradle-to-grave' or at least 'from cradle-to-factory gate'. The most advanced guidelines are those developed by the World Business Council for Sustainable Development (WBCSD) together with the World Resources Institute (WRI)[178]. The development of these guidelines has been accompanied by international congresses, education and training in order to gain experience by applying the draft guidelines to real-world studies on a global scale. It is to be hoped that these life cycle initiatives will lead to correct action for the reduction of the GHG emissions and help to reduce the – already inevitable – increase in global warming to a tolerable limit.[179]

---

176) British Standards Institution (2008), Sinden (2009), Finkbeiner (2009) and ISO (2013).
177) Valdivia *et al.* (2011).
178) WBCSD and WRI (2010, 2011).
179) Politically, this limit is presently set to an average increase of 2°(K) above the pre-industrial level.

**Table 4.9** Global warming potential ($GWP_{100}$) of some greenhouse gases (time horizon: 100 a).

| Chemical designation of the GHG | Life time in the troposhere (a) | $GWP_{100}$ (kg $CO_2$-equivalent per kg GHG)[a] |
|---|---|---|
| Carbon dioxide ($CO_2$) | Determined according to the Berne C-cycle[b] | 1 |
| Methane ($CH_4$) | 12.0 | |
| Fossil origin | — | 25[c] |
| Regenerative origin[e] | — | 23 |
| Nitrous oxide ($N_2O$) | 120 | 298 |
| HFC-23[d] ($CHF_3$) | 270 | 14 800 |
| HFC-32 ($CH_2F_2$) | 4.9 | 675 |
| HCF-125 ($C_2HF_5$) | 29 | 3 500 |
| HCF-134a ($CH_2FCF_3$) | 14 | 1 430 |
| HFC-152a ($C_2H_4F_2$) | 1.4 | 124 |
| HCF-143a ($CH_3CF_3$) | 52 | 4 470 |
| HFC-227ea ($CF_3CHFCF_3$) | 34.2 | 3 220 |
| HCF-236fa ($CF_3CH_2CF_3$) | 240 | 9 810 |
| HFC-245fa ($CHF_2CH_2CF_3$) | 7.6 | 1 030 |
| HFC-365mfc ($CH_3CF_2CH_2CF_3$) | 8.6 | 794 |
| HFC-43-10mee ($CF_3CHFCHFCF_2CF_3$) | 15.9 | 1 640 |
| Sulphur hexafluoride ($SF_6$) | 3 200 | 22 800 |
| Nitrogen trifluoride ($NF_3$) | 740 | 17 200 |
| PFC-14[d] ($CF_4$) | 50 000 | 7 390 |
| PFC-116 ($C_2F_6$) | 10 000 | 12 200 |
| PFC-218 ($C_3F_8$) | 2 600 | 8 830 |
| PFC-318 (cyclo-$C_4F_8$) | 3 200 | 10 300 |
| PFC-3-1-10 ($C_4F_{10}$) | 2 600 | 8 860 |
| PFC-5-1-14 ($C_6F_{14}$) | 3 200 | 9 300 |
| HFE-449sl[c] ($CH_3O (CF_2)_3CF_3$) | 5 | 297 |
| HFE-569sf2 ($CH_3CH_2O (CF_2)_3CF_3$) | 0.77 | 59 |
| HFE-347pcf2 ($CF_3-CH_2OCF_2CHF_2$) | 7.1 | 580 |

[a] $GWP_{100}$ data: selection from IPCC fourth assessment report.
[b] The average tropospheric residence time of $CO_2$ depends on a multitude of sources and sinks and thus cannot be described by a single value.
[c] This value (IPCC, 2007) includes indirect effects by increased formation of ozone and stratospheric water vapour; by the relatively short life time of methane the GWP value depends strongly on the selected time horizon: $GWP_{20}$: 63; $GWP_{500}$: 7.
[d] HFC: HydroFluoroCarbon; PFC: PerFluoroCarbon and HFE: HydroFluoroEther.
[e] By anoxic formation out of materials of vegetable origin (e.g. paper in a landfill).

### 4.5.2.3 Stratospheric Ozone Depletion

The second global impact category concerns the human caused depletion of the stratospheric ozone layer responsible for the shielding of short wavelength solar radiation below 290–300 nm from the earth's surface.[180] Ozone molecules of low concentration in the stratosphere but within a large layer (about 20 km) are in dynamic equilibrium of formation and decomposition (Chapman cycle; Equations 4.13 and 4.14). High energy UV radiation plays an important part in this dynamic process:

*Formation of stratospheric ozone:*

$$O_2 + h\nu \rightarrow 2O \quad \left(\lambda = \frac{c}{\nu} < 240\,\text{nm}\right) \tag{4.13}$$

$$2O + 2O_2 \rightarrow 2O_3$$

---

$$3O_2 \rightarrow 2O_3 \quad \text{Sum}$$

*Photolysis of ozone:*

$$O_3 + h\nu \rightarrow O + O_2 \quad \left(\lambda = \frac{c}{\nu} < 320\,\text{nm}\right) \tag{4.14}$$

$$O + O_3 \rightarrow 2O_2$$

---

$$2O_3 \rightarrow 3O_2 \quad \text{Sum}$$

Besides these, there are numerous ozone-forming and -degrading reactions in the stratosphere with trace components of the $HO_x$ and $NO_x$ 'family'.[181]

Already around the year 1970 and in connection with the development of supersonic air planes there had been fears of increased ozone depletion by nitrogen oxides emitted from aeroplanes into the lower stratosphere. An extension of this work[182] to the newly developed chlorine cycle by Rowland and Molina in 1974 and 1975 resulted in postulating causality between refrigerants, persistent spraying agents, and so on, (freons, CFC) and an additional ozone depletion based on plausible data and assumptions:

*Chlorine cycle of catalytic ozone depletion*[183]

$$Cl + O_3 \rightarrow ClO + O_2 \tag{4.15a}$$

$$ClO + O \rightarrow Cl + O_2 \tag{4.15b}$$

---

$$O + O_3 \rightarrow 2O_2$$

---

180) WMO (1999), Klöpffer and Meilinger (2001b) and Dameris et al. (2007).
181) Finlayson-Pitts and Pitts (1986).
182) Molina and Rowland (1974) and Rowland and Molina (1975).
183) Deutscher Bundestag (1991) and Möller (2010).

The chlorine atom in Equation 4.15a necessary for the initiation of the cycle, derives from photolysis of long-lived (persistent) chlorine compounds of anthropogenic origin which due to their persistence are capable of entering the stratosphere intact, particularly as a result of an extremely slow reaction with OH-radicals.[184] This is a very slow process – otherwise even more easily degradable compounds could enter into the stratosphere – and there is a temporal delay of several years between releases and the start of effectiveness.

The reaction to the work of Rowland and Molina was a direct and immense one. The few measurements of[185] on the concentration of freons in the atmosphere, which the authors had referred to, were confirmed and updated again and again. A trend of increasing concentrations of the freons or CFC was observed, above all for the most important[186] ones: trichlorofluoromethane (CFC 11, R11), dichlorodifluoromethane (CFC-12, R12) and 1,1,2-trichloro-1,2,2-trifluoroethane (CFC-113, R113). Besides, some chlorinated solvents, not considered as freons, were recognised as potential ozone-destroying substances and pursued by an analytical series of measurements for many years.

**4.5.2.3.1 Causing Substances** Freons were and still are partly used (in medicine) for the following:

- Propellant for sprays,
- Propellant for foam materials (e.g. polyurethanes),
- Refrigerant for refrigerators and smaller air conditioning systems (especially for cars),
- Cleaning agent (e.g. in the electronic industry),
- Smaller applications in medicine (asthma spray), analytic and spectroscopy (extraction agent, solvent for IR spectroscopy), and so on.

Already within a few years following the forecast of the effect, the use of CFC propellants was banned in the USA and several alternatives were quickly developed. A setback within the model calculations (simulating the ozone depletion) at the beginning of the 1980s indicated smaller ozone degradation rates as in earlier forecasts whereby the urgency of the measures seemed to have diminished.

**4.5.2.3.2 The 'Ozone Hole' and Legal Measures** After the wrong 'all-clear' signal, a totally unexpected discovery indicated the formation of an antarctic 'ozone hole' during the spring of the Southern hemisphere.[187] The formation of the ozone hole *is related, but not identical*, to the impact prognosticated by Rowland and Molina, of chlorine compounds.[188] It turns out that heterogeneous catalytic reactions on

---

184) Atkinson (1989) and Klöpffer and Wagner (2007a).
185) Lovelock, Maggs and Wade (1973) and Lovelock (1975).
186) Deutscher Bundestag (1991, 1992), Krol, van Leeuwen and Lelieveld (1998) and Bousquet et al. (2005).
187) Farman, Gardiner and Shanklin (1985).
188) McIntyre (1989).

strongly acid aerosol particles at the extremely low temperatures of the Antarctic stratosphere play an important part in mechanisms of reactions of the catalytic ozone depletion. As the Arctic stratosphere does not have such a deep cooling down potential compared to the Antarctic, the effect is less retractable but nevertheless measurable: in the Antarctic – and to a lesser extent – in the Arctic spring, there is a strong decrease of stratospheric ozone concentration depicted as the 'ozone hole'. Inspite of a measurable increase of ozone concentration in the course of the year, annually measured minimum concentrations have decreased. Measurements also show the decrease in the stratospheric ozone concentration in non-polar regions based on the impact of homogeneous catalysis predicted by Roland and Molina.[189] This decrease is less dramatic than the ozone holes, but continuous.

The discovery of the 'ozone hole' certainly accelerated the international political agreement under the patronage of the UN, particularly with regard to concrete measures (a Principle Declaration the 'Vienna Convention' dates before the discovery of the ozone hole):

- Vienna Convention for the protection of the ozone layer dated March 22, 1985 ('Vienna Convention'; Discovery of the ozone hole in autumn 1985).
- Montreal Protocol dated September 16, 1987 on substances inducing the decay of the ozone layer.[190]

Within an amazingly short period of time concrete lists and schedules for the production phase-out of substances causing the ozone depletion was provided. At the same time the development of chlorine free substitutes began for special areas of CFC applications. Unfortunately some of these substitutes are identical to those causing the greenhouse effect[191] (Table 4.9) resulting in counter-productive effects.

Both the interim report of the Enquète Commission as well as a statement by Rowland[192] showed the surprise following the emergence of the ozone hole that had not been predicted. This dramatically exemplifies that highly complex systems like the stratosphere are far from offering themselves to a complete computation. Unpredictabilities are always possible and therefore the precautionary principle is to be taken seriously, to act before full scientific evidence on an environmental damaging impact is provided. With a minimum concentration decrease the ozone hole continues to grow which has been verified by measurements. It now extends far beyond the continent[193] although the most important substances responsible for the ozone hole have been internationally banned for years causing a slow decrease of their concentration in the atmosphere.

**4.5.2.3.3 Impact Indicator and Characterisation Factors** The impact indicator for the category 'stratospheric ozone depletion' is the formation of chlorine (and

---

189) Rowland and Molina (1975) and Dameris et al. (2007).
190) Deutscher Bundestag (1988).
191) IPCC (1995a), Klöpffer and Meilinger (2001a) and IPCC/TEAP (2007).
192) Deutscher Bundestag (1988) and Rowland (1994).
193) Dameris et al. (2007).

bromine) atoms in the stratosphere by photolysis of volatile and persistent substances with chlorine or bromine as substituents. This definition of the impact indicator concerns both the ozone depletion in the homogeneous gaseous phase of the stratosphere predicted by Rowland and Molina (though it caught little public attention), as well as a spectacular but temporally and spatially limited depletion which relates to the ozone hole.

An *ODP* value was introduced for the quantification of a relative scale of ozone-harmful activities of substances. It literally reads[194]:

> *The ODP represents the amount of ozone destroyed by emission of a gas over the entire atmospheric lifetime (i.e. at steady state) relative to that due to emission of the same mass of CFC-11.*

Formally it is similar to the GWP and it is also handled alike. Numerical values originate from relatively simple model calculations. Values increase with persistence of a substance in the troposphere (i.e. with growing probability of entering the stratosphere) and with an increase of chlorine atoms per mass unit in the stratosphere. In the case of brominated halons an approximate 10-fold catalytic activity of bromine compared to that of chlorine is included. The reference factor is the ODP of CFC-11 or R11 which is arbitrarily set to one – in complete analogy to the GWP of carbon dioxide.

The ODP characterising factors for some important ozone-depleting substances are shown in Table 4.10.

The highest values (ODP $\gg$ 1) are due to brominated halons applied as fire-extinguishing agents. Other perhalogenated carbon compounds (molecules, only exhibiting chlorine and fluorine, but no H as substituent) figure between ODP = 0.5 and 1.1 R11-equivalents.

Dinitrogenmonoxide ($N_2O$) is a well-known ozone-depleting gas with a different impact mechanism: the persistent gas enters the stratosphere and is then converted by reaction with oxygen atoms into $NO_x$ ($NO + NO_2$). However this reaction in a complicated way depends on the site of the reaction (particularly in the proximity of the tropopause).[195] More recently, the potent GHG $N_2O$ has been proposed to become the most important ozone-depleting gas of this century (given the declining emissions of the CFC).[196] According to this recent paper, $N_2O$ reacts most efficiently in the mid stratosphere where the highest concentration of ozone is also found. The removal of $O_3$ contributes to the general decay (as originally proposed by Roland and Molina) but not to the formation of the yearly 'ozone hole'. Ravishankara, Daniel and Portmann (2009, loc. cit.) as well as Wuebbles (2009, loc. cit.) argue that this sink should not be neglected. Furthermore, an interim ODP for $N_2O$ has been calculated as 0.017 kg CFC 11 equivalents per kilogram $N_2O$ emitted.

---

194) WMO (1999).
195) Klöpffer and Meilinger (2001b).
196) Ravishankara, Daniel and Portmann (2009) and Wuebbles (2009).

**Table 4.10** ODP (based on mass) of some ozone-depleting gases according to WMO (1999); time horizon ∞ (stationary model).

| Compound | Residence time $\tau_R$ (a) | Life span $\tau_{OH}$ (a) | ODP (kg CFC11 eq/kg) |
|---|---|---|---|
| CFC-11, trichlorofluoromethane ($CCl_3F$) | 45 | <6400 | 1.0 |
| CFC-12, dichlorodifluoromethane ($CCl_2F_2$) | 100 | <6400 | 0.82 |
| CFC-113, trichlorofluoroethane ($CCl_2FCClF_2$) | 90 | — | 0.90 |
| CFC-114, 1,2 dichloro-1,1,2,2-tetrafluoroethane ($CF_2ClCF_2Cl$) | — | — | 0.85 |
| CFC-115, chlorine-1,1,2,2,2-pentafluoroethane ($CF_2ClCF_3$) | — | — | 0.40 |
| Tetrachloromethane ($CCl_4$) | 35 | >130 | 1.20 |
| Methyl chloride ($CH_3Cl$) | About 1.3 | 1.3 | 0.02 |
| HCFC-22, chlorodifluoromethane ($CHClF_2$) | 11.8 | 12.3 | 0.034 |
| HCFC-123, 2,2-dichloro-1,1,1-trifluoroethane ($CF_3CHClF$) | — | — | 0.012 |
| HCFC-124, 2-chlorine-1,1,1,2-tetrafluoroethane ($CF_3CHClF$) | — | — | 0.026 |
| HCFC-141b, 1,1-dichloro-1-fluoroethane ($CFCl_2CH_3$) | 9.2 | 10.4 | 0.086 |
| HCFC-142b, 1-chloro-1,1-difluoroethane ($CF_2ClCH_3$) | 18.5 | 19.5 | 0.043 |
| 1,1,1-trichloroethane ($CH_3CCl_3$) | 4.8 | 5.7 | 0.11 |
| Halon 1301, bromine trifluoromethane ($CBrF_3$) | 65 | — | 12 |
| Halon 1211, bromine chlorodifluoromethane ($CBrClF_2$) | 11 | — | 5.1 |
| Halon 2402, 1,2-dibromo-1,1,2,2-tetrafluoroethane ($CBrF_2CBrF_2$) | — | — | 6.0 |
| Methyl bromide ($CH_3Br$) | 0.7 | 1.8 | 0.37 (0.2-0.5) |
| Dinitrogenmonoxide ($N_2O$) | 120 | — | $0.017^c$ |

$^a$ Average tropospheric retention.[197]
$^b$ Average tropospheric life span, computed from the OH-reaction constant and an average concentration of OH-radicals in the troposphere.[197]
$^c$ Interim values according to Ravishankara, Daniel and Portmann (2009), see text above.

In this regard it should be noted, that about two-third of the total emission of $N_2O$ is natural and one-third (10.5 Mt a$^{-1}$) anthropogenic.

Lane and Lant[198] recommend that these findings should be included into the ozone depletion category of LCIA, if $N_2O$ emissions are significant. They propose to use the ODP calculated as interim characterisation factor for mid-point modelling. As a result, land use, agriculture and industries contributing to $N_2O$ emissions would also contribute to ozone depletion.

The release of $N_2O$ into the troposphere is essentially due to bacterial metabolism in the soil. It is enforced by fertilisers containing nitrogen in agriculture whereby the percentage of the nitrogen converted into $N_2O$ is possibly underestimated.[199] A practical aspect for the proposed inclusion is that $N_2O$ emissions have to be recorded anyhow for calculating the GWP within the impact category 'Climate change'.

197) Numerical values according to WMO (1999); for the definition of residence time and lifetime see Klöpffer and Wagner (2007a).
198) Lane and Lant (2012).
199) Crutzen et al. (2007).

Thetime scale, infinite in theory, for a computation of ODP values according to a flow equilibrium model of the World Meteorological Organization (WMO) is different to the calculation of GWP.[200] There are, however, also ODP values calculated for a relatively short period of time.[201] The objective of time dependence has also been discussed by the WMO.[202] Accordingly, stationary ODP values of relatively short-lived compounds (e.g. HCFC) are small because they are computed in relation to CFC-11 whose flow equilibrium will only be reached in centuries even presupposing releases. The brief impact of relatively short-lived compounds is therefore underestimated. If ODPs are calculated for a short time horizon, the values can be around an order of magnitude higher; however, because of smaller persistence, the values are still below those of CFC-11 and CFC-12. In LCAs with an expressly intended comparison of freons and substitution products in their goal definition, the time dependence should be explicitly considered (computation loc. cit. WMO, 1994). For LCAs not specifically concerned with this problem, stationary ODP values should be used. In Table 4.10 an excerpt of available ODP values is listed.

There has been a slight loss of importance for the assessment of ozone depletion because it is presumed that the Protocol of Montreal and its supplements have been adhered to. Even so, the effect will prevail for some decades and attention is therefore required, in LCA also! In preparing the LCI it should therefore be carefully investigated to what degree the process of substitution has occurred in the reference time of the LCA study. Furthermore, $N_2O$ should be included as potentially ozone degrading gas. This is especially important since there seems to be much more 'laughing gas' around in the atmosphere[203] and new sources in addition to agriculture are sought and found (e.g. nitric acid production). As this gas is equally important for the impact category 'Climate change', better LCI data are to be expected in the near future.

**4.5.2.3.4 Characterisation** The quantification (characterisation) of the impact category stratospheric ozone depletion results from ODP values as equivalence factors similar to those of the GWP:

$$\text{ODP} = \sum_i (m_i \times \text{ODP}_i) \qquad (4.16)$$

The load of the ozone-depleting gases per fU ($m_i$) can be taken from the inventory table and appropriate $\text{ODP}_i$ values from respective charts of the WMO.

ODP is a typical mid-point characterisation factor, based on the precautionary principle. There are other attempts to quantify the effects of stratospheric ozone depletion within 'endpoint' models. Evidently, an increase of UV radiation at the surface of the earth may damage both: human health and ecosystem quality.

---

200) Udo de Haes (1996).
201) Solomon and Albritton (1992).
202) WMO (1994).
203) Crutzen et al. (2007).

## 4 Life Cycle Impact Assessment

Because there are only data available for the human health effect, only this one has been quantified.[204]

### 4.5.2.4 Formation of Photo Oxidants (Summer Smog)

The photochemical smog (= photo smog), also called *summer smog* or *Los Angeles Smog* with a history of about 60 years to correlate to the air in California particularly in the region of Los Angeles.[205] High motorisation following a *de facto* removal of rail traffic, high intensities of solar radiation and geographical conditions hindering the exchange of air masses (inversion weather conditions) are the ideal basis for the formation of photochemical smog initiated by the following reaction sequence[206]:

$$NO_2 + h\nu \rightarrow NO + O \quad \left(\lambda = \frac{c}{\nu} < 405\,nm\right) \tag{4.17a}$$

$$O + O_2 \rightarrow O_3 \tag{4.17b}$$

As long as NO is present in sufficient concentration, ozone and NO react back to $NO_2$. Therefore secondary reactions with reactive hydrocarbons particularly with alkenes or carbon monoxide (CO) are also necessary for photo smog formation. These compounds bind NO through oxidation by the radical intermediate product $HO_2$ forming $NO_2$ again. Overall the concentration of NO is reduced resulting in a surplus of ozone which is part of human- and phyto-toxicity of the summer smog.

The reaction cycle of ozone formation is exemplified in Equation 4.18 for CO forming OOH:

$$CO + OH + O_2(+M) \rightarrow CO_2 + HO_2(+M) \tag{4.18a}$$

$$NO + HO_2 \rightarrow NO_2 + OH \tag{4.18b}$$

$$NO_2 + h\nu \rightarrow NO + O \left(\lambda = \frac{c}{\nu} < 405\,nm\right) \tag{4.18c}$$

$$O + O_2(+M) \rightarrow O_3(+M) \tag{4.18d}$$

$$net : CO + 2O_2 + h\nu \rightarrow CO_2 + O_3$$

(M: inactive impact partner).

Ozone, which is beneficial in the stratosphere for the absorption of short-wave UV radiation (see Section 4.5.2.3) is a threat to the environment and to human health if formed at ground level. In addition to the well-known toxic ozone, other human and ecotoxic substances are produced. Together they form the group of 'photooxidants'; hence the name of this impact category.

---

204) Struijs *et al.* (2010).
205) McCabe (1952).
206) Fabian (1992), Klöpffer, Potting and Meilinger (2001b) and Barnes, Becker and Wiesen (2007).

The basic mechanism of Equations 4.17 and 4.18 has long been known[207] but not as long as the 'London Smog'. This is because of the smoke and fog which originated at the time of the introduction of coal heating (sulphur content responsible for subsequent formation of sulphuric acid) and the open fireplaces of those days, and was feared for its health-damaging threats.[208] Although both phenomena are called *smog* the impacts are caused by different pollutants, different reaction mechanisms and refer to different impact categories (see also 'Acidification', Section 4.5.2.5).

For the formation of summer smog the following circumstances are thus necessary:

1. intense solar radiation with a high UV contribution
2. reactive nitrogen oxides $NO_x$ ($= NO + NO_2$)
3. reactive volatile organic compounds (VOC, especially alkenes) and/or CO.

Point 1

Intensities of necessary solar radiation have long been exaggerated: the radiation intensity as well as the spectral composition (UV + short-wave visible radiation) in central Europe is sufficient for the formation of summer smog, as has been known since the 1970s.[209] The effect increases downwind many kilometres off the formation of primary smog. In Europe, the metropolitan area of Athens, the capital of Greece, comes closest to the meteorological situation and radiation climate of Los Angeles.

Point 2

Reactive nitrogen oxides $NO_x$ are mostly released by car traffic (also by diesel engines, high contribution of trucks). Nitrogen dioxide provides oxygen atoms by photolysis in the short wavelength spectrum (red gas, absorption within the blue part of the solar spectrum), and in the near UV (see Equation 4.18). $NO_x$ also occurs in pure air areas in small but increasing concentrations in recent years.

Point 3

Unsaturated hydrocarbons mostly originate from traffic but also from industrial plants. The releases of motor vehicles have been reduced by means of a suitable catalytic converter (in California since the mid 1970s) but have not been completely avoided. In California, many years after the introduction of the catalytic converter, red smog has been observed again. There are also reactive natural hydrocarbons (terpenes) which react with traces of $NO_x$ to form a summer smog with particulate follow-up products (aerosol, *blue haze*) in sunny woods. The proverbial 'ozone' of the spicy air of woods and forests seems to have another (more real) meaning besides a metaphorical one.

For an efficient abatement of summer smog the reduction of volatile organic compounds (VOCs), of carbon monoxide *and* of $NO_x$ releases is indispensable.[210]

---

207) Finlayson-Pitts and Pitts (1986), Fabian (1992, Chapter 4.1) and Barnes, Becker and Wiesen (2007).
208) The word 'Smog' is an artificial word composed from smoke + fog. The definition by the Oxford dictionary (smog = fog intensified by smoke) is relevant only for the winter variant, see also Fabian (1992).
209) Becker *et al.* (1985) and Fabian (1992).
210) Finlayson-Pitts and Pitts (1986).

For a quantification of summer smog within an impact assessment (LCIA) it is not possible to consider continuously changing, climate related weather conditions. As is well known, LCA bears little relation to space and time, which proves to be a problem in this impact category. Accordingly there is low consensus among the LCA method developers: proponents of the causal chain are not amused if emissions of reactive organic compounds[211] are fully attributed to the summer smog; those supporting the precautionary principle will accept this attribution of emissions as the *worst case* being only a *relative one* in view of a *possible* contribution to summer smog. The fact that it only occurs at times and under unfavourable conditions is regarded as less important by this approach – according to the principle *'less is better'*.

A minimum requirement for quantification will therefore be a scale of the relative effectiveness (reactivity) of hydrocarbons and CO. A respective first scale has been elaborated as early as 1976.[212] Since the reaction of volatile hydrocarbons and CO with OH radicals is an important phase, the second order reaction rate constant of OH with the substance to be weighted is chosen (see also[213]);

$$k_{OH} = 10^{-10} \text{ to } 10^{-11} \text{ cm}^3 \text{molecule}^{-1} \text{s}^{-1} \text{(very reactive)}$$

This implies a very high reactivity (e.g. propene and terpenes). The reaction rate constant can be measured over many orders of magnitude. At the other end of the scale – for persistent compounds – it serves for a determination of a transition probability into the stratosphere:

$$k_{OH} = 10^{-15} \text{ to } 10^{-14} \text{cm}^3 \text{molecule}^{-1} \text{s}^{-1} \text{(inert)}$$

The formation of ozone at ground level serves as a mere *parameter of reference* for the noxiousness of the photo smog because $O_3$ is by no means the only pollutant by which summer smog is incompatible to human health and the environment. There are a number of other photo oxidants with human- and ecotoxicity such as peroxyacetyl nitrate and aldehydes like acrolein, a lachrymal gas. Less known are reaction products of OH and $NO_x$ with organic compounds such as

- trichloroacetic acid (TCA), for example, formed from of trichloro- and tetrachloroethane[214]
- nitrophenols (formed from benzene and toluene, BTX hydrocarbons) especially the extremely phytotoxic dinitrophenols; dinitro-o-cresol (DNOC) is a former in Germany non-authorised herbicide which is formed besides other nitrophenols and nitrocresols in the atmosphere).[215]

Both pollutants may play a role in the degradation of forests, which was attributed to acidification earlier (see Section 4.5.2.5). However, acidification alone cannot be the cause of these damages as they also occur on calcareous soils.

---

211) These releases are also designated as non-methane volatile organic compounds (NMVOC).
212) Darnall *et al.* (1976).
213) Klöpffer and Wagner (2007a).
214) Renner, Schleyer and Mühlhausen (1990).
215) Rippen *et al.* (1987).

#### 4.5.2.4.1 Indicators and Characterisation Factors
Impact indicator of the category 'summer smog' is the formation of photo smog mostly measured by the formation of the leading substance of ozone.[216] A list of characterisation factors is used for a quantification following similar considerations as the original list of Darnall et al. (1976, loc. cit), but on the basis of model calculations. The *photochemical ozone creation potential* (POCP) of ethene serves as reference, which is arbitrarily set to one.

Table 4.11 lists a selection of POCP values.[217] The calculation of the less recent values is done according to three scenarios valid for Europe and for a time horizon of 9 days.[218] The values by Labouze et al., 2004 (Table 1 'POCP$_{mean}$') refer to an average daily ozone concentration from 0 to 2.2 km height neglecting limit values for the environment and human health. Calculations considering those limits have also been accomplished but are not listed. The relative sequence of organic pollutant classes remains unchanged for various types of calculation. However, the POCP of NO$_x$ (as NO$_2$), according to Labouze et al. (2004, loc. cit.), strongly depends on the fundamentals of the calculations (POCP: 0.27–0.95 ethene equivalents). Values averaged in space and time are graded as complementary to the values by Derwent et al.[219] as they are independent of meteorological conditions and emissions at a given time; they are valid for Europe only. They are usually below the values of Derwent because they were deduced for average atmospheric conditions, not for conditions that promote smog formation.

The low value of methane in Table 4.12 is owing to the low reactivity with OH radicals. Because inventories often list aggregated values, factors like 'sum of hydrocarbons', 'volatile hydrocarbons' (VOC) or 'non-methane hydrocarbons' are of special practical importance. The POCP of the majority of reactive substances are within the range of 0.1–1 kg ethene equivalents per kilogram. This implies that the exact composition of the mixtures of VOCs is of minor importance for the result. However, methane should *not* be integrated into the VOC mixture because of its inertness.

The concept of *maximum incremental reactivity* (MIR)[220] is an alternative for a characterisation by POCP factors and was developed in California attempting to quantify ozone formation under 'optimum' conditions. It is not region-specific but simulates smog conditions with strong solar radiation and high pollution load. So-called incremental reactivity is defined as an increase of ozone concentration per C atom of a VOC. These values, however, depend on specific circumstances of a smog episode and cannot be directly used for ranking. Thus a maximum value was defined (MIR) (mg O$_3$/mg VOC). MIR values can be transformed into relative values by arbitrarily assigning the value one to a substance in complete analogy to POCP values and other characterisation factors.[221]

---

216) Potting et al. (2002), Norris (2002), Klöpffer et al. (2001a) and Klöpffer, Potting and Meilinger (2001b).
217) Klöpffer, Potting and Meilinger (2001b), Derwent, Jenkin and Saunders (1996), Derwent et al. (1998), Wright et al. (1997) and Labouze et al. (2004).
218) UNO (1991).
219) Derwent et al. (1998).
220) Carter (1994, 2003), Klöpffer (2002) and Potting et al. (2002).
221) Klöpffer et al. (2001)

**Table 4.11** POCP (kg of ethene equivalents per kg) of some substances after CML Udo de Haes (1996), update: www.leidenuniv.nl/cml/ssp/databases/cmlia/ index.html, Derwent, Jenkin and Saunders (1996), Derwent et al. (1998), Wright et al. (1997) and Labouze et al. (2004).

| Substance class | Emission (formula) | CML | Derwent et al.[a] | Labouze et al. |
|---|---|---|---|---|
| Alkanes | Methane ($CH_4$) | 0.007 | 0.034 | — |
| | Ethane ($C_2H_6$) | 0.082 | 0.14 | 0.021 |
| | Propane ($C_3H_8$) | 0.42 | 0.41 | — |
| | n-Butane ($C_4H_{10}$) | 0.41 | 0.60 | — |
| | n-Pentane ($C_5H_{12}$) | 0.41 | 0.62 | — |
| | n-Hexane ($C_6H_{14}$) | 0.42 | 0.65 | — |
| | Cyclohexane ($C_6H_{12}$) | — | 0.60 | — |
| | n-Heptane ($C_7H_{16}$) | 0.53 | 0.77 | — |
| | Average | 0.40 ($n=23$) | 0.60 ($n=25$) | 0.1 |
| Olefins (alkenes) | Ethene ($C_2H_4$) | 1 | 1 | 1 |
| | Propene ($C_3H_6$) | 1.03 | 1.08 | — |
| | 1-Butene ($C_4H_8$) | 0.96 | 1.13 | — |
| | Isoprene ($C_5H_8$) | — | 1.18 | 0.23 |
| | Styrene ($C_6H_5C_2H_3$) | — | 0.077 | — |
| | Average | 0.91 ($n=10$) | 0.91 ($n=12$) | 0.67 |
| Alkines | Acetylene ($C_2H_2$) | 0.17 | 0.28 | — |
| Aromatics | Benzene ($C_6H_6$) | 0.19 | 0.33 | — |
| | Toluene ($C_6H_5CH_3$) | 0.56 | 0.77 | — |
| | o-Xylene ($C_6H_4(CH_3)_2$) | 0.67 | 0.83 | — |
| | m-Xylene | 1.0 | 1.09 | — |
| | p-Xylene | 0.89 | 0.95 | — |
| | Ethylbenzene ($C_6H_5(C_2H_5)$) | 0.60 | 0.81 | — |
| | Average | 0.76 ($n=14$) | 0.96 ($n=16$) | 0.44 |
| Hydrocarbons | Average | 0.38 | — | — |
| Non-methane HC | Average | 0.42 | — | — |
| Alcohols | Methanol ($CH_3OH$) | 0.12 | 0.21 | — |
| | Ethanol ($C_2H_5OH$) | 0.27 | 0.45 | — |
| | Isopropyl alcohol ($C_3H_7OH$) | — | 0.22 | — |
| | Ethylene glycol ($CH_2OHCH_2OH$) | — | 0.2 | — |
| | Average | 0.196 | 0.44 ($n=9$) | — |
| Aldehydes | Acetaldehyde ($CH_3CHO$) | 0.53 | 0.65 | — |
| | Formaldehyde (HCHO) | 0.42 | 0.55 | 0.41 |
| | Average | 0.443 | 0.75 ($n=6$) | 0.063 |
| Ketones | Acetone ($CH_3COCH_3$) | 0.18 | 0.18 | — |
| | Average | 0.326 | 0.52 ($n=4$) | 0.067 |
| Organic acids | Acetic acid ($CH_3COOH$) | — | 0.16 | — |

Table 4.11 (Continued)

| Substance class | Emission (formula) | CML | Derwent et al.[a] | Labouze et al. |
|---|---|---|---|---|
| Halogenated hydrocarbons | Methyl chloride ($CH_3Cl$) | — | 0.04 | — |
|  | Methylene chloride ($CH_2Cl_2$) | 0.01 | 0.03 | — |
|  | Vinyl chloride ($C_2H_3Cl$) | — | 0.27 | — |
|  | Trichloroethene/Tri ($C_2HCl_3$) | 0.07 | 0.08 | — |
|  | Tetrachloroethene/Per ($C_2Cl_4$) | 0.005 | 0.04 | — |
|  | 1,1-Dichloroethene (VDC) | — | 0.23 | — |
|  | 1,2-Dichloroethane (EDC) | — | 0.04 | — |
|  | Average | 0.021 | 0.11 ($n = 9$) | — |
| Inorganic oxides | Nitrogen dioxide $NO_2$ | — | 0.028 | 0.95 |
|  | Carbon monoxide CO | 0.027 | 0.02 | — |
|  | Sulphur dioxide $SO_2$ | — | 0.048 | — |

[a] An extensive list of values by Derwent et al. (1998) can be found in Guinée et al. (2002).

Table 4.12 Relative and absolute MIR of some materials.

| Substance class | Substance (formula) | MIR (relative) (kg of ethene equivalents) | MIR (absolute) (mg $O_3$/mg VOC) |
|---|---|---|---|
| Alkanes | Methane ($CH_4$) | 0.002 | 0.0148 |
|  | Ethane ($C_2H_6$) | 0.034 | 0.25 |
|  | Propane ($C_3H_8$) | 0.066 | 0.48 |
|  | n-Pentane ($C_5H_{12}$) | 0.14 | 1.02 |
| Olefins (alkenes) | Ethene ($C_2H_4$) | 1 | 7.29 |
|  | Propene ($C_3H_6$) | 1.29 | 9.4 |
|  | 1-Butene ($C_4H_8$) | 1.22 | 8.91 |
|  | Iso-butene ($C_4H_8$) | 0.73 | 5.31 |
|  | Isoprene ($C_5H_8$) | 1.25 | 9.08 |
|  | α-Pinen ($C_5H_8$) | 0.45 | 3.28 |
| Alkines | Acetylene ($C_2H_2$) | 0.069 | 0.5 |
| Aromatics | Benzene ($C_6H_6$) | 0.058 | 0.42 |
|  | Toluene ($C_6H_5CH_3$) | 0.37 | 2.73 |
|  | m-Xylene ($C_6H_4(CH_3)_2$) | 1.12 | 8.15 |
|  | 1,3,5-Trimethyl benzene ($C_9H_{12}$) | 1.39 | 10.12 |
| Alcohols | Methanol ($CH_3OH$) | 0.077 | 0.56 |
|  | Ethanol ($C_2H_5OH$) | 0.18 | 1.34 |
| Aldehydes | Acetaldehyde ($CH_3CHO$) | 0.76 | 5.52 |
|  | Formaldehyde (HCHO) | 0.98 | 7.15 |
|  | Benzaldehyde ($C_7H_6O$) | 0 | −0.55 |
| Ketones | Acetone ($CH_3COCH_3$) | 0.077 | 0.56 |
| Inorganic oxides | Nitrogen dioxide $NO_2$ | Uncertain | Uncertain |
|  | Carbon monoxide CO | 0.0074 | 0.054 |

A selection of MIR factors (absolute and relative) has been listed in Table 4.13.

The data by Derwent et al. (1998) also include specifications for $NO_2$ as demanded by Finlayson-Pitts and Pitts (1986). Surprisingly $SO_2$ is also part of smog formation. As can be deduced from the data, POCP values calculated by different models do not exactly match. The total range of all known POCP value does however not exceed two to three orders of magnitude.

To be noted in Table 4.12 is the fact that absolute MIR values can also be negative in exceptional cases if a specific VOC inhibits the smog reactions. This is the case with benzaldehyde whose molecules react in the gasphase with $NO_x$ without radical formation and therefore interrupt the chain reaction forming ozone and other photooxidants.

**4.5.2.4.2 Characterisation/Quantification** In the context of a uniform impact category for the formation of photooxidants the quantification of the impact indicator is accomplished as follows with the help of the POCP characterisation factors:

$$POCP = \sum_i (m_i \times POCP_i) \quad \text{(kg ethene equivalents)} \quad (4.19)$$

where $m_i$ = load of the substance $i$ involved in summer smog formation per fU.

A very extensive data record (96 substances) of Derwent, Jenkin and Saunders (1996, loc. cit.) is in part reproduced in Table 4.12. The tropospheric ozone formation with average European climate conditions is characterised and can also be used for the quantification of a regional ozone formation as its own indicator for this impact (POCP = $POCP_{reg}$) as has been proposed by SETAC Europe.[222]

Alternatively the characterisation can be accomplished by MIR factors:

$$POCP_{loc} = \sum_i (m_i \times MIR_i) \quad \text{(kg ethene equivalents)} \quad (4.20)$$

The characterisation with the help of MIR factors (Table 4.13) is suitable for a quantification of summer smog in areas with particularly high solar radiation, unfavourable emission conditions and slowly varying weather conditions. They have rarely been used so far in LCIA practice, probably because of an altogether small relation to space in the classical LCA. In the following, attempts to include the space into the impact assessment of summer smog events are discussed.

**4.5.2.4.3 Regionalisation of the Impact Indicator** As already discussed, the formation of summer smog depends on regional and meteorological factors like 'background concentration' of relevant precursors. The RAINS (regional air pollution information and simulation) model, developed on behalf of the UNECE (United Nations Economic Commission for Europe)-convention with respect to extensive and transnational air pollution, calculates the spatially dissolved ozone formation for all of Europe and considers spatially varying meteorological conditions and tropospheric chemistry. In addition, the spatially resolved ozone concentrations are related to critical ozone limits for humans and the natural environment. This model was used by Potting[223] to obtain simple factors which relate the release of an

---

222) Klöpffer et al. (2001) and Potting et al. (2002).
223) Potting et al. (1998) and Hauschild and Potting (2001).

ozone precursor in a specific (emission) region to an overall impact in the entire impact region (here: Europe). Thus in principle with the help of POCP factors actual impacts on humans and the natural environment can be calculated. This of course presupposes knowledge of the emission site that is known in the case of a factory (foreground data), but not if generic data, emissions in other continents, and so on, are used. The results confirm the major contribution of $NO_x$ and the dependence of impacts on the site of emission. These facts are more important than small variations in POCP factors (see above).

The spatial differentiation was integrated into the official Danish impact assessment (EDIP2003).[224] Hauschild et al.[225] point out that the formation of ozone (reference substance of photo smog) not only occurs at ground level but also in the free troposphere. In this case apart from traces of $NO_x$, which are always present even in relatively clean air, CO and $CH_4$ are necessary. This tropospheric ozone is of great importance for atmospheric chemistry and meteorology and contributes to the greenhouse effect (see Section 4.5.2.2); however, due to smaller concentrations, contributes less to human toxicity and ecotoxicity. The main significance of the impact category is therefore, within a regionalised view, related to the damage of vegetation and of human health. Two subcategories are introduced to be able to separately acquire these impacts. Regionalised characterisation factors are calculated according to the RAINS-model,[226] which correlates the emissions (non-methane volatile organic substance, NMVOC and $NO_x$) of a European country to potential impacts in any other (European) country. The entire impact is the result of the sum of all relevant combinations between the model cells. Receptors are included into the model by mapped vegetation and population density. Site-dependent characterisation factors are computed and represented in a table. Since in 'European' LCAs many emissions of non-European countries occur or are of unknown origin 'site-generic'[227] characterisation factors have been suggested.

Conditions for an application of this method are:

- Sites of emissions and quantities allotted to the sites per fU for most NMVOCs[228] and $NO_x$ must be known, rare in the case of complex product systems.
- Characterisation factors and the model must be integrated into the software.
- It is accepted that two subcategories, vegetation and human health, are considered.
- The spatial resolution must be required by 'goal and scope'.

If this resolution is not necessary in order to achieve the goal of the study, the effort at present is not worthwhile yet. The method is further classified as *midpoint* with, however, a certain shift towards *endpoint*.

---

224) Hauschild et al. (2006).
225) See also Klöpffer, Potting and Meilinger (2001b).
226) Regionally Air Pollution Information and Simulation; Amann et al. (1999).
227) Site-generic as opposed to site-dependent.
228) CO is not designated separately.

### 4.5.2.5 Acidification

The inclusion of the impact category 'acidification' can be related to the following environmental problem areas:

- acidification of unbuffered waters
- damage to forests
- acidification of soils.

In the first case which has particularly been observed within the crystalline region of South Scandinavia,[229] a direct causal chain can be presumed between emissions and impact. In the south of Norway and Sweden, freshwater lakes on granite bedrock were transformed into diluted acids as a result of acid precipitation. Under the influence of acids, $Al^{3+}$ ions, which are toxic for aquatic organism, dissolve from aluminium silicates. These ions are absent at normal pH levels (about 5.5–6; unbuffered equilibrium with $CO_2$ of the troposphere). Aluminium ions, the acid itself and possibly further dissolved products extinguish most organisms of these usually shallow lakes. In Scandinavia a chemical analysis of precipitation at different times showed a relation between the direction of the wind and the acid load. Highest loads always occurred with winds from Great Britain and the Continent. Acidification was caused mainly by European power plants. A misleading 'policy of high chimneys' only aimed at a dilution of pollutants. Improvements in cleansing technology slowly improved the air quality, especially with regard to $SO_2$.

An acidification of waters that can be observed in Scandinavia is typical for all scarcely buffered surface waters, partly and indirectly also for groundwater of the crystalline rock, which come into contact with air masses from industrial areas. Part of the acid-forming gases also originates from agriculture. To these, belongs the base ammonia, which by oxidation is transformed into $NO_x$, which in the end reacts with water in an oxidising environment to become nitric acid. $NH_3$ and $NH_4^+$, respectively, which enter soil and waters, are oxidised by bacterial nitrification and contribute to acidification.

A second environmental area related to acidification is the so-called novel damage to the forest. While direct damage to vegetation by acid gases has been known for 150 a[230] – so to speak acute phytotoxic impacts by high concentration of acid gases – these novel damages have only been studied since around 1970. In Germany, a political issue of forest decline was initialised by an article in the magazine 'Der Spiegel' in 1980; only 3 a later a special report of a board of environmental experts on the subject of forest decline and air pollution was published.[231] A first hypothesis by Professor Ulrich[232] was similar to the one explaining the impacts of acidification of lakes: a discharge of toxic ions into the soils of the forests, implying a damage to mycorrhiza (symbiotic association between a fungus and the roots of plants), nitrogen over-fertilisation of low nutrient forest soils, and so on. This rather mono-causal interpretation could not be maintained as

---

229) Fabian (1992, Section 4.2).
230) Stoklasa (1923).
231) RSU (1983).
232) Ulrich (1984).

soon as forest damages also were detected on calciferous grounds. A summary of a long-termed research project of Austrian limestone Alps was published by 1998.[233] A simple cause-effect chain could not be deduced from this work either. As early as at the end of the 1980s, it was concluded that novel damages of forests[234] were a multi factorial illness caused by various stress factors, the most important of being air pollutants.[235]

- sulphur dioxide ($SO_2$) → oxidation to sulphuric acid ($H_2SO_4$)
- nitric oxides ($NO_x$) → oxidation to nitric acid ($HNO_3$)
- ammonia ($NH_3$) → oxidation to $NO_x$ and nitric acid
- hydrofluoric acid (HF)
- hydrogen chloride/hydrochloric acid (HCl)
- photo oxidants (ozone, peroxyacetyl nitrate, ...) (see Section 4.5.2.4)
- organic compounds which only form in the troposphere by reaction with OH and $NO_x$.

These compounds are deposited in and on trees by

- dry deposition
- wet deposition (precipitation)
- occult precipitation: surface of leaves and branches of trees and other plants act as collectors for condensed water vapour originating from clouds and fog.

The details of the corresponding impacts are so far unknown. A simple conversion factor for novel forest damage is not possible because of the complexity of the symptoms and little knowledge of the cause and effect relationship. Damaging impacts caused by acid gases including the base ammonia are handled by an *acidification potential* (AP).[236] Other contributions are covered in the category photo oxidants (see Section 4.5.2.4).

Acidification (current impact category) and over-fertilisation of low nutrition soils (see impact category eutrophication, Section 4.5.2.6) must be distinguished from one another. Further impacts caused by acidification are the washing out of nutritive substances (e.g. $K^+$, $Na^+$ and $Mg^{2+}$) plus the mobilisation of heavy metals. Both can induce damage to vegetation. Furthermore, heavy metals that are washed out can pollute ground waters.

For pollutants with regional impact – or more general 'stressors' – questions with regard to the selection of the best impact indicator arise, as already discussed in

---

233) Special issue of Environmental Science and Pollution Research (ESPR) Vol. 5, No. 1 (1998) ecomed, Landberg a.Lech.
234) In German: 'neuartige Waldschäden'; this rather vague expression was created after the realization that acidification alone could not be the reason; it slowly replaced the older term 'Waldsterben' (dying of the forests) used mainly in the popular press.
235) Papke et al. (1987).
236) Heijungs et al. (1992), Udo de Haes (1996), Udo de Haes et al. (1999a,b), Norris (2001, 2002) and Potting et al. (2002).

Section 4.5.2.4. Therefore a simple characterisation by means of an AP is discussed first.

**4.5.2.5.1 Impact Indicator and Characterisation Factors** Impact category 'acidification' has been chosen in ISO 14044 (2006b, loc. cit. Figure 4.3) as an example to describe an approach corresponding to standards:

- *LCI results*
  Example: $SO_2$, HCl, HF, and so on ($kg\, fU^{-1}$).
- *LCI results assigned to impact category (classification)*
  Emissions with an acidifying impact, for example, $NO_x$, $SO_2$, and so on, are assigned to the impact category acidification.
- *Category indicator, characterisation model*
  Release of protons ($H^+_{aq}$); calculation of AP equivalents (mostly as $SO_2$-eq).
- *Impact endpoints*
  Acid-related damages on aquatic ecosystems, forests, vegetation, buildings, works of art, and so on.

Quantification by an AP as proposed by Heijungs *et al.* (1992, loc. cit.) starts on top of a stressor-effect-chain and 'counts' protons per fU as $SO_2$ equivalents, occasionally also as mass or mole of protons. The impact indicator of this simple model is the acid formation from precursor compounds and a successive entry of acid through water whereby a total dissociation of the acid into protons and the respective anions is presumed. This is a very good approximation for strong acids but also weak acids can shift basic milieus (e.g. sea water) in direction of the neutral point (pH 7). This is often called *acidification* in spite of being a mere reduction of alkalinity (but even that is dangerous for many marine organisms whose shells start to dissolve around the neutral point).

The AP is a typical *midpoint* indicator which neither name nor model endpoints. The endpoints however as entirety are considered in the interpretation because the potential of acidification may cause numerous endpoints.

Characterisation factors are calculated according to the stoichiometry of the formation of acid from precursors. Sulphur dioxide is a precursor of the two-base acid $H_2SO_3$ (sulphurous acid) formed by solution in water of the gas $SO_2$. Because it can only produce one mol of protons (by dissolution + oxidation from $NO_x$), 1 mol $HNO_3$ thus corresponds to half a mole of sulphur dioxide, which produces 2 mol of protons following dissolution in water. It is not important that sulphurous acid is a relatively weak acid since in the environment it oxidises into very strong sulphuric acid ($H_2SO_4$).

It should be noted that the weak carbonic acid, formed by dissolution of $CO_2$ in surface waters, especially the oceans, has never been included in the list of acidifying substances. It actually is globally the most relevant acid, but only one negative consequence of this gas is included in the impact category 'climate change' and the acidification potential is not considered.

> **Sample Calculation**
>
> Conversion of 1 kg nitric acid to kilogram $SO_2$-equivalents:
> A molar ratio of $n(HNO_3)/n(H_3O^+) = 1/1$, a molar ratio $n(H_2SO_3)/n(H_3O^+) = 1/2$ and a molar ratio $n(H_2SO_3)/n(SO_2) = 1/1$ implies the following:
>
> $$m(SO_2) = \frac{m(HNO_3) \times M(SO_2)}{M(HNO_3) \times 2}$$
>
> ($M(HNO_3) = 63$ g mol$^{-1}$ and $M(SO_2) = 64$ g mol$^{-1}$).
> From $(HNO_3) = 1$ kg thus 0.51 kg $SO_2$-equivalents result.
> The conversion for 1 kg ammonia is similar; it produces protons by oxidation into $HNO_3$, in the atmosphere or by nitrification:
> A molar ratio of $n(NH_3)/n(HNO_3) = 1/1$, a molar ratio of $n(HNO_3)/n(H_3O^+) = 1/1$, a molar ratio $n(H_2SO_3)/n(H_3O^+) = 1/2$ and a molar ratio of $n(H_2SO_3)/n(SO_2) = 1/1$ implies
>
> $$m(SO_2) = \frac{m(NH_3) \times M(SO_2)}{M(NH_3) \times 2}$$
>
> ($M(NH_3) = 17$ g mol$^{-1}$ and $M(SO_2^-) = 64$ g mol$^{-1}$).
> For $m(NH_3) = 1$ kg thus result 1.88 kg $SO_2$-equivalents.

As an alternative to mass equivalents mol protons were proposed for characterisation, which is chemically better justified. To ensure consistency with other impact categories we propose as indicator the ability to segregate protons, and $SO_2$ kilogram equivalents as characterisation factor. These can be easily, and above all unambiguously, calculated according to the laws of stoichiometry. The most important characterisation factors are listed in Table 4.13. Very weak acids as, for example, carbonic acid ($H_2CO_3$ and its anhydride $CO_2$, respectively) are not included in the calculation of the AP in spite of a strong contribution to the acidification of the seas (see above). In view of decreasing pH-values of the oceans, characterisation factors for $H_2CO_3$ and $CO_2$ should also be included (separately calculated if meaningful).[237]

Organic acids, mainly weak acids, are currently also not assigned to the AP. Strong organic acids (e.g. TCA) should be assigned for the calculation of the AP in future.

**4.5.2.5.2 Characterisation/Quantification** The impact category is converted into an AP according to the values listed in Table 4.13 or to equivalence factors, which can easily be determined stoichiometrically for every defined (strong) acid, and is then aggregated:

$$AP = \sum_i (m_i \times AP_i) \quad (kg\,SO_2\text{-equivalents}) \tag{4.21}$$

where $m_i$ = load of the substance contributing to acidification per fU.

---

237) WBGU (2006).

**Table 4.13** Acidification potential (AP) of some gaseous emissions (Heijungs et al. (1992), Klopffer and Renner (1995), Hauschild and Wenzel (1998) and Norris (2001).

| Release (compound) | Formula | AP (kg $SO_2$-equivalents) |
| --- | --- | --- |
| Sulphur dioxide | $SO_2$ | 1 |
| Sulphur trioxide | $SO_3$ | 0.80 |
| Nitrogen monoxide | NO | 1.07 |
| Nitrogen dioxide | $NO_2$ | 0.70 |
| Nitrogen oxides (calculated as $NO_2$) | $NO_x$ | 0.70 |
| Nitric acid | $HNO_3$ | 0.51 |
| Ammonia | $NH_3$ | 1.88 |
| Phosphoric acid | $H_3PO_4$ | 0.98 |
| Hydrogen chloride (→ hydrochloric acid) | HCl | 0.88 |
| Hydrogen fluoride (→ hydrofluoric acid) | HF | 1.60 |
| Hydrogen sulphide | $H_2S$ | 1.88 |
| TRS[a] (calculated as S) | — | 2.0 |
| Sulphuric acid | $H_2SO_4$ | 0.65 |
| Organic acids | R–COOH | None at present[b] |
| Carbon dioxide (→ carbonic acid) | $CO_2$ | None at present[b] |

[a] Total reduced sulphur.
[b] See text.

The AP from the point of view of its simple and unambiguous determination is ideally suited for impact assessment. From the impact side, it may be doubtful whether, for example, acid gases released into oceans ($SO_2$ from crude oil of vessels!) are relevant in view of the alkaline buffering capability of the oceans. However, a recommendation not to consider emissions into the ocean as different from those released into the continental atmosphere is based on the precautionary principle (here: *less is better*): first, we know nothing of the impact of sulphur dioxide on stressed oceanic ecosystems; second, emission of gases by open sea vessels can drift for large distances along their routes, mostly along the coasts and third, an incentive to use purer oil should provided.[238] This would also diminish the high $SO_2$ load in harbours. Finally, it can also be presumed that during the incineration of low-quality bunker oil, numerous other pollutants are formed.

From the point of view of environmental politics it can be noted that the presumed correlation between acid gases and forest damage in the 1980s introduced significant efforts to flue gas purification, especially in power plants. This has led to a substantial overall reduction of the $SO_2$ load. It is more difficult to remove $NO_x$ from incineration gases. This is why the efforts to reduce the nitrogen oxides, which are not only acidifying but also eutrophying, toxic and additionally induce the formation of smog, show much slower success. $NO_x$ is not only released by chimneys of power plants high into the air, but has also sources near the surface, for example, car traffic.

---

[238] Workshop: 'Realeases of vessels at the coasts of Northern Germany', 12February, 2008, Hamburg http://www.aknew.org

**4.5.2.5.3 Regionalisation** Attempts for a regionalisation of non-global impact categories are based on *complex models*, which cannot be discussed here in detail. The following is meant to provide an overview on the state of the art and should entice further reading of the quoted primary literature.

José Potting has for the first time pointed out that a neglect of spatial dimensions in the impact assessment can imply wrong results for non-global impact categories.[239] Deficits of the *less-is-better* approach in the characterisation model 'acidification', which have already been discussed, have transformed this category into a test area for developments directed towards a more realistic spatial indicator model, which should also consider impact thresholds or critical loads.[240] Newer developments are directed to the calculation of (European) country-dependent characterisation factors for $SO_2$, $NO_x$ and $NH_3$. As a prerequisite for the application of these factors, an assignment to the European countries where the relevant emissions originated (per fU) is necessary. This is without doubt easier for stationary emitters, like power plants than for product systems. The same applies for all non-global impact categories. The resilience of ecosystems to be protected, a concept which is not considered in the definition of the AP, is introduced by means of *critical loads*[241] in divers regions. An atmospheric transport from emitting countries to the sites of impact is simulated and models of diverse complexity are applied. Within a critical evaluation of different approaches to the modelling[242] an exact but necessarily very complex model[243] was used to test options for simpler linear models. Advantages of simpler models in the context of impact assessment are self-evident. A decision towards a best suited characterisation factor has not yet been made. A useful characterisation factor independent of the model may be the average accumulated exceedance[244] of critical load. Since however the concept of critical load is *not* based on a dose-impact relation, from the authors' point of view further research is necessary whether critical loads may be used as surrogate. The underlying problem being of course that limit values are predominantly pragmatically defined and are not the result of precise scientific analyses.

Bellekom *et al.* (2006, loc. cit.) investigated the feasibility of the application of a site-dependent impact assessment for the impact category acidification within three existing LCAs (linoleum[245], rock wool[246], water pipe systems).[247] To achieve this, inventories, which usually do not list the origin of emissions, had to be extended.

---

239) Potting and Blok (1994) and Potting (2000).
240) Potting and Hauschild (1997a,b), Potting *et al.* (1998, 2002), Huijbregts *et al.* (2000a), Hauschild and Potting (2001), Krewitt *et al.* (2001), Hettelingh, Posch and Potting (2005), Bellekom, Potting and Benders (2006) and Sedlbauer *et al.* (2007).
241) Hettelingh, Posch and De Smet (2001).
242) Hettelingh *et al.* (2005).
243) The RAINS model used by Potting Amann *et al.* (1999) as well as the EcoSense model used by Krewitt were ranked by Hettelingh *et al.* (2005) as complex models.
244) Average Accumulated Exceedance Posch, Hettelingh and De Smet (2001) and Seppälä *et al.* (2006).
245) Gorree *et al.* (2000).
246) Schmidt *et al.* (2004).
247) Boersma and Kramer (1999 (NL)), quoted in Bellekom *et al.* (2006).

According to the authors this was done without great difficulty in all three studies. The site-dependent characterisation factors were those of EDIP 2003.[248] The RAINS (IIASA, Laxenburg) indicator model was chosen, geographical system boundary was EU15 + Switzerland, Norway. All further emissions outside the geographical system boundary or with unknown origin listed in the inventory have been assigned to a site-generic average characterisation factor. This also applies for non-European emissions. An analysis of the inventory and the used generic data required the maximum time.

The analysis of the linoleum LCA showed that the introduction of the site-specific characterisation factors implied no changes to the original statements of the study.

For the rock wool LCA, which was not completely comprehensible because of some data being confidential, conditions were favourable since the regarded emissions with the exception of $NO_x$ mostly originated from a single known production site. Thus regional assignment was accomplished with little difficulty. A comparison of site-specific (predominantly Denmark) to site-generic characterisation resulted in twice as high values for the former. Since the rock wool LCA was done without comparative analysis concerning comparator product systems it is impossible to say whether conclusions would have differed. However, a reduction of emissions at the production site seems to be among the first of possible improvements.

As the sole system among the systems of water pipes investigated by means of an LCA, the traditional copper pipe system was studied in detail. Here the difference between site-dependent and site-generic characterisation did not amount to more than 10%. The enrolment within the systems did not change, but the relative contribution of single unit processes to the total result did.

Seppälä et al. (2006, loc. cit.) proposed country specific characterisation factors on the basis of an accumulated exceedance of limit values.[249] This indicator was suggested as an alternative to an 'unprotected ecosystems surface' and was originally developed to calculate the reduction of burdens due to acidification and terrestrial eutrophication (next section) in Europe caused by reduction of the most important emissions ($SO_2$, $NO_x$ and $NH_3$) into the air. An investigation to find how much a specific reduction in one or many countries contributes to the reduction of the overall load in Europe was conducted. Only the exceeding of critical regional loads was considered.[250] The calculation of characterisation factors was accomplished according to an 'exact model' (see above).[251] Factors for 35 European countries and 5 oceanic areas calculated for the year 2002 and estimated for 2010 are provided in a table. A comparison of results with an alternative model based on the indicator 'unprotected ecosystems surface' and calculated with RAINS (see above) is yet to be accomplished.

---

248) Hauschild et al. (2006); EDIP2003, Guidelines from the Danish EPA, to be published.
249) Seppälä et al.; see above and Hettelingh et al. (2005).
250) This concept implies that almost unspoiled areas – existing so far in Europe – may be 'filled' to the limit value without a negative impact in the calculations. In LCA jargon this is called *only above*, see also Hogan et al. (1996) and Seppälä et al. (2006); for the use of limit values in the impact assessment see also Sections 4.5.3.2 and 4.5.3.3.
251) The model is based on the EMEP model of the United Nations Economic Commission for Europe.

A recommendation for one or the other indicator model cannot be provided at present.

#### 4.5.2.6 Eutrophication

Eutrophication can best be translated as over-fertilisation or excess supply of nutrients. The impact category eutrophication is listed in every LCA but at closer look poses some difficulties.[252]

The substances that cause eutrophication cannot be generally referred to as *pollutants* as such, rather as plant nutrients. Its surplus implies as a first impact a forced photosynthetic increase of biomass (growth of plants, especially algae). A change of this supply causes changes in the spectrum of species in an ecosystem.

An important secondary impact in water bodies is the consumption of oxygen by means of bacteriological degradation of dead biomass. A strong increase of growth of algae induces more extinct biomass at the bottom of the water body with subsequent decay and can completely change the character of, for example, a lake or estuary: a formerly clean lake with drinking water quality can evolve into water with an anoxic (free of oxygen) depth layer. The reduction of the oxygen content changes the composition of species. In extreme cases an anaerobic ecosystem that is not desirable evolves.

The impact assessment can be differentiated between aquatic eutrophication (eutrophication in the original sense) and terrestrial eutrophication or over-fertilisation.[253] Gases like $NO_x$ and $NH_3$ (terrestrial eutrophication) and substances in the effluent of waste water sewage plants, which have not been completely decomposed or removed, are considered as well as untreated waste water entries into water bodies (aquatic eutrophication).

**4.5.2.6.1 Aquatic Eutrophication** An entry of nutrients into water bodies can occur both by the water path and by air.

The most important nutrients for plants are the elements *phosphor and nitrogen* in a resorbable nutrition compound, primarily as water soluble salts. Gaseous nitrogen from the air can only be used by some 'specialised' plants like legumes in symbiosis with nodule bacteria. The impact assessment only considers compounds suited for uptake by plants and only these must be integrated into the inventory. Other elements important for plant nutrition like potassium, copper (high concentrations are toxic!) and other trace elements are not integrated into the impact assessment. Phosphor is the limiting[254] element in most fresh water bodies (surface water like lakes and rivers as well as groundwater) whereas in seawater generally nitrogen is the limiting element. Estuaries and brackish waters can be either P- or N-limited. Terrestrial ecosystems are mostly N-limited. These differences can only be

---

252) Klöpffer and Renner (1995), Udo de Haes (1996), Udo de Haes *et al.* (1999a,b, 2002), Finnveden and Potting (1999, 2001), Guinée *et al.* (2002), Potting *et al.* (2002), Norris (2002), Seppälä *et al.* (2006) and Toffoletto *et al.* (2007).
253) 'Fertilisation' should not be misunderstood in an agricultural (intended) context, although agricultural wastes, increased run-off, and so on can (unintended) contribute to eutrophication.
254) If a shortage of a limiting element occurs even an overdose of other substances necessary for growth cannot produce biomass. The same applies for special amino acids in the food of animals.

considered in the impact assessment with geographically highly resolved inventory data. Without doubt phosphor and nitrogen are the highest contributors to the impact category eutrophication.

The former impact category 'chemical oxygen demand' has been integrated as an indicator for nutrients into the impact category 'eutrophication'. As a matter of fact, organic compounds that are biologically degradable under consumption of oxygen are better characterised by the *biochemical* oxygen demand (BOD) but the data availability of COD is much better and the BOD is included in the COD. As a result of aerobic bacterial degradation of organic compounds (particular or dissolved) the oxygen concentration decreases, $CO_2$ concentration and the concentration of inorganic nutrient salts increase. The impact of an increased entry of these organic compounds into the waters is therefore comparable to an over-fertilisation with P- and N-containing compounds followed by increased growth of algae.

Heat released by power plants into waters can have a qualitatively similar impact: bacterial degradation processes of organic matter are speeded up. Furthermore different species compared to those of cooler waters are favoured. Keywords like eutrophication, BOD and heat have been comprised into one impact category by SETAC Europe[255] (see Table 4.2).

An attempt can be made to comprise eutrophication and BOD or COD as one indicator; however, heat cannot be included. It has to be assigned and assessed separately, if relevant, in the study under consideration, as for instance for thermal power plants. Experience shows that heat is never an impact category of its own nor nearly ever integrated as indicator into the eutrophication category.

Following the precautionary principle for a calculation of the aquatic eutrophication potential (EP) or *nutrification potential* (*NP*) it is presumed that every *unintended* nutrient supply to the environment, contrary to targeted fertilisation in the technosphere, can imply over-fertilisation. Neither the local situation nor pre-existing pollution loads of the site are considered. As a basic idea behind the scenes it is presumed that loads caused by humans, no matter whether by nutrient or pollutant overload *can* have a damaging or at least an unintended impact on the environment. This is in accordance with experience: eutrophic lakes, unrestricted growth of algae in the estuaries and over-fertilisation of forests in case of terrestrial eutrophication (forest soils are generally nutrient-poor). Oceans are also very nutrient-poor. Only in upwell regions where nutrient rich water from the deeper layers drifts to the surface, under natural circumstances and based on an increased growth of algae (primary production) high amounts of biomass are observed at all trophic levels.

Also in the marine ecosystems, an increased entry of nutrients can imply undesirable, and above all (in contrary to many limnic ecosystems), uncontrollable changes. Many eutrophic lakes of anthropogenic origin could be remediated by strict regulations with respect to discharge of waste water and the implementation of waste water sewage plants (i.e. prevention of the input).

---

255) Udo de Haes (1996).

### 4.5.2.6.2 Indicator and Characterisation Factor

The impact indicator for aquatic eutrophication is the undesired formation of biomass in limnic (lakes, ponds and rivers) and marine ecosystems (estuaries, brackish water seas and open ocean) by an entry of fertilising substances into the environment.

A specification of a relative fertilisation effect of P and N is at the centre of the definition of an equivalent or characterisation model. In the table of CML[256] (Table 4.15) equivalence factors similar to those of the AP are made according to simple stoichiometric calculations. Derivations of factors are based on an average composition of algae biomass according to the relation $C : N : P = 106 : 16 : 1$, named Redfield-relation[257] due to its revelation by Alfred Redfield (1890–1983), valid to a surprising accuracy in Deep Oceans:

$$C_{106}H_{263}O_{110}N_{16}P$$

A question arises with respect to the contribution of a nutritive substance (X) to the formation of algae biomass by photo synthesis if it contains the limiting element and if all other elements in biologically available form are presumed to be abundantly present.

X + substrate, trace elements, $h\nu \rightarrow$

$$\eta(\text{algae biomass} = C_{106}H_{263}O_{110}N_{16}P) \tag{4.22}$$

With this definition of EP values of all P- and N-containing compounds can unambiguously be stoichiometrically calculated. This directness is the charm of the method which disregards all local restrictions and derives a potential impact – similar to an AP – from the chemical formula alone. Of course it has to be reassessed with scientific prudence whether the regarded compound that can actually provide the nutritive element, is biologically available! There is no taking into account of regional composition of the water bodies or of observed discrepancies in the Redfield-relation.

---

**Sample Calculation**

If $X = P$ (a molecule or an ion with a bioavailable P-atom), $\eta = 1$, thus 1 mol of P ($M = 31 \text{ g mol}^{-1}$) causes the formation of 1 mol alga biomass of an average composition $C_{106}H_{263}O_{110}N_{16}P$ ($M = 3550 \text{ g mol}^{-1}$).
A molar ratio $n(P)/n(\text{algae biomass}) = 1/1$ results to

$$m(\text{algae biomass}) = \frac{m(P) \times M(\text{algae biomass})}{M(P)}$$

For 1 kg P thus 114.5 kg algae biomass is calculated.

---

256) Heijungs et al. (1992), Klöpffer and Renner (1995), Hauschild and Wenzel (1998) and Guinée et al. (2002).
257) Redfield (1934), Redfield, Ketchum and Richards (1993) and Samuelsson (1993); http://de.wikipedia.org/wiki/Redfield-relation

If X = N (a molecule or an ion with a bioavailable N-atom), $\eta = 1/16$, thus 1 mol of N ($M = 14\,\mathrm{g\,mol^{-1}}$) enables the formation of 1/16 mol of alga biomass of the above composition.

A molar ratio $n(N)/n(\text{algae biomass}) = 16/1$ results in

$$m(\text{algae biomass}) = \frac{m(N) \times M(\text{algae biomass})}{M(N) \times 16}$$

For 1 kg N thus 15.8 kg algae biomass is calculated.

For the sake of descriptiveness the EP refers to 1 kg $PO_4^{3-}$ (Table 4.14). This definition on phosphate equivalents is arbitrary, like the choice of $SO_2$ as reference substance for an AP or the choice of $CO_2$ for the GWP of the greenhouse effect.

**Table 4.14** Eutrophication potential (EP) of important emissions (Heijungs et al. (1992), Lindfors et al. (1994, 1995) and Klopffer and Renner (1995).

| Emission (entry path) | Formula | Eutrophication potential (EP) (kg $PO_4^{3-}$-equivalent) |
|---|---|---|
| Nitrogen monoxide (air) | NO | 0.20 |
| Nitrogen dioxide (air) | $NO_2$ | 0.13 |
| Nitrogen oxides (air) | $NO_x$ | 0.13 |
| Nitrate (water) | $NO_3^-$ | 0.1 |
| Ammonium (water) | $NH_4^+$ | 0.33 |
| Nitrogen | N | 0.42 |
| Phosphate | $PO_4^{3-}$ | 1 |
| Phosphor (water) | P | 3.06 |
| Chemical oxygen demand (COD) | As $O_2$ | 0.022 |

### Sample Calculation

Nitrogen and phosphor enter the environment as compounds. There are however data in the inventory where emissions are calculated as N or P as common, for example, in sewage engineering.

1. Conversion of '1 kg P' in (kg of phosphate equivalents):
   A molar ratio $n(P)/n(PO_4^{3-}) = 1/1$ results to

   $$m(PO_4^{3-}) = \frac{m(P) \times M(PO_4^{3-})}{M(P)}$$

   ($M(P) = 31\,\mathrm{g\,mol^{-1}}$ and $M(PO_4^{3-}) = 95\,\mathrm{g\,mol^{-1}}$). For 1 kg P thus 3.06 kg phosphate is calculated.

2. Conversion of '1 kg N' in (kg of phosphate equivalents):
   A molar ratio in the algae biomass of $n(P)/n(N) = 1/16$ and a molar ratio of $n(P)/(PO_4^{3-}) = 1/1$ results to

   $$m(PO_4^{3-}) = \frac{m(N) \times M(PO_4^{3-})}{M(N) \times 16}$$

   ($M(N) = 14 \text{ g mol}^{-1}$ and $M(PO_4^{3-}) = 95 \text{ g mol}^{-1}$)
   For 1 kg N thus 0.42 kg phosphate equivalents are calculated.

3. Conversion of 1 kg NO in (kg of phosphate equivalents):
   A molar ratio in the alga biomass of $n(P)/n(N) = 1/16$, a molar ratio of $n(N)/n(NO) = 1/1$ and a molar ratio of $n(P)/n(PO_4^{3-}) = 1/1$ results to

   $$m(PO_4^{3-}) = \frac{m(NO) \times M(PO_4^{3-})}{M(NO) \times 16}$$

   ($M(NO) = 30 \text{ g mol}^{-1}$ and $M(PO_4^{3-}) = 95 \text{ g mol}^{-1}$) For 1 kg NO thus 0.2 kg phosphate equivalents are calculated.

Similarly aquatic EPs (as phosphate equivalents) of arbitrary nutrients containing P and N can be calculated. Specification of the EP as phosphate equivalent have been largely accepted (for an over fertilisation of soil, see below). For the most important emissions the EP values are shown in Table 4.14.

Besides compounds containing P and N only the over-fertilisation effect according to the COD of organic compounds is integrated into the characterisation. The COD is a concentration related sum parameter often applied in waste water analysis and therefore often available in the inventory (LCI). A prerequisite for the use of these data for LCA is that the determination of COD loads is possible. All substances contained in water that can be oxidised with potassium dichromate under defined conditions are considered. The result is given as the mass of oxygen (mg l$^{-1}$) necessary for the oxidisation of the substances in the water if oxygen were the oxidising agent. Both biologically degradable and biologically non-degradable organic materials as well as some inorganic materials are included. As the stoichiometry of the oxidation reaction is well-known, the COD can be calculated according to the known formula of the substance to be oxidised. Because biologically non-degradable organic substances as well as inorganic substances included within the COD do not contribute to eutrophication, the BOD would be better suited for this impact category. Here the mass of oxygen (in mg l$^{-1}$), which is utilised by bacteria in the course of the use of organic compounds as nutrition within a defined time period is indicated. The BOD is therefore a measure for biologically degradable substances in the water and simulates processes that induce the decrease of oxygen concentration in the water. However, many more data for the COD are provided from waste water analysis. The BOD has to be determined separately by experiment. If exclusively aerobic degradable organic substances are present in the water the COD is equal to the BOD, otherwise the COD is always larger. Therefore a consideration of the COD in the impact category, eutrophication, under precautionary criteria always refers to the maximum possible value.

The following consideration correlates the COD with the EP:

Because the damaging impact of biodegradable carbon compounds is the result of an 'oxygen depletion'[258] the oxygen demand is taken for a definition of the EP-value (kg of phosphate equivalents). To completely oxidise a molecule of the model biomass it is presumed that 138 molecules of oxygen are additionally necessary besides the 110 oxygen atoms already present in the molecule (Equation 4.24). The chemical species considered for N and P following the oxidation are $NO_3^-$ and $HPO_4^{2-}$ which prevail in aerobic water with usual pH levels:

$$C_{106}H_{263}O_{110}N_{16}P + 138O_2 \rightarrow 106CO_2 + 122H_2O$$
$$+ 16NO_3^- + 1HPO_4^{2-} + 18H^+ \text{[259]}$$

---

**Sample Calculation**

A molar ratio of $n(P)/n(O_2) = 1/138$ results in

$$m(PO_4^{3-}) = \frac{m(O_2) \times M(PO_4^{3-})}{M(O_2) \times 138}$$

($M(O_2) = 32$ g mol$^{-1}$ and $M(PO_4^{3-}) = 95$ g mol$^{-1}$)
For 1 kg COD calculated as $O_2$ thus 0.022 kg phosphate equivalent results.

---

Thus, somewhat artificially, the correlation of the impact of eutrophication by oxidisable organic compounds with oxygen depletion is quantitatively achieved. The COD of persistent ('refractory') compounds should actually not be included in the calculations, as these materials do not contribute to oxygen consumption. The COD is therefore a *worst case* approximation of the BOD.

For the calculation of the terrestrial EP mainly emissions into the air are considered, for the aquatic EP direct and indirect entries into the waters.[260] The majority of continental emissions into the air reach the soil by dry and wet deposition. The emissions into water and the settling on soil influence ecosystems which can be over-fertilised. Soil and water are interconnected: Washing-off surplus fertilisers from agricultural surfaces and penetration into the groundwater (nitrate pollution). Over-fertilisation of agriculturally used grounds (technosphere, not ecosphere) is not considered here because by definition the *inadvertent* eutrophication including the unavoidable sequel impacts is discussed exclusively. This is because of the restriction of LCA towards environmental pollution. Clearly, in the case of

---

258) Classical eutrophication due to P and N induces oxygen depletion as a frequent secondary effect.
259) Kummert and Stumm (1989).
260) Mauch and Schäfer (1996).

agriculture, the connection between the technosphere and environment is close and the necessary distinction is often delicate.

**4.5.2.6.3 Characterisation/Quantification** By simply equating the EP and the phosphate equivalents according to Table 4.14 the following is valid:

$$\text{EP} = \sum_i (m_i \times \text{EP}_i) \quad (\text{kg PO}_4^{3-}\text{-equivalents}) \tag{4.23}$$

where $m_i$ = load of the substance $i$ contributing to eutrophication per fU.

**4.5.2.6.4 Terrestrial Eutrophication** The characterisation according to Equation 4.23 does not differentiate between aquatic and terrestrial eutrophication and therefore represents the simplest method. The advantage of clarity of this computation is a result of strong simplifications on the impact side. Therefore, attempts have been made to sacrifice part of the simplicity for an approach to reality, for example, by the 'Nordic Guidelines'.[261] A division of the impact category into aquatic and terrestrial seems to be the most promising.[262] As most emissions into the inland air are deposited on soils, emissions into air of compounds containing N mostly as $NO_x$ and $NH_3$ are the most important input to the soil and can therefore be assigned to terrestrial eutrophication. For water bodies relevant emissions into water (phosphate, ammonium, COD/BOD, etc.) should be considered.

If this division is made, the EP of the soil nutrients are often expressed as nitrate equivalents (EP nitrate = 1). This formal conversion merely serves as a better distinctness and indicates the fundamental contribution of nitrate as the most frequent limiting element in soil. Phosphate however can also be used as reference.

For a calculation of terrestrial eutrophication only emissions to air with an over-fertilising impact are considered. For a separate assessment of this subcategory, the eutrophication of water is quantified only by emissions into water including COD. For the classification step in impact assessment this has to be kept in mind! According to Table 4.14 mainly the nitrogen oxides NO, $NO_2$ and their sum $NO_x$ (calculated as $NO_2$) are considered for terrestrial eutrophication calculation. Ammonia ($NH_3$) is also part of terrestrial eutrophicating emissions into the air, but in the air is quickly transformed into $NO_x$ or in water into the ammonium ion ($NH_4^+$). The amount of nitrogen that reaches surface-near air by way of $NH_3$ should not, however, be neglected.

Characterisation of terrestrial eutrophication is also calculated according to Equation 4.23 but separately assigned.

**4.5.2.6.5 Regionalisation** A discussion of regionalisation is closely linked to the discussion of the impact categories 'formation of photo oxidants (summer smog)' and 'acidification'. The basic set of problems remains the same. Transport by air can in principle be calculated with identical models (RAINS, EMEP, etc.). Transport by water abides to different rules. The most important entry of fertilisers into

---

[261] Lindfors *et al.* (1994, 1995), Finnveden and Potting (2001) and Guinée *et al.* (2002).
[262] Lindfors *et al.* (1994, 1995) and Udo de Haes (1996).

surface waters by means of run-off depends definitely on the local site formation. As these difficulties do not occur for terrestrial eutrophication by air, which additionally is partly caused by the same pollutants as the acidification, country specific characterisation factors were at first only deduced for this subcategory.[263]

The model is similar to the one introduced in Section 4.5.2.5. From the inventory data only $NO_x$ and $NH_3$ are considered, however the site of emission must at least be approximately known. This should be more easily acquirable with large stationary emission sources than with highly distributed ones. The characterisation factors are calculated for 35 European countries and 5 oceanic regions for the years 2002 and are estimated for 2010 and listed in a table. They apply to the indicator 'accumulated exceedance' of limit values as already discussed.

### 4.5.3
### Toxicity-Related Impact Categories

#### 4.5.3.1 Introduction

In this section two impact categories are discussed where traditionally little harmony between LCA research and application, and the groups of societies involved has been observed: *human toxicity* and *ecotoxicity*.[264] The reason might be a problem of the application of scientific knowledge and feasibility, with little resolution of time and space offered by the inventory data. The most radical proposal to solve the problem by neglect can only exceptionally be adhered to, because

- first, most people consider protection of health as an important, in the US most important,[265] aspect of environmental protection and therefore also of environmental assessment tools
- second, LCA, *Environmental LCA*[266] is an ecological evaluation of product systems or more generally of *human activities*[267] which includes the protection of human health and of ecosystems as fundamental to human existence for every definition chosen. Whether, in addition, nature has its own rights, implying the renunciation of the anthropocentric point of view, does not have to be discussed here.[268]

Therefore, the impact categories human toxicity and ecotoxicity cannot be neglected even if in some LCA studies – depending on the Goal and Scope definition – the interrelated issues of the protection of resources, of energy savings and of global impacts (above all 'Climate change') are in the lead. However, a scientifically approvable elaboration on toxicity-related categories implies new challenges for the inventories, which have not been met yet. This is especially true for organic pollutants that may be emitted over the entire life cycle of a product.

---

263) Huijbregts and Seppälä, 2001, Seppälä, Knuutila and Silvo (2004) and Seppälä *et al.* (2006).
264) Klöpffer (1996a); more recently, a major breakthrough has been achieved by a UNEP/SETAC working group, see Hauschild *et al.* (2007) and Rosenbaum *et al.* (2008, 2011).
265) Bare *et al.* (2002).
266) Klöpffer (2008).
267) SETAC (1993).
268) Beltrani (1997) (this author represents a moderate anthropocentric point of view: humans with regard to their own species should not destroy the fundamentals of life!).

Considerable progress in the area, mostly in the context of the EU-projects OMNITOX[269] and UNEP/SETAC Life Cycle Initiative,[270] are discussed in the two subsequent Sections 4.5.3.2 and 4.5.3.3. First we take a look at the 'simpler' indicators based on the principle *less-is-better*, which can also be used with non-regional inventory data and be applied without knowledge of special impact mechanisms. The method IMPACT2002+ is of special importance for a characterisation of the impact categories human toxicity and ecotoxicity, and was developed by an international working group chaired by Olivier Jolliet.[271] It interconnects 14 *mid-point* categories with four *damage categories*:

- human health
- quality of ecosystems
- climate change
- resources.

These '*Damage Categories*' are also called *Safeguard Subjects*, *Areas of Protection* and, most ambiguously, *Endpoints*.[272] The damage category called *climate change* in IMPACT2002+ should not be confused with the impact category of the same name even though they are closely connected.

### 4.5.3.2 Human Toxicity

**4.5.3.2.1 Problem Definition** According to Table 4.4 this impact category was designated by DIN/NAGUS[273] as 'toxic hazard to humans', by SETAC Europe[274] as *human toxicological impacts;* a similar definition has been given in the Nordic Guidelines.[275] In a second working group of SETAC Europe for impact assessment in LCA (WIA-2)[276] the expression *human toxicity* is applied.

The main difficulty of this category, more than those already discussed, is owing to the fact that a strictly scientific composite indicator 'upstream' in the impact hierarchy that is closer to the emissions (mid-point) does not exist. Toxic molecules do not have a common attribute which corresponds to the acidic function in the case of AP or to the P or N content in the case of the EP respectively. Neither measurable physical or chemical parameters exist according to which an impact potential, like GWP or ODP could be calculated using theoretical models. There are too many different impact mechanisms leading to diseases or groups of diseases

---

269) Operational Models and Information tools for Industrial applications of eco/TOXicological impact assessment (OMNITOX). Special edition Int. J. LCA Vol. 9, No. 5 (2004); Larsen *et al.* (2004) and Molander *et al.* (2004).
270) Jolliet *et al.* (2004); *http://lcinitiative.unep.fr*; *http://www.uneptie.org/pc/sustain/lcinitiative/home/htm*; Hauschild *et al.* (2008) and Rosenbaum *et al.* (2008, 2011).
271) Jolliet *et al.* (2003, 2004).
272) Assignable damages are called endpoints in (eco) toxicology. For these test procedures or research methods are developed. Known endpoints for example are probabilities of mortality such as LD50 or LC50.
273) DIN/NAGUS (1996).
274) Udo de Haes (1996).
275) Lindfors *et al.* (1994, 1995).
276) Udo de Haes *et al.* (2002).

and too little is known on the causal and quantitative relations with the chemical or other noxes that are part of the inventory.

This multitude of impact mechanisms can be ordered into groups but show a large diversity concerning dose-response relationships. For many substances, toxicological impact thresholds, which are the basis of limit or indicative value calculation, can be identified. Of these only those effect thresholds investigated and measured by experiments can be considered. The level of scientific knowledge of possible impacts is a prerequisite for an experimental examination. Furthermore, for the specification of effect thresholds the metrological conditions for the quantification of a defined impact is essential. Ideally the basis for a limit or indicative value is the highest dose where no (adverse) effect has been observed (NO(A)EL[277]). However, below this dose impacts which have not (yet) been targeted in experiments or can neither be measured nor quantified are possible. The regulatory toxicology is concerned with possibilities and limitations to derive safe limit or indicative values.[278] Important points of discussion are combination effects, chronic impacts in the low dose range, a definition of what can be considered as adverse effect as well as the handling of safety factors. This discussion targeted risk assessment considering defined exposure conditions are not regarded here.

The derivation of indicative values for carcinogenic or mutagenic substances without effect threshold may refer to an acceptable upper-bound excess lifetime cancer risk (e.g. US EPA: unit risk method).

Many approaches to the treatment of the impact category human toxicity are based on limit or indicative values which have been derived within diverse explanation contexts.

### 4.5.3.2.2 Simple Weighting Using Occupational Exposure Limit or Indicative Values

For a weighting of toxic emissions into the air documented in the inventory a highly detailed chart with limit or indicative values deduced scientifically by a uniform method would be useful. The absolute amount of these values is not decisive for a relative ordering system in the impact assessment as the substances are arranged by their risk potential alone (see Sections 4.1 and 4.4.3.2).

In Germany, for example, for hazardous working materials and chemical substances toxicologically substantiated limit values of a maximum working site concentration (MAK[279]) are derived on behalf of the German research foundation.[280],[281] These values are the German occupational exposure limits (OELs) and serve as example here. The limits are published by the German Federal Institute for Occupational Safety and Health in Technical Rules for Hazardous Substances (TRGS 900[282]) and are updated every year. The DFG list is rather long and both

---

277) No observed (adverse) effect level: the highest level of a dose where no (adverse) effect has been observed.
278) Reichl and Schwenk (2004).
279) Maximale Arbeitsplatzkonzentration.
280) DFG (2007).
281) For an overview on occupational exposure limits in other countries see: http://www.ilo.org (International Labour Organization)
282) *German*: Technische Regeln für Gefahrstoffe 900: Occupational exposure limits.

organic and inorganic compounds and chemically badly characterised airborne substances like dust are listed.

As a relative ordering system of substances concerning human toxicological impact, OEL-values are basically suited as *mid-point* characterisation factors: They are available for many substances without the necessity of a breakdown into individual damage impacts (illness patterns, endpoints), which would imply more subcategories. They are deduced by a uniform method with consideration of scientific literature.

The characterisation according to this method results in a human toxicity potential (HTP)[283] of type 'c.V. human toxicity' (Equation 4.24), which is acceptable here because only the environmental problem field 'human toxicity' is discussed. The ecotoxicological impacts and other impact categories contrary to being jointly addressed in the BUWAL method[284] are handled separately.

$$\text{HTP} = \sum_i \left( \frac{m_i}{\text{MAK}_i} \right) \quad (m^3 \, fU^{-1}) \tag{4.24}$$

HTP = human toxicological potential. $m_i$ = mass of the substance $i$ released into the air, for which a $\text{MAK}_i$ value was deduced per fU:

The unit $(m^3 \, fU^{-1})$ results for a load (mg per fU) and a MAK value $(mg \, m^{-3})$. The HTP can also be normalised to a reference substance (e.g. 1,4-dichlorobenzene (DCB)) Guinée et al., which is arbitrarily assigned to a HTP of one; this could, however, lead to the impression of a uniform impact indicator, which in this category would be even less adequate than for those already discussed.

The HTP defined in Equation 4.24 only maps the risk potential of emissions weighted according to the MAK-values assessed in the inventory. The quantification according to Equation 4.24 therefore provides an aggregation by weighting according to MAK-values which have been uniformly deduced by a DFG expert team. These values have the exclusive function of a **relative toxicity scale** and are not to be applied for a weighting of an actual exposure to hazard at the working site, the more since the working site at the centre of the technosphere is out of the scope of an LCA.[285]

The experts of MAK-commission in DFG also proposed the ranking of carcinogenic or suspected carcinogenic chemicals into groups of varying carcinogenic impact probability for humans that are also applicable for a relative weighting.[286]

An objection against MAK-values or OELs of other countries as starting point for weighting is based on the fact that the numbers of limit and indicative values vary by nation within a certain range. Here a scope for a discretion margin can be perceived on the part of those boards deducing these values. In view of the geographical system boundary, German MAK-values could be used for studies in

---

283) Heijungs et al. (1992).
284) BUWAL (1990).
285) This is controversially discussed, particularly by colleagues from the Scandinavian countries Poulsen et al. (2004); in our opinion hazards at the working site is part of a 'product related social assessment (Societal LCA, SLCA)' as part of a sustainability valuation of products. See Klöpffer and Renner (2007) and Klöpffer (2008, Section 6.3.3).
236) See also unit risk concept of US-EPA.

Germany. If the EU is the geographical boundary, averages from EU states could be applied, for international studies average from the OECD countries. (OELs of little industrialised states may be inspired by those of the industrialised countries). A worldwide labour protection organisation (ILO) based in Geneva lists those values from all over the world. Thus a data base of international scale were given which at least for substances with effect threshold values would allow a relative weighting. A summary of international OELs (D, EU; USA; GUS) has been published by Sorbe.[287] In this work which also lists other toxicological limit values, an overall of 18 000 substances are listed.

All lists should be critically reviewed as to whether limit or indicative values for the considered substances have been deduced according to uniform methods (scientific data base, considered impacts, duration of exposure, target groups, security factors, acceptable risk and other boundary conditions). If the explanatory framework of listed substances varies considerably, the lists do not meet the criteria of a reliable ranking of toxicity of substances relative to each other. As such, the limit value for a substance A within an explanatory framework 'working site protection' deduced for an exposure of healthy employees for five days a week 8 hours a day cannot be compared with that of a substance B with respect to the in-house air in apartments deduced for a continuing exposure including sensitive population groups, in a single list for a relative weighting of toxicity of A and B. An adaption of multiple lists of those values into one without a critical reflection of the explanatory framework is therefore prohibited.

Table 4.15 exemplifies inconsistencies concerning the mix of limit and indicative values from various explanatory frameworks for three well-known substances: MAK-values are juxtaposed indicative values for substances of the indoor air pollution. With regard to their use for weighting in the impact assessment the absolute values are not the subject in this discussion. This example shows that values can vary according to varying explanatory frameworks and the relative order of substances can change.

A method comparison in the US[288] introduces a characterisation equivalent to the 'MAK method' within the group 'comparison of toxicity' but proposes the use of Acceptable Daily Intake (ADI) values for characterisation:

$$\text{TBS} = \sum_i \left( \frac{Q_i}{Q_{\text{ref}}} \right) m_i \qquad (4.25)$$

TBS, *Toxicity-based scoring*
$Q_i$ 1/$\text{ADI}_i$ (kg body weight × d/mg); reciprocal ADI of the substance $i$

---

287) Sorbe (1998).
288) Hertwich, Pease and McKone (1998).

**Table 4.15** Indicative- and limit values for varying explanatory frameworks.

| Substance | Indicative values for indoor air[a] | | | MAK values[b] |
|---|---|---|---|---|
| | BGA[c] RW[e] | IRK/AOLG[d] RW I[f] | RW II[g] | DFG (2007) |
| Toluene | — | 0.3 mg m$^{-3}$ (1996) | 3 mg m$^{-3}$ (1996) | 190 mg m$^{-3}$ |
| Formaldehyde | 125 µg m$^{-3}$ (1977) | — | — | 0.37 mg m$^{-3}$ Carcinogen (category 4[h]) |
| Pentachlorophenol | — | 0.1 µg m$^{-3}$ (1997) | 1 µg m$^{-3}$ (1997) | No value established Carcinogen (category 2[i]) |

[a] Reichl and Schwenk (2004).
[b] DFG (2007).
[c] German Federal Office of Health (BGA – Bundesgesundheitsamt).
[d] Indoor air hygiene commission of the German Federal Environmental Agency (IRK – Innenraumlufthygienekommision) and working groups of Highest Regional Authorities (AOLG).
[e] Indicative value (RW – Richtwert).
[f] Indicative value I: Concentration of a substance in the indoor air whereby according to present knowledge even for lifelong exposition no health demanding impacts are to be expected.
[g] Indicative value II: concentration of a substance which if reached or exceeded requires an immediate call for action as it is expected especially for sensitive people in case of a prolonged exposition to have a health threatening impact.
[h] Carcinogenic category 4: substances with carcinogenic impact where genotoxic impacts are of no or of minor importance. For an adherence to MAK and BAT[289] values no considerable contribution to a risk for cancer can be expected.
[i] Carcinogenic category 2: substances regarded as carcinogenic for humans where due to long-term animal experiments and epidemiological investigations a considerable contribution to a risk for cancer can be expected.

$Q_{ref}$ 1/ADI$_{ref}$ (kg body weight × d/mg); reciprocal ADI of the reference substance
$m_i$ released mass of the substance $i$ per fU.

The reference substance can be chosen arbitrarily.

If in the impact category, human toxicity, a weighting is accomplished by selected limit or the respective indicative values, toxic substances that are not considered in appropriate lists have to be determined separately and may be verbally interpreted and evaluated.

**4.5.3.2.3 Characterisation with Supplementary Exposure Estimation** The neglect of exposure by the simple weighting method according to Equation 4.24 is a limitation. Thus acute highly toxic substances are over-estimated by their low OELs like MAK-values because for an exposure over environmental media, the acute toxicity – contrary to the working site – is of minor importance for the following two reasons:

---

289) *German:*'Biologosche Arbeitsplatz Toleranzwerte', biologically permissive values at the working site.

- Usually large dilutions occur for an exposure via environmental media, thus the effect thresholds are usually not reached.
- Acutely toxic substances (not always, see polychlorinated dibenzo-*p*-dioxin and polychlorinated dibenzofuran, PCDD/F) are often reactive and therefore, by trend rapidly degradable (abiotically or biotically)[290] contrary to non-reactive (persistent) substances; strong toxics as, for example, phosphine ($PH_3$) are therefore not relevant in the environment except in the case of accidents. Even then long-term damaging consequences seldom occur because of dilution and degradation. Short-term acute impacts are part of the impact category casualties, which is hardly ever used.

Contrary to acute often short-lived poisons (in the narrow sense of the term) persistent substances can be ranked as potentially environmentally toxic even for minor toxicity and minor concentrations. This is especially true for ecotoxicity[291], for an exposure by environmental media and by the nutrition chain, but also for human toxicity.

All quantification procedures exceeding a simple weighting or the formation of groups by means of limit values must introduce the investigation of uptake pathways and exposure analysis[292], a usual procedure in toxicology and in chemical risk assessment. A general treatment of human toxicity according to SETAC[293] formally corresponds to ecotoxicity (see Section 4.5.3.3) but the protection goals differ:

- *Human toxicity*: personal health of every individual (also the unborn).
- *Ecotoxicity*: the functionality of the system characteristics of entire ecosystems as well as the biodiversity, not however, with few exceptions[294], single individuals.

In anticipation of the section on ecotoxicity and as delimitation from human health and its significance (human toxicity) it is already noted here that the expression '*ecosystem health*' is not undisputed.[295] For ecosystems that are not organisms, as they, for example cannot multiply and have no strict boundaries in space, no illnesses can be defined, also the absence of illness, that is, health, practically cannot be defined. The pleasant metaphor of an animated earth as a super organism (Gaia[296]) used by J. Lovlock's also belongs to that issue. A descriptive definition of the unique characteristics of organisms has been provided by Monod.[297]

It is true for the above distinction that for human toxicity the individual and with ecotoxicity the ecosystem is the primary *safeguard subject*. The species (animals and plants) is at the centre of these two extremes. Species protection is a declared safeguard subject in an ecosystem that is not of absolute importance: another species can (often) adopt the same function (occupy the same ecological niche)

---

290) Klöpffer (1996b) and Klöpffer and Wagner (2007a,b).
291) Klöpffer (1989, 1994a, 2001) and Scheringer (1999).
292) Mackay (1991), Trapp and Matthies (1996, 1998) and TGD (1996, 2003).
293) Udo de Haes (1996) and Udo de Haes *et al.* (2002).
294) Exceptions are species threatened with extinction with only single specimen left and especially beautiful old trees ('natural monuments').
295) Suter (1993).
296) Lovelock (1982, 1990).
297) Monod (1970).

without a collapse of the ecosystem. Nature has coped with the extinction of many species. This cannot be, however, a license for the present-day practice of extinction of species by humans, because of totally different time scales! Protection of species is therefore a very important short time goal; in the long term, however, the ecosystem aspects prevail. Referring to Lovelock again one could say 'Gaia's ability for learning has to be sustained', create new species and thus biodiversity is sustained as a long-term result. This problem field is closely related to time scales ('time ecology').[298]

Without restriction to any method proposed in literature the working group 'Impact Assessment of Human an Ecotoxicity in Life Cycle Assessment' of SETAC Europe[299] has proposed the following general formula for treatment of human toxicity

$$S_i^{nm} = E_i^m F_i^{nm} M_i^n \qquad (4.26)$$

$E$ effect factor (EF)
$F$ fate (distribution and degradation)
$M$ mass (load per fU).

The score ($S$) of a substance $i$ for the environmental compartment m is at the left of the Equation 4.26. The original emission was released into compartment n (n = m is the only case considered for a simple weighting). It is attempted to integrate exposure and effect into one equation which is the basic principle for a risk assessment of chemicals. The first expression on the right side is the EF weighting the considered adverse effect in compartment m. This factor can be very similar to usual weighting factors, for example:

$$E_i^m = \frac{1}{\text{NEC}_i^m} \qquad (4.27)$$

NEC = *no effect concentration* or NOEC = *no observed effect concentration* of substance $i$ in compartment m (e.g. a volatile chemical in compartment air).

With such a definition of the weighting factor for the effect Equation 4.26 – except for notation – differs from Equation 4.27 only by the factor F. Still not included is the summation of the results for all toxic substances quantified in the inventory.

The second expression on the right in Equation 4.26 is the *fate and exposure factor*[300] *of substance i*, which has been emitted into compartment n and transferred into compartment m (e.g. by evaporation, deposition, etc.) considering degradation processes and accumulation. It can be observed that this factor can only be determined by modelling or by estimations with knowledge of physical and chemical properties of the molecule. Such calculations are part of the risk assessments of chemicals, but however, of little reliability because of multiple simplifying assumptions and low quality of input data.[301]

---

298) Held and Geißler (1993, 1995) and Held and Klöpffer (2000).
299) Jolliet et al. (1996): Impact Assessment of Human and Eco-toxicity in Life Cycle Assessment, in: Udo de Haes (1996, S. 49–61).
300) Jolliet et al. (1996).
301) Klöpffer (1996a, 2002, 2004) and Klöpffer and Schmidt (2003).

The third expression provides the entry into the primary compartment n (air, water, soil). This entry can be taken from the inventory (LCI).

The result of the characterisation (S) is the sum of all substances $i$ considering all compartments m and n:

$$S = \Sigma_i \Sigma_m \Sigma_n S_i^{nm} \tag{4.28}$$

A simple HTP or TBS Equations 4.24 and 4.25 with a weighting by OEL or ADI values is a special case of Equation 4.28 with the assumption $n = m = $ air, $NEC_i^m = OEL_i$ or $ADI_i$ with a neglect of the fate factor which signifies the degradation and transfer between the environmental media.

An attempt to assess various toxicities individually or at least by groups is further complicated. A category human toxicity and a characterising model HTP or S necessarily produces a range of subcategories which correspond to the selected toxicity endpoints.

*Attempts for a Specification of Exposure Factors* Difficulties with respect to exposure factors for human toxicity are somehow similar to those of the impact category ecotoxicity (see Section 4.5.3.3). A first determination of exposure factors has been made by Guinée and Heijungs.[302] The HTP definition by the authors considers the intake of pollutants by air (respiratory) and by nutrition (orally). The dermal absorption was so far neglected. The exposure via an environmental medium is estimated by a Mackay-III model that describes the flow equilibrium between the media air, water, soil and sediment within a global *Unit-world*[303]-*Box-Model*. This model considers degradation processes. These are however only adequately specified for a number of substances which is especially valid for biological degradation. Quantification of abiotic degradation in the air however is near to appropriate.[304]

For an application of the Mackay model – a similar application is valid for other distribution models – the so-called flow/pulse problem occurs: in this model the mass input is considered as a continuous mass flow, for example x kg/d into compartment air. The inventory however supplies a load per fU (kg fU$^{-1}$) with an unspecified dispersion in space and time. This load can be approximated as pulse of uncertain characteristic, the target medium being known from the inventory. As a work-around, Guinée and Heijung proposed to refer exposure and effect to an arbitrarily selectable reference substance which would eliminate conversion (load/flow) and result in a dimensionless toxicity potential HTP. The HTP of the reference substance is set equal to one. Although this approach seems logical it is not convincing due to its highly artificial character which does not reflect the diverse toxic effects. Therefore, those rather complicated calculations will not be discussed in detail here. Heijungs and co-workers were later able to demonstrate that pulses in Mackay-like models, like, for example in the Dutch model USES,

---

302) Guinée and Heijungs (1993) and Guinée et al. (1996a).
303) Mackay (1991) and Klöpffer (1996a, 2012b). Four media or compartments (air, water, soiland biota) are designated as boxes or compartments with a possible reference to subcategories in more complex models.
304) Klöpffer and Wagner (2007a).

can be regarded as flows.[305] This model was the starting point for an expansion into an European model 'EUSES' (European Union System for the Evaluation of Substances) which can be purchased from the European Chemicals Bureau at the Joint Research Centre, Ispra. In Guinée et al. (1996b), eight solutions of the flow/pulse problem are presented which come to the same conclusion. In the same paper, equivalence factors for human toxicity for 94 chemicals are listed and a strategy for deriving new not yet calculated substances is provided. Equivalence factors for the HTP refer to the substance pDCB in the compartment air.

**4.5.3.2.4 Harmonised LCIA Toxicity Model** Analyses on the depth of detail which could be obtained in LCIA (contrary to, e.g. chemical risk assessment) marked the beginning of a more recent development of toxicity evaluation in LCA.[306] By Hertwich, Pease and McKone (1998) simple methods (see Section 4.5.3.2.2) were integrated into the analyses of toxicity potentials or -equivalents called *toxicity-based-scoring*. For LCA a *Human Toxicity Potential* (*HTP*) according to logic of GWP and similar LCIA-characterisation factors is recommended. It is however noted within the model calculations that no quality criteria were developed for physical and chemical parameters, which are generally measurable with larger accuracy than toxicities. Even in Europe these requirements to data quality are only statutorily regulated for 'new substances'.[307] The situation is slowly improving due to the European Community Regulation on chemicals REACH (Registration, Evaluation, Authorisation and Restriction of Chemical substances). But errors can still occur becuse of missing or faulty substance data.

The simple toxicity evaluation according to Section 4.5.3.2.2 has not been further developed, but the methodology outlined in Section 4.5.3.2.3 was elaborated. A solution for multiple toxicity endpoints was shown by Hofstetter[308] the concept of a 'disability adjusted lost life years (DALYs)' elaborated on behalf of the WHO.[309] With this concept and based on the assumption of a linear and cumulative dose-effect relationship all partial effects are converted into an approximated 'lost years of life'. By conversion with fate and effect factors of inventory data as emitted pollutants per functional unit this results to an (DALY/fU). The equations necessary are principally the same as Equations 4.26 and 4.27. A quantitative and unambiguous

---

305) Heijungs (1995), Guinée et al. (1996a,b) and Wegener Sleeswijk and Heijungs (1996). USES: Uniform System for the Evaluation of Substances.
306) Hertwich, Pease and McKone (1998), Hertwich et al. (2001), Huijbregts et al. (2000b, 2005a), Guinée et al. (2002), Udo de Haes et al. (2002), Pant, Christensen and Pennington (2004) and Molander et al. (2004) both in OMNITOX Special Issue, International Journal Life Cycle Assessment Vol. 9, No. 5; Jolliet et al. (2003, 2004).
307) 'New substances' according to the European Union chemicals legislation are those which were circulated after the taking effect of the chemical law (1981). A remaining of approx. 100 000 substances are called 'Old substances'. This difference has now been removed within the new European regulation REACH
308) Hofstetter (1998), WHO (1996) and Müller-Wenk (2002a).
309) 'Years of life lost' by premature death due to illness per capita can be computed from statistics and be calculated as an impairment of quality of life by diseases. This requires factors that were determined by an international Panel. A factor scarcely below one means slight impairment, a factor close to zero, very high impairment.

reference to the fU is important for the impact assessment within LCA. The DALY concept has, however, not been generally adopted.

Should toxicological effect data be present for a sufficiently large number of substances, a HTP and the score $S$ respectively (Equation 4.27) can be computed if fate factors for the same sequence of substances can be computed. Here and by international comparison of methods the largest difficulties occur because multimedia models developed for various purposes, particularly for risk assessment of chemicals provide strongly deviating results.[310] This can be due to the structure of the models or the input data and simplifying assumptions. It is to the credit of the globally acting 'UNEP-SETAC Life Cycle Initiative'[311] that the most important model developers and users established a revised methodology 'USEtox'.[312] by model simplifications, comparative investigations and an inclusion of toxicological effect data . For the first time it seems possible that a uniform sequence of several thousands of recommended characterisation factors will be available particularly for LCIA. Starting point for the development of USEtox were the following seven LCIA and multimedia models:

- CalTOX (USA)[313]
- IMPACT 2002 (Switzerland)[314]
- USES LCA (the Netherlands)[315]
- BETR (Canada, the USA)[316]
- EDIP (Denmark)[317]
- WATSON (Germany)[318]
- EcoSense (Germany).[319]

Independent of LCIA issues nine multimedia models had been comparatively investigated by a team of experts of the OECD[320] upon their suitability for the computation of persistence and long-distance transportation potential. Starting from these and in the context of OMNITOX (2004) conducted model comparisons the most important elements of the model system were identified and implemented into a basic model by consensus.[321] It was thus possible to eliminate the largest deviations of the indicator results. Since multimedia models are better discussed in the context of ecotoxicity these aspects will be included in Section 4.5.3.3. Typical for *human toxicity* is the calculation of human exposure on the basis of a fate factor (in conjunction with ecotoxicity), and the intake fraction as well as the human EF. These factors are calculated for the most important intake pathways based on,

---

310) http://se.setac.org/files/setac-eu-0248-2007.pdf, http://www.lcacenter.org/InLCA2007/presentations/97pdf
311) Specially: 'Task Force on Toxic Impacts', http://lcinitiative.unep.fr
312) USEtox: The UNEP/SETAC toxicity model; Rosenbaum et al. (2008, 2011).
313) McKone, Bennett and Maddalena (2001) and Hertwich et al. (2001).
314) Pennington et al. (2005).
315) Huijbregts et al. (2005b).
316) McKone, Bennett and Maddalena (2001).
317) Hauschild and Wenzel (1998).
318) Bachmann (2006).
319) EU (1999, 2005).
320) Fenner et al. (2005).
321) Rosenbaum et al. (2008).

strictly speaking, false[322] assumptions of a linear dose-response relationship for all types of toxic impacts. This results in a toxicity factor of approximately $0.5/ED_{50}$ for every disease endpoint and intake pathway, where $ED_{50}$ refers to a daily dose causing an effect with a probability of 50% of human life time. Up to four EFs are calculated:

- Cancer by intake exposure
- No cancer illness by intake exposure
- Cancer by inhalation exposure
- No cancer illness by inhalation exposure.

The human toxicity factors are of dimension comparative toxic units for human health ($CTU_h$) or number of diseases/kilgram substance intake. A reference to mass ensures a linkage to released quantities of a substance and to the fU. Since calculations can be automated with the help of software and integrated databases, the focus shifts towards the inventory, which by then should also include emissions of a multitude of organic substances. In the most recent data set, over 3000 organic chemicals are included (2500 with freshwater EFs). For 'metals' (mostly ionic), dissociating organic compounds and amphoteric compounds (e.g. surfactants) only interim characterisation factors,[323] have been determined since their *fate* factors are usually difficult to calculate by multimedia models.[324]

The toxicological information needed for the determination of the $ED_{50}$-values was taken from extensive data collections, for example, US EPA. The same is valid for the physico-chemical data of the compounds necessary for the computation.

Detailed information necessary for a practical *application* of this methodology can be obtained in the USEtox special issue[325] including both human toxicity and ecotoxicity (next section).

### 4.5.3.3 Ecotoxicity

**4.5.3.3.1 Protected Objects** Some problems of the impact category 'ecotoxicity' have already been addressed within the category human toxicity. Protected objects in the category ecotoxicity are primarily ecosystems, from small-scale ecosystems to the macro ecosystem Earth including the atmosphere (Lovelock's 'Gaia').[326] Interdependence between biotic and abiotic factors within the complex structure of producers, consumers, decomposers and the physical environment is typical of ecosystems. Biotic factors are organisms at various trophic levels. Usually primary producers, consumers and destructors are differentiated. From dead biomass the destructors generate nutrients which are needed by producers.

---

322) A more realistic assumption would make a reference to the functional unit, necessary for LCA, impossible.
323) Henderson *et al.* (2011).
324) Classical multi-media models were developed for non-dissociating organic molecules without a surface-active impact, see Mackay (1991) and Klöpffer (1996b, 2012b).
325) Jørgensen and Hauschild (2011).
326) Gaia was the old (pre-olympic) earth goddess and primeval mother, earth itself; Lovelock (1982, 1990).

Examples for abiotic factors are radiation (intensity and spectral composition), temperature, pH value, currents (air and water), chemical composition of water, soil and air and the cycles and rhythms of these factors. Ecosystems are open systems implying an energy transfer with the surrounding environment exists and they are interconnected by material transfer over the atmosphere or the water as well as by an exchange of organic materials with energetic content.

Ecotoxicity examines harmful changes of structures and functions of ecosystems caused by an anthropogenic entry of substances. Because of the complexity of interactions in ecosystems the examined system is usually strongly simplified in practice and an issue under investigation is selected within defined boundary conditions: often only a small number of selected test organisms (fish, daphnia, algae or earthworms, etc.) are examined for selected effects and under defined laboratory conditions. An examined effect in the simplest case is acute toxicity of a substance on the test organism, measured by the concentration of the substance in the water whereby 50% of the test organisms die ($LC_{50}$). Tests for the determination of chronic toxicity, reproduction or carcinogenic effect are more elaborate. In some cases model ecosystems (so-called mesocosms) are also examined in the laboratory or in a field study; frequently small ponds are used. From these results it is difficult, if not impossible, to draw conclusions on the damage of the impacting structures on larger ecosystems.[327]

**4.5.3.3.2 Chemicals and Environment** The handling of chemicals as substances in the sense of the chemical laws by humans requires a knowledge of whether the regarded substance is a poison, originally for reasons of worker and consumer protection (see impact category human toxicity). A frequently used measure for the acute toxicity of a substance is $LD_{50}$ experimentally determined on animals. Ecotoxicology is the application of this confined terminology on poisons to environmental issues, as perceived by admission authorities: usually the $LC_{50}$-value is determined in water, for example, for the water flea (daphnia magna), a small water crustacean suitable as a test organism. It is presumptuous to extrapolate from single species tests with a few organisms to ecotoxicity – even if safety factors are considered – for the following reasons:

1. *Ecological systems can react more sensitively than individuals of a species*
   Chemicals can intervene with the function of ecosystems with no toxic effects at all on cells, such as by a disturbance of the chemical communication system.[328] Stratospheric ozone depletion which is separately addressed in the impact assessment is another example. A combination of ecotoxicological test systems that would have been successful in identifying this impact can hardly be imagined. Freons (used e.g. as refrigerants) are practically non-toxic and therefore, not ecotoxic either, strictly speaking. $LC_{50}$- values cannot determine either a direct eco-toxic impact like the damage of the endocrine system of

---
327) (Klöpffer (1989, 1993, 1994c).
328) Stumm (1977) and Klaschka (2008, 2009).

animals living in the wild by endocrine disruptors or potential side-effects of GMOs.[329]

2. *Ecological systems can also react less sensitively than individuals of a species*
The disappearance of only one species in an ecosystem will generally not destroy the complex interdependency: another species with similar environmental requirements will replace the extinct one and provide the function of the former species (occupy an ecological niche). If it is not a remarkable species, only specialists will be able to perceive the difference.

Ecosystems go through a development from youth to maturity or climax. A permanent biocoenosis develops, according to regional climates and local soils, water and topographic conditions. Climax ecosystems are therefore highly different from one another, for example, tropical rain forests, oak and beech mixed woodlands, lakes in the high mountains or savannah. These spacious ecological systems are called *biomes*. In relation to stages of youth and growth, characteristics of climax systems are:

- huge variety in spatial and functional structuring;
- often interlaced food chains;
- closed nutrient cycles;
- good stability towards small and small stability towards large disturbances from the outside;
- small net primary production. This means that biomass cannot be extracted without a major disturbance of the system.

In relation to the stage of youth, stability against small disturbances is based on a larger flexibility of the complex interdependency in the climax system. With large disturbances the system needs a very long time to again develop into a climax ecosystem. If, for example, in a region with tropical rain forest the relatively thin humus layer due to clear cutting and violent rainfalls is washed away, for a very long period the tropical rain forest will not develop because of the now changed local soil and water conditions.

*Resumee* It is not possible from single species tests to deduce statements concerning ecosystems, it is however done taking into account 'safety factors'. The results are NECs, NOECs or PNECs. The values are derived from measured values (the lowest measured effect concentration; at best only $LC_{50}$ values are available on several organisms or for longer durations to be able also to determine chronic effects). They are usually defined for water as a test environment. For the compartments soil and sediment, there are much less values available; air in ecotoxicity is only concerned as the recipient and transportation medium. Phytotoxic effects can be transferred directly by air. In LCA these effects are partly considered by the impact categories acidification and summer smog. Also global atmospheric impacts, which in a wider sense are part of ecotoxicity as well, are addressed in separate impact categories. Scientific ecotoxicology, similarly to human toxicology,

---

[329] Klöpffer *et al.* (1999) and Klöpffer (1998a, 2001).

comprises a thorough study of damaging impacts on diverse species caused by damaging substances and set up mechanisms. A transmission of these single findings on all possible species and particularly on ecosystems (see above) is at present not possible and, in the case of the ecosystems, is probably not possible at all.

The actual protection goal in the impact category ecotoxicity can be described by the prospering, quality and sustainability of the natural ecosystems, but cannot be satisfactorily defined. It has been tried, mostly in the US, to introduce the descriptive metaphor *ecosystem health* as the actual protection goal. This was strongly contradicted by Suter[330] in an analysis of the term and its consequences (see Section 4.5.3.2.3). The expression health suggests an understanding of the functioning of ecosystems, which is not true[331], and that sick ecosystems could be healed. A more suitable expression for the higher goal of protection suggested by Suter is quality or *sustainability* of ecosystems, and examples of the necessary investigations and data for characterisation are provided.

### 4.5.3.3.3 Simple Quantification of Ecotoxicity without Relation to Exposure

Similar to the impact category human toxicity there is a 'zeroth approximation'.[332] for ecotoxicity; a primary loaded compartment ($m = n$) is considered in Equation 4.26 and the fate factor is not regarded ($F = 1$). Under this boundary conditions a large NEC list provided by, for example, US EPA,[333] can be used for weighting in the regarded compartment (ecotoxicity potential, ETP, according to Equation 4.29). As discussed for the use of the OELs like MAK- or ADI values for human toxicity, the NEC values are exclusively applied for a relative weighting of emissions into water or soil as determined in the inventory (LCI). By application of this simple method a division into two indicators, one for water and one for soil, cannot be avoided, if ecotoxicity in the soil compartment is not neglected for lack of data.

$$\mathrm{ETP} = \sum_i \left( \frac{m_i}{\mathrm{NEC}_i} \right) \quad \text{(l water or kg soil per fU)} \tag{4.29}$$

with $m_i$ = mass of substance $i$ released into water and/or soil with a documented $\mathrm{NEC}_i$ value, per fU. ETP is usually separated into an aquatic (ETPA) and a terrestrial ecotoxicity potential (ETPT). The unit of the ETP (l water; kg ground) results from the units of NEC values, common units are mg l$^{-1}$ for water and mg kg$^{-1}$ dry weight ('ppm') for soil. The load per fU from the inventory should be used accordingly in milligrams.

Another option would be an ETP or ETPA or ETPB in Equation 4.29 by quotient formation on any reference substance (e.g. 1,4-dichlorobenzene), with its ETP arbitrarily set to one. The indicator result then reads 'kg of DCB equivalents' per fU for ecotoxicity in water or in soil. The only argument against such units is the same as for HTPs: it assumes a similar mechanism for all aggregated emissions whereas in reality there may be a totally different impact context.

---

330) Suter (1993).
331) Schmid and Schmid-Araya (2001).
332) The expression 'zeroth approximation' originates from quantum physics and designates a theoretical problem solution, which excludes interactions between subsystems.
333) Heijungs *et al.* (1992) and Klöpffer and Renner (1995).

If the reference of OELs leads to a relative weighting of the HTP of substances, a formal relative weighting concerning the documented NECs is similarly obtained. The difference is that usually a large data background is consulted for the derivation of limit and indicative values for the protection of the human health. NEC and NOEC values on the other hand usually bear no relation to the complex interdependencies in ecosystems, (see discussion on the protection goal above) but were determined for single organisms. Nevertheless it is a first step towards a relative weighting of the ecotoxicity. A further disadvantage of the simple weighting is the neglect of persistence and bioaccumulation of compounds, which should be determined in an extra score or at least in a list from the inventory. Otherwise the most toxic persistent environmental chemicals often without a high acute toxicity would by aggregation not (or not sufficiently) be considered.[334] Substances with a very high ecotoxicity ($LC_{50} < 1\,\mu g\,l^{-1}$) dominate the result. Contrary to human toxicity, which only refers to humans, biologically speaking to a single species gives rise to a predicament, as described in Sections 4.5.3.3.1 and 4.5.3.3.2 for ecotoxicity, as to the exact objective of the protection goal and to the choice of the impact indicator, which describes the damaging impacts. Simple characterisations according to Equation 4.31 will therefore only be recommendable for simple LCAs and must be justified within the 'goal and scope'.

### 4.5.3.3.4 Inclusion of Persistence and Distribution into Quantification

An inclusion of substance properties, which describe transport, distribution and degradation, requires more subtle structured impact indicators. These must include a fate factor ($F$) according to the formula described in the paragraph on human toxicity (Equation 4.30):

$$S_i^{nm} = E_i^m F_i^{nm} M_i^n \tag{4.30}$$

$S_i^{nm}$ score of a substance $i$ for compartment $m$, the release originating in compartment $n$
$E$ effect factor (for (target) compartment $m$)
$F$ fate (distribution from compartment $n$ to $m$ and degradation)
$M$ mass (entry of substance $i$ into compartment $n$ per fU).

Reciprocal NOEC, which are known for the regarded compartment $m$ (water, soil and sediment) and preferably have been evaluated, can be used. Values that have been determined for multiple species and have been averaged are more meaningful than those of a single species.[335]

Modelling necessary for the computation of fate factors are state of the art[336], the problem being boundary conditions like the size of compartments (global or local), selected transfer and degradation models as well as numbers for degradation and

---

334) Stephenson (1977), Frische *et al.* (1982), Klöpffer (1989, 1994a, 2001), Müller-Herold (1996), Scheringer (1999) and Klöpffer and Wagner (2007a,b).
335) Larsen and Hauschild (2007a,b).
336) Mackay (1991), Trapp and Matthies (1996, 1998), Fenner *et al.* (2005) and Klöpffer (2012b)

transfer constants.[337] A list of a minimum of data required for an ecotoxicity model is integrated into the OMNITOX method description[338]:

- Dissociation constants (acid/base)
- Reaction constant of second order for the reaction with OH in the gaseous phase
- Half-life of hydrolysis in water
- Henry constant or air/water distribution coefficient
- Fusion point
- Molar mass
- Octanol/water distribution coefficient
- Particle gas distribution coefficient (e.g. after Junge)
- Distribution coefficient (steady state) between water and sediment
- Distribution coefficient (steady state) between water and soil
- Vapour pressure
- Water solubility
- Acute lethal toxicity for fresh water fish
- Acute toxicity for invertebrates
- Alga growth inhibition as degradation of the growth rate
- Alga growth inhibition as reduction of the biomass
- Ready biodegradability by different end points
- Inherent biodegradability, biological degradability after adaptation of the degrading micro-organisms or their enzyme system to the substrate

Some, but not all of these data must be declared by the producer for assessment of the potential risk of chemicals. This unsatisfactory situation applies even for the most progressive chemical law of the world, REACH,[339] as work-around quantitative structure–activity-relationships (QSARs) are used to obtain an estimate of the characteristics of substances with a minimum of information, often only a structural formula. Even measured physico-chemical data often scatter within a wide range. This was shown by the example of the well-known environmental chemical dichloro-diphenyl-trichloroethane (DDT) and its transformation product dichlorodiphenyldichloroethylene (DDE).[340] The only possible way-out is an evaluation of the data available and a normative definition on which data should be used for exposure calculations. A first step has been made in the model USEtox[341] already discussed in Section 4.5.3.2.4 in the context of the human toxicity indicator. As an ecotoxicological indicator in the USEtox model, so far only the *aquatic ecotoxicity* has been worked out, presumably because a maximum of data are available in this context, mostly for daphnia, fish and algae. So far, characterisation factors (fresh water ecotoxicity) for about 2500 substances have been provided.[342] The factors

---

337) Hertwich, Pease and McKone (1998).
338) Guinée *et al.* (2004); see also Klöpffer (1996b, 2012b), Schwarzenbach, Gschwend and Imboden (2002) and Klöpffer and Wagner (2007a).
339) EC (2006) and Scheringer and Hungerbühler (2008); see also Klöpffer (1996c, 2002, 2004).
340) Eganhouse and Pontolillo (2002).
341) Rosenbaum *et al.* (2008, 2011), Jørgensen and Hauschild (2011) and Hauschild, Jolliet and Huijbregts (2011).
342) Rosenbaum *et al.* (2008) and Henderson *et al.* (2011).

extend over a range of ten orders of magnitude which is why a classification by an order of magnitude of the individual values is judged as being sufficient and the deviations of the individual models from each other (to maximally three orders of magnitude) are regarded as tolerable.

A restriction of the model according to the authors (Rosenbaum et al., 2008, loc. cit.) is the modelling of oceans as sinks, and ecotoxicity only referring to fresh water organisms. Moreover, averaged sensitivities of tested groups of organism, not those of the most sensitive species are integrated into the determination of the effect factors. This procedure is justified, however, by the fact that in LCA – contrary to chemical risk assessment – no risk is measured but comparative statements are to be made.[343] Thus, average values are more useful than extrema.

A comprehensive treatment of USEtox including examples from the praxis of LCIA is now available (special issue of The International Journal Life Cycle Assess. edited by Jørgensen and Hauschild (2011, loc. cit.). The model developed for the purpose of LCIA (as opposed to chemical risk assessment) has been described earlier by Hauschild.[344]

### 4.5.3.4 Concluding Remark on the Toxicity Categories

For toxicity-related impact categories, like for all others, the inclusion of these categories into the impact assessment has to be announced in 'Goal and Scope' and the method used has to be defined. Surely, LCAs on chemicals as such or as a substantial part of products, for example, detergents, drugs, solvents, agricultural chemicals, and so on must address those two impact categories, human toxicity and ecotoxicity, in order to give convincing and credible results. Besides, simpler impact indicators can further be employed for materials or products where chemicals are only released in small quantities into the environment. This is not the case if the use phase is dominated by cleansing or maintenance.

For procedures including persistence and long-range transfer, great progress has been made in recent years and more is to be expected (For a recent survey see.[345] Toxicological data like, for example, no observed effect level (NOEL) or NOEC and derived values are, however, always at the basis to span a first relative scale.

A systematic inclusion of human toxicity and ecotoxicity into the impact assessment increases requirements in the inventory:

1. The table of emissions should list as many individual chemicals as possible.
2. Data should not only include generic data records on energy and transportation processes but be determined by a careful specific process analyses.
3. Data asymmetries (particularly frequent within toxicity) must be avoided in comparative LCAs.

Data asymmetries within these impact categories can result in gross false estimations because human and ecotoxicity of chemicals covers a range of many orders of magnitude. Thus a particularly 'poisonous' substance may even though

---

343) Larsen and Hauschild (2007a,b).
344) Hauschild et al. (2008).
345) Klöpffer (2012b)

released in a small quantity per functional unit dominate the impact assessment. If a comparable product is less carefully investigated the product system with superior investigation will be discriminated, that is, that sluggards (who do not thoroughly investigate) or swindlers (if releases of toxic substances were consciously suppressed) will even be 'rewarded'. Therefore special care has to be taken if the results of the toxicity indicator are used for comparative assertions, and a particular effort is necessary for a critical review of results.

Hauschild et al.[346] brought the situation to the point in writing:

> Without an appropriate link to the inventory, the impact assessment is bound to do a poor job.

### 4.5.4
### Nuisances by Chemical and Physical Emissions

#### 4.5.4.1 Introduction

Nuisances within the impact assessment are those which do not directly lead to diseases or heavy damages in ecosystems but are considered by humans as disturbing, annoying or as reducing the quality of life. These include above all smell and noise. The latter is regarded by the population as a very highly ranked environmental problem and noise applied at continuously high levels can actually make a person ill. The traditional designation 'nuisance' is therefore only justified for small doses. Noise can also be considered as physical emission perceived by physiological-sensory means. As such it would be on even level with probably the most precarious physical emission of hard radiation (see Section 4.5.5), which is however not directly perceived as sensorial, and can even in small doses on a long-term basis cause damages.

In view of their impact radius, noise and smell are assigned to local range. In industrialised countries both are however pervasive through an abundance of sources.

#### 4.5.4.2 Smell

Smelling nuisances have as a starting point many human activities in industry, trade and in the agriculture. An obvious differentiation into good and bad smells is practically hardly feasible as the exhaust air of an odorous substance factory (without sophisticated purification) is rarely less troubling than that of a field fertilised with liquid manure. It can be assumed that each inadvertent (man-made) smell is considered an environmental exposure. As weighting factors odour threshold values (OTVs) are suitable.[347] Although they can only be determined quite inaccurately by 'smelling panels' some useful lists exist (e.g. Heijungs et al., 1992; Guinée et al., 2002, loc. cit.); however, a single one, sanctioned by a team of experts would be better. With the help of these values as a weighting tool, a type of

---

346) Hauschild et al. (2011).
347) Heijungs et al. (1992), Klöpffer and Renner (1995) and Guinée et al. (2002).

critical air volume related to the nuisance by smells is obtained:

$$\text{Smell potential} = \sum_i \left(\frac{m_i}{\text{OTV}_i}\right) \quad \left(\text{m}^3\,\text{fU}^{-1}\right) \tag{4.31}$$

with $m_i$ = mass of the substance $i$ released into the air, for which an $\text{OTV}_i$ was determined, per fU.

Here also a conversion into kg equivalent to a reference substance could be obtained that would also make sense (same mechanism of smell perception).

Smells are extremely specific sensory perceptions, which can react to the smallest chemical structural differences. It is therefore a problem, if an inventory only lists cumulative values without reference to defined substances. Which is an odour smell threshold value of 'VOC'and/or 'HC', and so on? The same is also valid for toxicities.

The smell should be quantified only with product systems where smelling nuisances particularly for a comparison of products are concerned. A listing of all smells by life cycle is not very informative.

#### 4.5.4.3 Noise

Noise by the judgement of a majority of the population, subjectively belongs to the most disturbing of environmental factors. It is to a large extent traffic-dependent, but stationary plants and services can also be sources of noise.

In contrast to most other emission categories, where emissions are chemical substances, noise as a release is a physical emission and from the receiver side (like all sensory impressions) a physiological–psychological effect. Sound pressure is quantified by a relative (logarithmic) scale (decibels). This unit is unfortunately not suited to represent the total amount as a sum of partial amounts. In a direct comparison of products (e.g. car, mowing machine) a standard distance can be specified to measure the noise level of the device.

It is more difficult to include diffuse noises of multiple sources as for instance traffic noise. For transport by truck as a first approximation the transport distance per fU (vehicle-kilometres) can be used as a proxy instead.[348]

In model calculations for the highway from Milano to Bologna an equivalence factor for cars and trucks was deduced.[349] A 'disturbed time period' of residents is calculated and the limits of noisy loads that can be assigned to a nuisance are adapted from regulations. Performances of transport are determined in the inventory as passenger kilometres (car, rail, …) and tonne-kilometres (truck, rail, vessel, …) (see Section 3.2.5). For the special case studied by Lafleche and Sacchetto (1997) the following factors were calculated for the first time:

- number of disturbed human hours/passenger kilometre (car) = 0.000688
- number of disturbed human hours/tonne-kilometre (truck) = 0.000747.

---

348) Schmitz, Oels, and Tiedemann (1995).
349) Lafleche and Sacchetto (1997).

**Table 4.16** Health damage in DALYs per 1000 km.

|  | Motor vehicle type 1 (passenger car, etc.) DALY (a) | Motor vehicle type 2 (truck, etc.) DALY (a) |
|---|---|---|
| Communication disturbance (during the day) | 0.00013 | 0.0013 |
| Sleep disturbance (at night) | 0.0027 | 0.026 |

Simplified, the number of disturbed human hours can be calculated for one passenger kilometre (pkm) or (metric) tonne-kilometre (tkm) as follows:

$$1\,\text{tkm} \approx 1\,\text{pkm} \approx 7 \times 10^{-4}\ (\text{disturbed human hours})$$

For these factors to be suitable for a generalisation in future, average resident densities have to be determined and integrated into the calculations. Besides, point sources and rail transports should be integrated.

Whereas Lafleche and Sacchetto (1997, loc. cit.) regard noise as a mere nuisance, Müller-Wenk[350] advocates a quantitative relationship between street traffic noise and health damages which mainly comprises sleep and communication disturbances. The calculations or respective estimations were accomplished for 1000 additional vehicle kilometres (car and truck) on the Swiss road system and therefore includes overland traffic as well as traffic within townships. The year of reference was 1995. The evaluation was done according to the DALY method[351] and was supplemented by the interviews with experts. The results for the damaging impacts 'communication disturbance' (during daytime) and 'sleeping disturbance' (during night) are presented in Table 4.16 in DALY per 1000 km. For a conversion into common LCA specifications as passenger kilometre (car) and tonne-kilometre (truck) per fU, further assumptions are necessary. The figures are valid for Switzerland and can be used for other countries of average noise load. The authors recommend a division by 2 for countries with a minor noise load (in Europe, e.g. Finland, Denmark and Sweden) and a multiplication with 2 for countries with a major noise load (in Europe, e.g. Spain and Slovakia).

For the evaluation, assumptions of countries concerned have to be made and information on whether transports occur during the day or at night has to be obtained. Transport by rail has not yet been considered in the analysis. In view of the broader grid structure only a smaller fraction of the population is concerned. Müller-Wenk (2004) outlines how to analyse rail traffic similar to road traffic. An assignment to flight-kilometre is problematic as noise only occurs for starting and landing of air planes near the air port. Especially for comparative LCAs these restrictions must be kept in mind in the case where transport systems differ qualitatively (e.g. truck vs. rail).

---

350) Müller-Wenk (1999a, 2002b, 2004)
351) WHO (1996) and Hofstetter (1998); Disability Adjusted lost Life Years, see Section 4.5.3.2.4.

More recent work by Althaus *et al.*[352] extends the pioneering work of Müller-Wenk (2004). First of all, the authors analysed and evaluated five methods proposed using the following set of criteria.[353] The method has to be applicable to

1. both generic and specific transports
2. different modes of transport
3. different vehicles within one mode
4. transports in different geographic contexts
5. different temporal contexts
6. and be compatible with the ISO LCA standards.

The result was that none of the methods fully complied with all criteria.

On the basis of the results of part 1 of the study, a new framework for the inclusion of traffic noise has been developed.[354]

Among the three methods identified as best suited for further development, the 'Swiss EPA method', developed by Müller-Wenk[355], was chosen for a deeper analysis. It should be noted that the study by Althaus *et al.*[356] deals predominantly with the LCI-aspects of the problem and that the development of an appropriate impact assessment method for noise (including, e.g. consequences of sleep disturbance) is still on the agenda.

A completely different approach for a characterisation of noise by the *fuzzy-sets-method*[357] is only indicated since it is unclear as to how the selected indicator can be related to the fU.

### 4.5.5
### Accidents and Radioactivity

#### 4.5.5.1 **Casualties**
This impact category has been integrated into the category list by the first SETAC working group for impact assessment[358] but no method for a characterisation or quantification was proposed. It belongs to the categories to be considered if two product systems strongly differ in this respect. A pre-study accomplished on behalf of the UBA Berlin[359] clarifies the concept and provides a framework for a methodical development. It remains to be clarified, whether (similar to toxic effects in the workplace) a 'product-related social life cycle assessment (SLCA)' still under development within a framework of a sustainability analysis[360] (see Chapter 6) is the right place for casualties as an impact category. This discussion is of special

---

352) Althaus, de Haan and Scholz (2009a,b).
353) Althaus, de Haan and Scholz (2009a).
354) Althaus, de Haan and Scholz (2009b).
355) Müller-Wenk (2002b, 2004).
356) Althaus, de Haan and Scholz (2009b).
357) Benetto, Dujet and Rousseaux (2006); to the use of Fuzzy sets (a numeric expert system) in LCA see also Thiel *et al.* (1999), Weckenmann and Schwan (2001) and Güereca *et al.* (2007).
358) Udo de Haes (1996).
359) Kurth *et al.* (2004).
360) Klöpffer and Renner (2007).

importance for rare invents with a very high potential of harm, as, for example, in nuclear electricity production.

### 4.5.5.2 Radioactivity

Without this impact category and the category 'casualties' nuclear power would do quite well in the impact assessment. The greenhouse effect and most chemical emissions related to the energy output are small compared to thermal power generation with fossil sources of energy. The risk of a maximum credible accident remains that is hardly to be calculated according to the classical 'insurance formula'

$$\text{Risk} = \text{extent of damage} \times \text{probability of occurrence}$$

because

- the extent of damage is extremely large but cannot be quantified;
- the probability of occurrence is >0 but very small, and cannot be verified statistically.

Statistical material at the basis of insurance mathematics is therefore missing. As a substitute an attempt is made using general technical knowledge including modelling as a basis for reasoning by analogy to make risk calculations, which are however very controversial. Furthermore, the problem of the final disposal of radioactive waste is not yet solved (anywhere in the world).

Incidents and leakages of nuclear power and reprocessing plants can be considered by impact assessment. Emissions under normal operation conditions should be considered in the toxicity impact categories. A first quantification attempt of the impact category radioactivity is based on the number of radioactive decays per time unit[361] originating from the emissions. The SI unit of radioactivity[362] is the Becquerel named in honour of the discoverer A. H. Becquerel (1852–1908). It signifies the number of radioactive decays per second. The conversion into the former unit Curie is done according to

$$1 \text{ Curie (Ci)} = 3.7 \times 10^{10} \text{ Becquerel (Bq)}$$

The unit Becquerel has the disadvantage that further important information concerning the impact of nuclear radiation like radiation type ($\alpha, \beta, \gamma$)[363], the energy per particle or quantum and the half-life of radio active atomic nuclei is not considered. For aggregation the non-weighted inventory data *Bq per fU* can be chosen.

A characterisation beyond this non-weighted aggregation has been described in literature.[364] It takes into account that hard radiation can not only harm human health but also the environment and that radionuclides can accumulate in the environment with a well-known possibility of transfer to humans, for example, by contaminated nutrition.

---

361) Suter and Walder (1995) and Guinée *et al.* (2002).
362) ISO (1981) and Deutsche Normen (1978a,b).
363) $\alpha$-radiation consists of $He^4$-nuclei, $\beta$-radiation is high-energy electron radiation, $\gamma$-radiation is extremely short-wave electromagnetic radiation; the three radiation types mainly differ by their energy content and particularly by their ability to penetrate matter, which increases from $\alpha$ to $\gamma$.
364) Solberg-Johansen, Clift and Jeapes (1997).

For a deduction of an indicator model an equation essentially the same as Equation 4.30 is chosen, combining mass, impact and fate. The so-called *environmental increments* (EIs)[365] have been proposed as EFs for an environmental load. Such factors have been deduced for the most important radionuclides in radioactive waste. From experience it is known that even in natural surroundings a very small exposure of organisms occurs, and species and ecosystems obviously can cope with it.[366] EIs of individual nuclides are determined, not completely without certain arbitrariness, because of their natural variability of occurrence in ecosystems (minimum area 1 ha, minimum time period 1 a). They serve as a proxy similar to 'No Effect Concentration' in ecotoxicology.[367] For artificial radionuclides EI values with more or less plausible assumptions were applied.

For an application in the impact assessment a media related indicator model must be developed for terrestrial, air-related and aquatic exposition.

The simplest application of the EI method (fate factor = 1) formally resembles a 'c.V.' (Section 4.2):

$$C_i = m_i \left( \frac{1}{EI_i} \right) \quad (4.32)$$

where

$m_i$ mass of the released nuclide $i$ (into the regarded medium), for which an $EI_i$ was determined, per fU
$C_i$ possible contribution of nuclide $i$
$M_i$ mass $i$ per functional unit
$EI_i$ *Environmental Increment* of $i$ ((Bq kg$^{-1}$ soil), (Bq m$^{-3}$ water) or (Bq m$^{-3}$ air)).

These partial amounts can, as with the chemicals, be summed to an overall potential.

In order to also include the life time and distribution of the nuclides (fate) simple exposure models can be used. The human toxicity of radioactive emissions is quantified according to Solberg-Johansen *et al.* (1997, loc. cit) by non-weighted Bq but not characterised.

For radioactivity also, as in the case of other rarely applied impact categories (see Section 6.3.1), advancement in methodology, harmonisation of different variants and testing in real LCAs should have a high priority.

## 4.6
## Illustration of the Phase Impact Assessment by Practical Example

Requirements of the impact assessment are divided into mandatory and optional elements. For an elaboration of optional elements clearly more degrees of freedom

---

365) Amiro (1993).
366) The method was developed for emissions from nuclear plants and nuclear waste and is therefore closer to a threshold-thinking than to the 'less is better' of classical LCA, Amiro and Zach (1993).
367) Solberg-Johansen *et al.* (1997).

exist. The following illustration of the phase impact assessment on the basis of a case example[368] is done as outlined in Section 4.3.

*Mandatory elements*:

1. Selection of impact categories, – indicators and characterisation factors (Section 4.6.1):
   These specifications according to ISO 14040/44 are to be determined in the first phase, 'definition of goal and scope', as data procurement in the inventory must guarantee that the required data for the selected impact categories are available (see Chapter 2). The detailed discussion takes place here, as the scientific background of the consideration of impact categories was discussed in Section 4.5. A fundamental examination is necessary whether all inputs and outputs that have been quantitatively considered are able to map to the selected impact categories. If this is not the case there are two possibilities: either the inventory has to be revised or impact categories with insufficient data quality have to be discarded. Because of the absence of a list of impact categories in ISO 14044 mandatory for all LCAs (see Section 4.3.2.1), the selection in each study has to be comprehensible and transparent and must be justified. For many impact categories, indicator models with impact indicators have been established. For others like, for example, 'human or ecotoxicity', quite diverse models are used by different working groups (see Section 4.5.3). Therefore indicator models used in the study should also be comprehensible and transparent.
2. Classification (Section 4.6.2):
   The classification in each study is accomplished for the selected impact categories. The inventory data are ordered according to their scientifically established contribution to the selected impacts categories: they are ordered in classes (hence the name 'classification').
3. Characterisation (Section 4.6.3):
   By the selected indicator models for the considered impact categories the inventory data assigned by classification are transferred into impact indicators. This is done by means of characterisation factors.

*Optional elements*:

4. Normalisation (Section 4.6.4):
   If a normalisation is accomplished the reference quantities must be defined.
5. Grouping (Section 4.6.5):
   If the normalised data are further ordered (sorted or ranked) according to their relevance, the ordering criteria must be represented with transparency. That is particularly important as value-based elements start to enter here.

---

368) IFEU (2006).

6. Weighting (Section 4.6.6):
   Because weighting includes value choices, specifications for the basics of weighting is indispensable for the credibility of an LCA.
7. Additional analysis of data quality:
   Each step of the impact assessment should be accompanied by a critical reflection of the quality of the data base. As described in Section 4.3.3.4, the analysis of the data quality is explicitly required in the interpretation (see Chapter 5).

## 4.6.1
### Selection of Impact Categories – Indicators and Characterisation Factors

In the 'definition of goal and scope' of the illustrative study example the following reasons are quoted for the selected impact categories:

> The impact assessment in this study is based on the following impact categories[369]:
> A) Resource-related categories:
>    - Demand for fossil resources,
>    - Land use (forest).
> B) Emission-related categories:
>    - Greenhouse effect,
>    - terrestrial eutrophication,
>    - acidification,
>    - summer smog
>    - aquatic eutrophication.
>
> The separation of the impact category eutrophication into aquatic and terrestrial eutrophication is in view of the different mechanisms of effect within soil and water.
> The mechanisms of effect for all categories considered (with the exception of land use forest) are scientifically founded and they are usually convertible from the inventory data. This is confirmed by their widespread use in national and international LCAs. The general acceptance of these impact categories can thus be presumed.[370] They can be considered as standard in the common practice of LCA.

---

369) In the examined systems ozone destructive substances were not released in relevant quantities, hence, for economic reasons this impact category was not considered.
370) In LCA practice it is hardly possible to make a complete estimate of all environmental issues. In the present study by pre-selection of specific environmental issues a restriction is already made. A desirable broad examination of as many environmental issues as possible is frequently not done due to the different quality of the available inventory data and a differing scientific acceptance of individual impact models.

> Regarding the evaluation of land use there are different approaches in LCA. The scientific discussion is among other things concerned with how to ecologically evaluate a given land use.
>
> The impact categories human and ecotoxicity are also among the 'standard categories' of LCA. Here too, different approaches are used for a consideration within the impact assessment. Points of criticism vary to an extent where no direct harmonisation is expected in the near future.[371] Besides, there are even problems at the level of the inventory, for example, incomplete inventory data, which in the end can infer misinterpretations. Human and ecotoxicity have for these reasons not been evaluated in the context of the impact assessment of this study.

In the example study the assignment of inventory parameters to the selected impact categories and the respective impact indicator models are already represented in the phase 'definition of goal and scope' (see Table 2.1).

In the supplement of the example study, the reasons for the selection of impact indicators as well as the indicator models as basis for the impact assessment are described in more detail. For every one of the selected impact categories equivalence factors of the selected indicator model for all assigned inventory parameters are presented in a table. There are impact categories like, for example, 'Greenhouse effect'[372] for which a detailed presentation in every LCA seems to almost be exaggerated because of an existing consent with respect to the indicator model. Other impact categories on the other hand are discussed in scientific literature (e.g. photo-oxidant formation) with their controversies, and the indicator model selected for the special study must be transparent and comprehensibly described.

### 4.6.1.1 (Greenhouse) Global Warming Potential

> The greenhouse effect (global warming and climate change) as impact category is the negative environmental effect of heating of the terrestrial atmosphere caused by anthropogenic activity. It has already been described in detail in the appropriate references.[367] The most frequently applied indicator so far in LCAs is the radiative forcing[368] and is indicated as $CO_2$-equivalent value (GWP). The characterisation method is generally accepted.
>
> Substances and their $CO_2$-equivalence values, which can be found in calculations of the greenhouse potential as 'global warming potential (GWP)', are listed in Table 4.17:

---

371) This LCA study was performed in 2006.
372) Now called *climate change*.
373) IPCC (1995a, 2001).
374) CML (1992) and Klöpffer 1995

**Table 4.17** Global warming potential of substances considered in the context of this project.

| Greenhouse gas | $(GWP_i)_{100}$ ($CO_2$-equivalents) |
|---|---|
| Carbon dioxide ($CO_2$) | 1 |
| Methane ($CH_4$)[a] | 25.75 |
| Methane ($CH_4$), regenerative | 23 |
| Dinitrogenmonoxide ($N_2O$) | 296 |
| Tetrachloromethane | 1 800 |
| Tetrafluoromethane | 5 700 |
| Hexafluoroethane | 11 900 |

[a] In Houghton et al. (2001) indirect impacts like oxidation from $CH_4$ to $CO_2$ are not included in the GWP values. For methane from fossil sources the GWP value increases if formed (fossil) $CO_2$ is considered.
Houghton et al. (2001).

The contribution to the greenhouse effect is the sum of the products of released quantities of the individual greenhouse-relevant pollutants ($m_i$) and their respective GWP ($GWP_i$) according to the following formula:

$$GWP = \sum_i (m_i \times GWP_i)$$

### 4.6.1.2 Photo-Oxidant Formation (Photo Smog or Summer Smog Potential)

The gases and their POCPs, which could be procured in the context of this LCA are listed in Table 4.18.

**Table 4.18** Ozone formation potential of substances considered in the context of this project.

| Noxious gas | $POCP_i$ (ethene equivalents) |
|---|---|
| Ethene | 1 |
| Propene | 1.123 |
| Methane | 0.006 |
| Hexane | 0.482 |
| Formaldehyde | 0.52 |
| Ethanol | 0.399 |
| Aldehydes (average) | 0.443 |
| Benzene | 0.22 |
| Toluene | 0.637 |
| Xylene | 1.1 |
| Ethylbenzene | 0.73 |
| Hydrocarbons NMVOC from diesel releases | 0.7 |
| NMVOC (average) | 0.416 |
| VOC (average) | 0.377 |

CML (1992), Guinée et al. (2002) and Klöpffer (1995).

Only individual substances with a defined equivalence value to Ethene were considered. For hydrocarbons, often quoted in literature, which are not precisely defined as substances, an average equivalence value from CML (1992) is used.

The POCP was determined according to the following formula:

$$POCP = \sum_i m_i \times POCP_i$$

### 4.6.1.3 Eutrophication Potential

As an indicator for the calculation of an unwanted nutrient supply the EP is selected in units of phosphate equivalents.[375] Pollutants or nutrients with their respective characterisation factor in the context of this project are listed in Table 4.19.

Table 4.19  Eutrophication potential of substances considered in the context of this project.

| Pollutant | $PO_4^{3-}$-equivalents ($NP_i$) |
|---|---|
| Eutrophication potential (soil) | |
| Nitrogen oxides ($NO_x$ as $NO_2$) | 0.13 |
| Ammonia ($NH_3$) | 0.327 |
| Eutrophication potential (water) | |
| Phosphate | 1 |
| Phosphorus compounds calculated as P | 3.06 |
| Chemical oxygen demand (COD) | 0.022 |
| Ammonia/ammonium $NH_3/NH_4^+$ | 0.327 |
| Nitrate ($NO_3^-$)/$HNO_3$ | 0.095 |

Guinée et al. (2002) and Klöpffer (1995).

In a simplified assumption it is presumed that all nutrients released by air constitute an over-fertilisation of the soil and all nutrients (including COD substances) of waters lead to eutrophication of water bodies. Since a nutrient entry into water by air is small compared to that by waste water, this assumption represents no considerable error.

To distinguish between nutrient entry into soil and into water the contribution to the EP is calculated as the sum of the products of NP and the released quantities of individual pollutants.

---

375) CML (1992) and Klöpffer (1995).

For the eutrophication of soil (S) the following applies:

$$NP_S = \sum_i (m_i \times NP_i)$$

For the eutrophication of water (W) the following applies:

$$NP_W = \sum_i (m_i \times NP_i)$$

### 4.6.1.4 Acidification Potential

Acidification because of acid-forming substances can occur in both terrestrial and aquatic systems.

An impact indicator AP as described[376] is regarded as adequate. Therefore, no specific properties of the loaded land and water systems are required. The calculation of the AP is usually done in units of $SO_2$ equivalents. Pollutants considered in this study are listed with their AP as $SO_2$ equivalent in Table 4.20.

**Table 4.20** Acidification potential of the substances in the context of this project.

| Pollutant | $SO_2$-equivalents ($AP_i$) |
|---|---|
| Sulphur dioxide ($SO_2$) | 1 |
| Hydrogen sulphide ($H_2S$) | 1.88 |
| Sulphuric acid | 0.65 |
| Hydrogen chloride (HCl) | 0.88 |
| Hydrogen fluoride (HF) | 1.6 |
| Hydrogen cyanide | 1.6[a] |
| Nitrogen oxides ($NO_x$) | 0.7 |
| TRS (total reduced sulphur) calculated as S | 2.0 |
| Carbon disulphide | 1.68 |
| Ethane thiol | 1.03 |
| Mercaptan | 0.84[b] |
| Ammonia | 1.88 |

[a] Assumption like HF.
[b] Assumption: Propanthiol.

Guinée et al. (2002) and Klöpffer (1995).

The contribution to the AP is calculated as the sum of the products of AP and released quantities of individual pollutants according to the following formula:

$$AP = \sum_i (m_i \times AP_i)$$

---

376) CML (1992) and Klöpffer (1995).

4.6.1.5  **Resource Demand**

For an evaluation of the resource demand within the impact assessment, the scarcity of resources is usually used as criterion. Despite an alleged good accessibility of the environmental aspect 'resource demand' some fundamental aspects will have to be clarified in future. This is of particular concern for a meaningful classification of resource types and for the definition of scarcity.

Because of a pre-selection of priority impact categories in this study only the categories energy resources and land use are described.

4.6.1.5.1  **Energy Resources**

In this study an aggregation of the resource 'energy' is twofold: On the one hand the concept of its evaluation as cumulative primary energy demand (CED) is applied, on the other hand the finiteness of fossil primary energy sources is considered.

**Cumulative Energy Demand**

The CED is not an impact parameter but an inventory figure. Nevertheless it is used in the study as important information in the interpretation and is therefore specified here.

CED is applied as a category for the evaluation of primary energy.

It is an inventory parameter and represents the sum of the energy content of all primary energy sources, which can be traced back to the system boundaries. 'CED fossil' is the sum of exclusively fossil primary energy sources. The consumption of uranium is assessed by 'CED nuclear'. The computation of 'CED nuclear' is done by an efficiency mark-up of 33% of the electricity generated by nuclear power used by the investigated systems. Besides, the 'CED water power', 'CED renewable' and 'CED other' as well as the 'CED sum' of all CED values as a result of the inventory is listed. 'CED other' is assessed as an energy demand related to data records with no information concerning the type of energy production. 'CED water power' is based on an efficiency of 85%.

**Scarcity of Fossil Fuels**

According to the method of the UBA the static range of energy sources serves as an indicator for the scarcity of fossil fuels.[371] The static range is derived from the data of the world reserve and current consumption of the respective resource. The scarcity is converted into crude oil equivalence factor (ROE) (UBA, 1995).

---

377) The reliability of the static range as scarceness indicator is impaired by uncertainties concerning the state of known and economically exploitable resources.

**Table 4.21** ROE: crude oil equivalence factors for fossil sources of energy according to UBA Berlin[a].

| Raw materials in deposits | Static range (a) | Heat value, $H_u$ (MJ kg$^{-1}$) or (MJ Nm$^{-3}$) | ROE$_i$ |
|---|---|---|---|
| Crude oil | 42 | 42.622 | 1 |
| Natural gas | 60 | 31.736 | 0.5212[b] |
| Hard coal | 160 | 29.809 | 0.1836 |
| Brown coal | 200 | 8.303 | 0.0409 |

[a] The static ranges of hard and brown coal used by the UBA significantly deviate from recent values (Table 4.7). The former UBA values (UBA, 1995) probably referred to Germany while the values in Table 4.7 were calculated from worldwide reserves and production figures. The energy contents are average values of unknown provenance and should only be used in the context of ROE calculation according to UBA. Their exaggerated accuracy is probably due to conversions from the outdated unit calorie (cal). The ROE values are rounded; on the other hand not all decimal digits are consistent.
[b] The ROE for natural gas of the UBA table quoted refers to the heat value in MJ m$^{-3}$ (31 736 MJ Nm$^{-3}$). 31 736 MJ m$^{-3}$ divided by 60 a results in an ROE of 0.529 (not 0.5212) related to volume. With a natural gas density of 0.78 kg m$^{-3}$ a heat value of 40.69 MJ kg$^{-1}$ results plus an ROE of 0.678 referred to mass. The use of the value of 46.1 MJ kg$^{-1}$ in accordance with Table 3.5 for natural gas results in an ROE referred to mass for a static range of 60 a of 0.77. In the practical example (see also Section 4.6) an ROE referred to mass for natural gas of 0.6202 was used.

The ROEs (ROE: Röhöläquivalenzfaktor) considered by UBA are defined according to Equation 4.33 and listed in Table 4.21. Crude oil is assigned to one.

$$\text{ROE}_i = \frac{\text{NCV}_i}{\text{static range}_i(a)} \tag{4.33}$$

ROE$_i$: crude oil equivalence factor of fossil resource $i$
NCV$_i$: Net calorific value of resource $i$ (MJ kg$^{-1}$).

Deviating from Table 4.21 a ROE for natural gas of 0.6202 is used in the sample study (see Table 4.22).

**Table 4.22** Fossil resources consumption (ROE): raw oil equivalents/fU.

| Fossil resources | Inventory result (see Table 3.13) (kg) | Characterisation factor factor in kg raw oil equivalent/kg | Impact indicator value |
|---|---|---|---|
| Brown coal | 5.48 | 0.0409 | 0.22 kg ROE |
| Natural gas | 8.03 | 0.6202 | 4.98 kg ROE |
| Oil | 12.22 | 1 | 12.22 kg ROE |
| Hard coal | 1.27 | 0.1836 | 0.23 kg ROE |
| **Sum ROE: crude oil equivalence value** | | | 17.66 kg ROE |

The following applies for the calculation of the ROE:

$$\text{ROE} = \sum_i (m_i \times \text{ROE}_i)$$

**4.6.1.5.2 Land Use**

Considering the ecological capacity of an area implies taking all area-related environmental impacts into account like, for example, the decrease of biological diversity, land erosion, impairment of the landscape, and so on. In contrast to the terms area or surface it seems to be appropriate to circumscribe with 'natural space' all its inherent natural correlations.

For this purpose, a method for the impact assessment was developed in an LCA on graphic papers on behalf of the UBA[378] which is based on a description of 'grades of naturalness' (hemerobic levels) of natural space.[379] In the present study hemerobic levels II–VI are considered.

It is particularly stressed in the study that the impact assessment is not a risk analysis (see also Section 4.4.3.2):

It is explicitly pointed out that the impact assessment represents an analysis instrument in the context of an LCA. The results are partly based on model assumptions and previous knowledge of certain impact relations and they are to be regarded in the general context. Under no circumstances are forecasts, for example, of concrete impacts, threshold values or dangers, which are caused by the examined product systems, being made or provided.

## 4.6.2
## Classification

In Section 3.7 the inventory of the packaging '1-l-cardboard with closure' for fruit juices was presented as an example. Data were already marked, which were later transferred into impact categories based on definitions presented in Section 4.6.1. This is the phase of classification.

## 4.6.3
## Characterisation

Using characterisation factors specified for the study, the classified data are transferred into impact indicator values.

Classification and characterisation are integrated into the most relevant Software tools. A plausibility check is however always recommended, as designations of the

---

378) UBA (1998).
379) Klöpffer and Renner (1995).

**Table 4.23** Assessed value: CED (cumulative energy demand) (no characterisation factors)/fU.

| Energy as CED | Inventory result (see Table 3.14) (kJ) |
|---|---|
| CED fossil | 9.62E+05 |
| CED nuclear | 3.17E+05 |
| CED other | 1.23E+03 |
| CED water power | 9.76E+04 |
| CED renewable | 6.58E+05 |
| *CED sum* | *2.04E+06* |

**Table 4.24** Assessed value: demand on natural space (no characterisation factors)/fU.

| Demand on natural space | Inventory result (see Table 3.17) ($m^2$) | Addition of classed areas (Class 2–6) | |
|---|---|---|---|
| Space requirement Class 2 | 4.74E−01 | | |
| Space requirement Class 2 (FRG) | 3.15E−02 | | |
| Space requirement Class 2 (north)[380] | 1.04E+00 | | |
| | | Area 2 | 1.55E+00 $m^2$ |
| Space requirement Class 3 | 4.77E+00 | | |
| Space requirement Class 3 (FRG) | 2.83E−01 | | |
| Space requirement Class 3 (north) | 1.19E+01 | | |
| | | Area 3 | 1.70E+01 $m^2$ |
| Space requirement Class 4 | 1.15E+01 | | |
| Space requirement Class 4 (FRG) | 1.36E−01 | | |
| Space requirement Class 4 (north) | 3.16E+01 | | |
| | | Area 4 | 4.32E+01 $m^2$ |
| Space requirement Class 5 | 1.79E+00 | | |
| Space requirement Class 5 (FRG) | 2.22E−02 | | |
| Space requirement Class 5 (north) | 7.33E+00 | | |
| | | Area 5 | 9.14E+00 $m^2$ |
| **Space requirement (entire)** | | | *7.09E+01 $m^2$* |

same elementary flow for different imported data records can be quite different (see Section 3.7.6). It must therefore be made sure that the software is capable of correctly assigning all designations of inputs and outputs in the study.

In Tables 4.22–4.28 the selected inventory parameters for the impact categories are listed in accordance with Section 3.7 in the column 'inventory result'. The impact indicator values can be calculated by use of the characterisation factors in accordance with Section 4.6.1. The result as a sum in each impact category refers to the fU of the packing system, here, supply of 1000 l fruit juice/nectar in 1-l-beverage carton with closure at the point of sale.

---

380) Skandinavia.

**Table 4.25** Global warming potential (GWP)$_{100}$: $CO_2$-equivalents/fU.

| Greenhouse effect | Inventory result (see Table 3.18) (kg) | Characterisation factor in kg $CO_2$-equivalents/kg | Impact indicator value GWP$_{100}$ (kg $CO_2$-equivalents) |
|---|---|---|---|
| $C_2F_6$ | 2.12E−05 | 11 900 | 2.52E−01 |
| $CF_4$ | 2.43E−04 | 5 700 | 1.39E+00 |
| $CH_4$ | 1.55E−01 | 25.75 | 4.00E+00 |
| $CH_4$, renewable | 6.90E−02 | 23 | 1.59E+00 |
| $CO_2$, fossil | 6.07E+01 kg | 1 | 6.07E+01 |
| $N_2O$ | 1.13E−03 | 296 | 3.34E−01 |
| Carbon tetrachloride | 1.36E−11 | 1 800 | 2.45E−08 |
| **Global warming potential (total)** | | | *6.83E+01* |

**Table 4.26** Summer smog potential (POCP): ethene equivalents/fU.

| Summer smog | Inventory result (see Table 3.18) (kg) | Characterisation factor in kg ethene equivalents/kg | Impact indicator value POCP (kg of ethene equivalents) |
|---|---|---|---|
| Aldehydes, unspecific. | 1.27E−07 | 0.443 | 5.63E−08 |
| Benzene | 7.53E−05 | 0.22 | 1.66E−05 |
| Ethanol | 2.55E−08 | 0.399 | 1.02E−08 |
| Ethene | 1.08E−05 | 1 | 1.08E−05 |
| Ethylbenzene | 7.21E−11 | 0.73 | 5.26E−11 |
| Formaldehyde | 3.76E−04 | 0.52 | 1.96E−04 |
| Hexane | 7.53E−06 | 0.482 | 3.63E−06 |
| Methane | 2.24E−01 | 0.006 | 1.34E−03 |
| NMVOC of diesel emissions | 3.53E−03 | 0.7 | 2.47E−03 |
| NMVOC, unspecific including TOC[381] calculated as NMVOC[382] | 1.39E−02 | 0.416 | 5.78E−03 |
| NMVOC (hydrocarbons) | 1.22E−02 | 0.416 | 5.08E−03 |
| Propene | 8.02E−06 | 1.123 | 9.01E−06 |
| Toluene | 2.35E−07 | 0.637 | 1.50E−07 |
| VOC, unspecific | 5.49E−02 | 0.377 | 2.07E−02 |
| VOC (Hydrocarbons) | 3.35E−03 | 0.377 | 1.26E−03 |
| Xylene | 3.17E−07 | 1.1 | 3.49E−07 |
| **Summer smog potential (total)** | | | *3.69E−02* |

---

[381] Total organic carbon.
[382] Assuming an average molar ratio n (H)/n (C) of 3/1 in NMVOC, the TOC value is divided by 0,8 and added to NMVOC unspecific.

**Table 4.27** Acidification potential (AP): $SO_2$-equivalents/fU.

| Acidification | Inventory result (see Table 3.18) (kg) | Characterisation factor in kg $SO_2$ equivalent/kg | Impact indicator value AP (kg $SO_2$-equivalents) |
|---|---|---|---|
| Ammonia | 4.04E−03 | 1.88 | 7.60E−03 |
| Carbon disulphide | 6.15E−12 | 1.68 | 1.03E−11 |
| Hydrogen chloride | 1.04E−03 | 0.88 | 9.15E−04 |
| Hydrogen cyanide | 2.41E−09 | 1.6 | 3.86E−09 |
| Ethanethiol | 1.35E−12 | 1.03 | 1.39E−12 |
| Hydrogen fluoride | 1.02E−03 | 1.6 | 1.63E−03 |
| Mercaptan | 5.77E−08 | 0.84 | 4.85E−08 |
| Sulphurdioxide | 1.53E−01 | 1 | 1.53E−01 |
| Carbondisulphide | 1.89E−10 | 1.68 | 3.18E−10 |
| Sulphuric acid | 1.04E−13 | 0.65 | 6.76E−14 |
| Hydrogen sulphide | 2.11E−04 | 1.88 | 3.97E−04 |
| Nitrogen oxides ($NO_x$ calculated as $NO_2$) | 1.79E−01 | 0.7 | 1.25E−01 |
| TRS (total reduced sulphur) | 7.79E−04 | 2 | 1.56E−03 |
| **Acidification potential (total)** | | | 2.90E−01 |

**Table 4.28** Eutrophication potential (NP): $PO_4^{3-}$ equivalents/fU.

| Eutrophication | Inventory result (see Table 3.18 and 3.19) (kg) | Characterisation factor (kg $PO_4^{3-}$ equiv/kg) | Impact indicator value NP (kg $PO_4^{3-}$-equiv) |
|---|---|---|---|
| Ammonia (A)[383] | 4.04E−03 | 0.347 | 1.40E−03 |
| $NO_x$ calculated as $NO_2$ (A) | 1.79E−01 | 0.13 | 2.33E−02 |
| **Eutrophication potential (terrestrial)** | | | 2.47E−02 |
| Ammonia (W)[384] | 1.87E−05 | 0.327 | 6.11E−06 |
| Ammonium (W) | 1.67E−04 | 0.327 | 5.46E−05 |
| Ammonium calculated as N | 6.73E−06 | 0.42 | 2.83E−06 |
| COD (W) | 3.31E−01 | 0.022 | 7.28E−03 |
| Nitrate | 2.66E−02 | 0.095 | 2.53E−03 |
| Nitrate calculated as N | 1.25E−08 | 0.42 | 5.25E−09 |
| N-compounds, unspecific (W) | 1.22E−05 | 0.42 | 5.12E−06 |
| N-compounds calculated as N (W) | 4.22E−03 | 0.42 | 1.77E−03 |
| Phosphate (W) | 1.08E−07 | 1 | 1.08E−07 |
| P calculated as $P_2O_5$ (W) | 2.94E−08 | 1.338 | 3.93E−08 |
| Phosphorus | 8.50E−08 | 3.06 | 2.60E−07 |
| P-compounds calculated as P (W) | 1.33E−03 | 3.06 | 4.07E−03 |
| Nitric acid | 8.75E−07 | 0.128 | 1.12E−07 |
| **Eutrophication potential (aquatic)** | | | 1.57E−02 |
| **Eutrophication potential (total)** | | | 4.04E−02 |

383) Emission into air.
384) Emission into water.

**Table 4.29** Tabular presentation of the results of the impact assessment after the stage of characterisation.

| Indicator | Result | Unit |
|---|---|---|
| Sum: raw oil equivalents | 17.66 | kg ROE |
| Sum CED | 2.04E+06 | kJ |
| Space requirement (total) | 7.09E+01 | m$^2$ |
| Global warming potential (GWP$_{100}$) | 6.83E+01 | kg CO$_2$-equivalents |
| Summer smog potential (POCP) | 3.69E−02 | kg of Ethene equivalents |
| Acidification potential (AP) | 2.90E−01 | kg SO$_2$- equivalents |
| Eutrophication potential (NP aquatic) | 1.57E−02 | kg PO$_4^{3-}$- equivalents |
| Eutrophication potential (NP terrestrial) | 2.47E−02 | kg PO$_4^{3-}$- equivalents |

As the final result of the classification the inventory data are bundled and thus prepared for the interpretation: Many data of the inventory are however not transferred into impact indicators. If in a study the categories human- and eco-toxicity are considered, further data of the inventory could be consulted. As described above in the sample study these impact categories were not included because of controversial discussions in professional circles (see also Section 4.5). Table 4.29 shows the summation of results of impact categories considered and additionally, the two inventory figures 'CED' and 'space requirement'.

In Section 2.3.1 the goals outlined in the sample study were indicated. In order to be able to redeem these goals the data are further prepared by Normalisation for the interpretation (see Section 4.6.4).

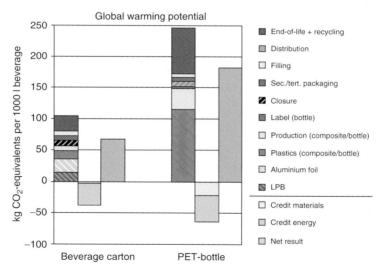

**Figure 4.4** Sectoral analysis of the product system variants referring to the impact category climate change.

For a deduction of optimisation potentials a sectoral analysis on the level of the impact indicators following characterisation is very useful. Figure 4.4 shows the sectoral analysis of the variants 1-l-carton system and 1-l-PET-system using as example the impact category greenhouse effect (now climate change) related to the fU. A prerequisite of a sector analysis is the modelling in the inventory in a way, that environmental loads can be assigned to individual phases of the life cycle. In view of an optimisation analysis during the interpretation (see Chapter 5) such information is very valuable.

Figure 4.4 serves exclusively as an illustration of the use of a sectoral analysis on the level of impact indicators. It will be integrated into the discussion of the component interpretation.

### 4.6.4
### Normalisation

In every study the selected bases of normalisation are to be described. In the example study[383] the normalisation is based on the specific contribution and resident equivalents (see Section 4.3.3.1).

> For the normalisation accomplished here, the impact related aggregated environmental loads are represented as 'specific contribution' by means of resident equivalents. These indicate an average contribution to the respective impact category per inhabitant in a given geographical reference area per annum. Thus information on relevance of individual categories can be obtained.
>
> For the normalisation accomplished here on selected examples the environmental load of Germany and Western Europe is consulted as reference value. The data consulted for the normalisation are provided in Table 4.30. Total load values and a quantity per inhabitant – corresponding to one resident equivalents (REQs) – of Germany and Western Europe respectively are specified.
>
> Results of the classification, which initially refer to the fU as defined in the goal and scope definition, are scaled by the total consumption of the regarded beverages in Germany with respect to Western Europe. The basis for Germany is an annual consumption of fruit juices and fruit nectars of 4555 million litres. The source for the derivation of these values was (Tetra Pak, 2005, 2006).
>
> The results of the described derivations are specific contributions of the examined options to the respective categories. For a representation of REQs no sectoral analyses is applied to the results of the selected categories because of a higher interest in the overall contributions of the impact categories (see Figure 4.5).

---

383) IFEU (2006).

## 4 Life Cycle Impact Assessment

**Table 4.30** Data for the determination of the specific contribution (IAV) (from IFEU, 2006); IAV = inhabitant average value.

| | Load per annum | | | REQ | | |
|---|---|---|---|---|---|---|
| | Germany | Western Europe | | Germany | Western Europe | |
| **Inhabitants** | | | | | | |
| Inhabitants | 82 532 000 | 397 404 900 | a,b | | | |
| **Resources** | | | | | | |
| Brown coal | 1 547 000 | 2 239 394 | c,d | 18 744 | 5 635 | MJ |
| Natural gas | 3 025 000 | 15 552 120 | c,d | 36 652 | 39 134 | MJ |
| Crude oil | 5 478 000 | 26 042 733 | c,d | 66 374 | 65 532 | MJ |
| Hard coal | 1 920 000 | 6 823 563 | c,d | 23 264 | 17 710 | MJ |
| Total area | 35 703 000 | 400 000 000 | e,f | 4 326 | 10 065 | m² |
| **Emissions (air)** | | | | | | |
| Ammonia | 601 000 | 3 540 000 | g,h | 7.28 | 8.91 | kg |
| Arsenic | 33 | 193 | i,h | 0.0004 | 0.0005 | kg |
| Benzene | 42 900 | 306 000 | j,h | 0.52 | 0.77 | kg |
| Benzo (a)pyrene | 13.76 | 209 | k,h | 0.0002 | 0.0005 | kg |
| Cadmium | 11 | 133 | i,h | 0.00013 | 0.00033 | kg |
| Hydrogen chloride | | 730 000 | h | | 1.84 | kg |
| Chromium | 115 | 646 | i,h | 0.0014 | 0.0032 | kg |
| Dioxin (TCDD/F) | 1.25 | 3.55 | l,m | 15.15 | 8.93 | kg |
| Nitrogen dioxide | 205 000 | 1 300 000 | g,h | 2.48 | 3.27 | kg |
| Hydrogen fluoride | 124 000 | | n | 1.5 | | kg |
| Carbon dioxide, fossil | 865 000 000 | 3 390 000 000 | g,h | 10 481 | 8 530 | kg |
| Carbon monoxide | 4 155 000 | 42 800 000 | g,h | 50.34 | 107.7 | kg |
| Methane | 3 582 000 | 20 265 000 | g,h | 43.4 | 50.99 | kg |
| Nickel | 159 | 1 580 | i,h | 0.0019 | 0.004 | kg |
| NMVOC | 1 460 000 | 14 000 000 | g,h | 17.69 | 35.23 | kg |
| Nitrogen oxides (as NO₂) | 1 428 000 | 106 | g,l,h | 17.3 | 0.00027 | kg |
| PCB | 43.6 | 12 220 000 | l,g,h | 0.00053 | 30.75 | kg |
| Sulphur dioxide | 616 000 | | | 7.46 | | kg |

## 4.6 Illustration of the Phase Impact Assessment by Practical Example

| | | | | | | |
|---|---|---|---|---|---|---|
| Dust (PM10) | 224 930 | | 1 350 000 | [h] | 2.73 | 3.4 | kg |
| **Emissions (water)** | | | | | | | |
| Phosphorus | 33 000 | g | 224 000 | [h] | 0.39984 | 0.56366 | kg |
| Nitrogen | 688 000 | g | 1 370 000 | [h] | 8.33616 | 3.44737 | kg |
| **Aggregated values for the impact assessment** | | | | | | | |
| Crude oil equivalents | 189 702 096 | | 878 621 435 | t ROE-equivalents | 2298.53 | 2210.9 | kg |
| Greenhouse effect | 1 017 916 500 | | 4 296 623 750 | t $CO_2$-equivalents | 12334 | 10812 | kg |
| Acidification | 2 943 880 | | 29 317 600 | t $SO_2$-equivalents | 35.67 | 73.77 | kg |
| Eutrophication (terrestrial) | 395 990 | | 3 059 000 | t $PO_4$-equivalents | 4.8 | 7.7 | kg |
| Eutrophication (aquatic) | 389 940 | | 1 260 840 | t $PO_4$-equivalents | 4.72 | 3.17 | kg |
| Eutrophication (total) | 785 930 | | 4 319 840 | t $PO_4$-equivalents | 9.52 | 10.87 | kg |
| Summer smog (POCP) | 630 290 | | 8 200 000 | t Eth-equivalents | 7.73 | 20.63 | kg |
| Area | 357 033 | | 400 000 | $km^2$ | 0.43 | 1.01 | ha |

[a] Federal Statistical Office Germany 2004 (December 31, 2003).
[b] Eurostat (1 January 2005).
[c] Data to the environment (UBA Germany) 2005, year of reference 2000.
[d] Eurostat, from 'energy balances – data 2002–2003, detailed tables, 2005 edition'.
[e] Data to the environment (UBA Germany) 2005, referred to the year 2001, source: Federal Statistical Office 2003.
[f] Internet June 2004.
[g] Data to the environment (UBA Germany) 2005, year of reference 2003.
[h] Reference releases Western Europe, 1995, data from CML (April 2004).
[i] Data to the environment (UBA Germany) 1996, year of reference 1995.
[j] German parliament's Enquête Commission on 'substance chains and material flows'1993, p. 146.
[k] IFEU study 'POP Emissions in Germany', year of reference 1994.
[l] Information UBA Germany.
[m] European dioxin inventory – Stage II – year of reference 2000.
[n] Data to the environment (UBA Germany) 92/92, year of reference 1991.

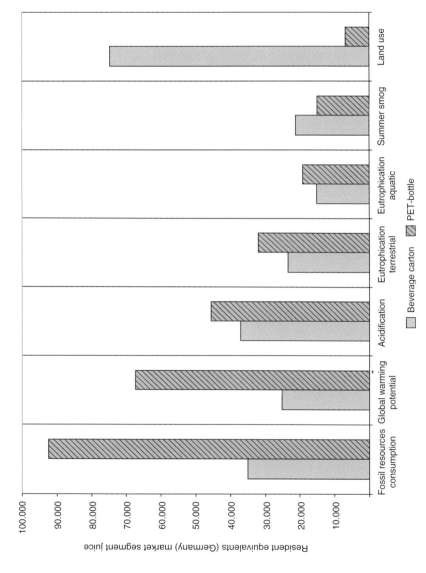

**Figure 4.5** Resident equivalents of the plotted impact indicators for the beverage carton and the PET bottle based on 1-l-beverage carton and the 1-l-PET-bottle for fruit juices with reference to Germany.

On the basis of the data in Table 4.30 and the impact indicator result for the greenhouse effect, the following specific contribution and resident equivalents (REQs) for the example system carton packaging results:

---

**Sample Calculation: Calculation for the total consumption Germany for beverage carton; REQ normalised impact indicator result:**

| Impact category: | Global warming potential: |
|---|---|
| Global warming potential per fU | 68.3 kg $CO_2$-Eq/1000 l (see Table 4.29) |
| Total consumption in Germany | $4.555 \times 10^6$ l |
| Global warming potential total | $68.3 \times 4.555 \times 10^3$ kg $CO_2$-Eq $= 311 \times 10^6$ kg $CO_2$-Eq |
| Total greenhouse potential, Germany | $1\,017\,916\,500 \times 10^3$ kg $CO_2$-Eq (see Table 4.30) |
| Inhabitants in Germany | 82 532 000 (see Table 4.30) |
| REQ Global warming potential | 12 334 kg $CO_2$-Eq/inhabitant (see Table 4.30) |
| *Specific contribution product system* | $3.06 \times 10^{-4}$ |
| *REQ product system* | 25 223[a] |

$$\text{Spec. contribution of product system} = \frac{\text{total GWP of the product system}}{\text{total GWP of Germany}}$$

$$\text{REQ of product system} = \frac{\text{total GWP of the product system}}{\text{REQ GWP of Germany}}$$

[a] All data specified in this section are rounded for clarity. A REQ of 25212 as the result of the study is obtained if all decimals are considered.

---

For the normalisation of the impact indicator, land use, the respective total areas of Germany or Western Europe were considered. This corresponds to the assessment in other impact categories with reference to German or Western European total values.

The geographical frame of references values (e.g. for GWP Germany) and causing emissions (GWP of carton system) are not necessarily identical since for instance the GWP due to the production of raw cardboard originates in the Nordic countries. Generally this has little impact on the results of normalisation.

Normalised indicator results are usually plotted. Figure 4.5 shows as an example a comparison of the packing systems 'beverage carton' and 'PET bottle' for fruit juice and fruit nectars, normalised with reference to the total consumption in Germany.

### Results of the Normalisation

The diagram of indicator results normalised by resident equivalents for selected scenarios show which impact categories contribute more ore less to the related total values in Germany. This implies that within impact categories with the highest specific values a reduction of environmental loads of regarded packaging systems would have a particularly strong effect on an environmental improvement in Germany.

The reading of differences between the scenarios measured in REQ is exemplified by the example of the greenhouse effect in Figure 4.5.

With reference to Germany the beverage carton obtained 25212 REQ and the PET system 67358 REQ. Assuming that the total annual consumption of fruit juices and fruit nectars in Germany were exclusively packed in 1000 ml carton or 1000 ml PET-bottles, a saving of GWP, equivalent to 42146 statistical inhabitants, would result if only the carton alternative is considered.

## 4.6.5
## Grouping

In the stage, grouping, value-based elements are integrated (see also Section 4.3.3.2). The example study refers to the ranking of environmental problem fields concerning their ecological priority which was compiled by the federal environmental office (UBA).[384]

This study does not work out in an own grouping system. As an alternative, reference is made to the ranking of impact categories according to the classification developed by the environmental protection agency (UBA Germany) and used for beverage packaging LCAs by UBA (Table 4.31).

**Table 4.31** Classification of ecological priorities developed by environmental protection agency (UBA Germany) and used by UBA in beverage packaging LCAs UBA (2000).

| Impact category | Ecological priority (UBA, 2000) |
|---|---|
| Greenhouse effect | Very large ecological priority |
| Fossil resource demand | Large ecological priority |
| Eutrophication (terrestrial) | Large ecological priority |
| Acidification | Large ecological priority |
| Summer smog (~ surface-near ozone formation) | Large ecological priority |
| Eutrophication (aquatic) | Average ecological priority |
| Land use, forest | Average ecological priority |

---

384) Schmitz and Paulini (1999).

In Figure 4.5 REQ normalised results of considered impact categories are represented.

The highest normalised indicator results for Germany occur for impact categories fossil resource demand, greenhouse effect and land use. There is a particularly large margin with regard to the remaining impact categories.

Taken by themselves the normalised results suggest that the results of scenarios with respect to the impact categories fossil resources demand, greenhouse effect and natural utilisation of space should be adequately considered within an interpretation of the results. A special relevance of these categories for a comparison of scenarios is also owing to large absolute differences in REQs.

### 4.6.6
### Weighting

'Weighting' for LCAs including comparative assertions to be disclosed to the public is not permissible in accordance with ISO 14040/44 and therefore not included in this study.

## References

Alcamo, J., Doll, P., Henrichs, T., Kaspar, F., Lehner, B., Rosch, T., and Siebert, S. (2003) Development and testing of the Water GAP 2 global model of water use and availability. *Hydrol. Sci. J.*, **48** (3), 317–337.

Althaus, H.J., de Haan, P., and Scholz, R.W. (2009a) Traffic noise in LCA. Part 1: state–of-science and requirement profile for consistent context-sensitive integration of traffic noise in LCA. *Int. J. Life Cycle Assess.*, **14** (6), 560–570.

Althaus, H.J., de Haan, P., and Scholz, R.W. (2009b) Traffic noise in LCA. Part 2: analysis of existing methods and proposition of a new framework for consistent, context-sensitive LCI modeling of road transport noise emission. *Int. J. Life Cycle Assess.*, **14** (7), 676–686.

Amann, M., Bertok, I., Cofala, J., Gyrfas, F., Heyes, C., Klimont, Z., and Schöpp, W. (1999) *Integrated Assessment Modelling for the Protocol to Abate Acidification, Eutrophication and Ground-level Ozone in Europe*, Publications Series Air and Energy, vol. 132, Netherlands Ministry for Housing, Spatial Planning and the Environment, The Hague.

Amiro, B.D. (1993) Protection of the environment from nuclear fuel waste radionuclides: a framework using environmental increments. *Sci. Total Environ.*, **128**, 157–189.

Amiro, B.D. and Zach, R. (1993) A method to assess environmental acceptability of releases of radionuclides from nuclear facilities. *Environ. Int.*, **19**, 341–358.

Atkinson, R. (1989) Kinetics and mechanisms of the gas-phase reactions of the hydroxyl radical with organic compounds. *J. Phys. Chem. Ref. Data Monogr.*, **1**, 1–246.

de Baan, L., Alkemade, R., and Koellner, T. (2013) Land use impact on biodiversity in LCA: a global approach. *Int. J. Life Cycle Assess.*, **18** (13) Special issue, 1216–1230.

Bachmann, T.M. (2006) Hazardous substances and human health: exposure, impact and external cost assessment at the European scale, in *Trace Metals and other Contaminants in the Environment*, vol. 8, Elsevier, Amsterdam.

Baitz, M. (2002) Die Bedeutung der funktionsbasierten Charakterisierung von Flächen-Inanspruchnahmen in industriellen Prozesskettenanalysen. PhD thesis. University of Stuttgart, *www.shaker.de* (accessed 31 October 2013) (German).

Bare, J.C., Norris, G.A., Pennington, D.W., and McKone, D.W. (2002) TRACI – the tool for the reduction and assessment of chemical and other environmental impacts. *J. Ind. Ecol.*, **6** (3/4), 49–78.

Bare, J., Pennington, D.W., and Udo de Haes, H.A. (1999) Life cycle impact assessment sophistication. *Int. J. Life Cycle Assess.*, **4** (5), 299–306.

Barnes, I., Becker, K.-H., and Wiesen, P. (2007) Organische verbindungen und der photosmog. *Chem. Z.*, **41** (3), 200–210.

Barnthouse, L., Fava, J., Humphreys, K., Hunt, R., Laibson, L., Noesen, S., Norris, G., Owens, J., Todd, J., Vigon, B., Weitz, K., and Young, J. (eds) (1998) *Life-Cycle Impact Assessment: The State-of-the-Art. Report of the SETAC Life-Cycle Assessment (LCA) Impact Assessment Workgroup*, 2nd edn, Society of Environmental Toxicology and Chemistry, Pensacola, FL.

Becker, K.H., Fricke, W., Löbel, J., and Schurath, U. (1985) Formation, transport and control of photochemical oxidants, in *Air Pollution by Photochemical Oxidants* (ed. R. Guderian (Hrsg.)), Springer-Verlag, Berlin, pp. S.1–125.

Bellekom, S., Potting, J., and Benders, R. (2006) Feasibility of applying site-dependent impact assessment of acidification in LCA. *Int. J. Life Cycle Assess.*, **11** (6), 417–424.

Beltrani, G. (1997) Safeguard subjects. The conflict between operationalization and ethical justification. *Int. J. Life Cycle Assess.*, **2** (1), 45–51.

Benetto, E., Dujet, C., and Rousseaux, P. (2006) Fuzzy-sets approach to noise impact assessment. *Int. J. Life Cycle Assess.*, **11** (4), 222–228.

Berger, M. and Finkbeiner, M. (2010) Water footprinting: how to address water use in life cycle assessment? *Sustainability*, **2**, 919–944. doi: 10.3390/su2040919

Berger, M. and Finkbeiner, M. (2012) Methodological challenges in volumetric and impact-oriented water footprints. *J. Ind. Ecol.* doi: 10.1111/j.1530-9290.2012.00495.x

Bösch, M.E., Hellweg, S., Huijbregts, M., and Frischknecht, R. (2007) Applying cumulative exergy demand (CExD) indicators to the ecoinvent database. *Int. J. Life Cycle Assess.*, **12** (3), 181–190.

Boulay, A.M., Bulle, C., Bayart, J.B., Deschênes, L., and Margni, M. (2011a) Regional characterization of freshwater use in LCA: modelling direct impacts on human health. *Environ. Sci. Technol.*, **45** (20), 8948–8957.

Boulay, A.-M., Bouchard, C., Bulle, C., Dechênes, L., and Margni, M. (2011b) Categorizing water for LCA inventory. *Int. J. Life Cycle Assess.*, **16** (7), 639–651.

Bousquet, P., Hauglustaine, D.A., Peylin, P., Carouge, C., and Ciais, P. (2005) Two decades of OH variability as inferred by an inversion of atmospheric transport and chemistry of methyl chloroform. *Atmos. Chem. Phys. Discuss.*, **5**, 1679–1731.

BUWAL (1998) Brand, G., Scheidegger, A., Schwank, O. and Braunschweig, A. *Bewertung in Ökobilanzen mit der Methode der ökologischen Knappheit. Ökofaktoren 1997*, Schriftenreihe Umwelt Nr. 297 Ökobilanzen, Bundesamt für Umwelt, Wald und Landschaft (BUWAL), Bern.

Brandao, M. and Milà i Canals, L. (2013) Global characterisation factors to assess land use impacts on biotic production. *Int. J. Life Cycle Assess.*, **18** (6), 1243–1252.

Breedveld, L., Lafleur, M., and Blonk, H. (1999) A framework for actualising normalisation data in LCA: experiences in the Netherlands. *Int. J. Life Cycle Assess.*, **4** (4), 213–220.

Brentrup, F., Küsters, J., Lammel, J., and Kuhlmann, H. (2002a) Impact assessment of abiotic resource consumption: conceptual considerations. *Int. J. Life Cycle Assess.*, **7** (5), 301–307.

Brentrup, F., Küsters, J., Lammel, J., and Kuhlmann, H. (2002b) Life cycle impact assessment of land use based on the hemeroby concept. *Int. J. Life Cycle Assess.*, **7** (6), 339–348.

British Standards Institution (2008) Publicly Available Specification (PAS) 2050:2008. Specification for the Assessment of the Life Cycle Greenhouse Gas Emissions of

Goods and Services, British Standards Institution.

Brühl, C. and Crutzen, P.J. (1988) Scenarios of possible changes in atmospheric temperatures and ozone concentrations due to man's activities, estimated with a one-dimensional coupled photochemical climate model. *Clim. Dyn.*, **2**, 173–203.

Brunner, P.H. and Rechberger, H. (2004) *Practical Handbook of Material Flow Analysis*, Lewis Publishers (CRC Press), Boca Raton, FL. ISBN: 1566 706 041.

Bundesamt für Umweltschutz (BUS), Bern (Hrsg.) (1984) *Oekobilanzen von Packstoffen*, Schriftenreihe Umweltschutz, Nr. **24**, Bundesamt für Umweltschutz (BUS), Bern.

Bundesverband der Deutschen Industrie e.V. (BDI) (1999) *Die Durchführung von Ökobilanzen zur Information von Öffentlichkeit und Politik*. BDI-Drucksache Nr. 313, Verlag Industrie-Förderung GmbH, Köln. ISSN: 0407–8977.

BUWAL (1990) Ahbe, S., Braunschweig, A. and Müller-Wenk, R. *Methodik für Oekobilanzen auf der Basis ökologischer Optimierung*, Schriftenreihe Umwelt, Nr. **133**, Bundesamt für Umwelt, Wald und Landschaft (BUWAL, Hrsg.), Bern.

BUWAL (1991) Habersatter, K. and Widmer, F. *Oekobilanzen von Packstoffen. Stand 1990*, Schriftenreihe Umwelt, Nr. **132**, Bundesamt für Umwelt, Wald und Landschaft (BUWAL, Hrsg.), Bern.

Carson, R. (1962) *Silent Spring*. Penguin Books, Harmondsworth, Middlesex (UK) 1982, 1st edn, Houghton Mifflin.

Carter, W. (1994) Development of ozone reactivity scales for volatile organic compounds. *J. Air Water Manage. Assess.*, **44**, 881–889.

Carter, W.P. (2003) Documentation of the SAPRC-99 Chemical Mechanism for VOC Reactivity Assessment. Final Report to California Air Resources Board, Contract No. 92–329 and 95–308.

Ciambrone, D.F. (1997) *Environmental Life Cycle Analysis*, Lewis Publishers, Boca Raton, FL.

CML (1992) Environmental Life Cycle Assessment of Products, Guide and Backgrounds, Center of Environmental Science (CML), Netherlands Organisation for Applied Scientific Research (TNO), Fuels and Raw Materials Bureau (B&G), Leiden.

Crowson, P. (1992) *Minerals Handbook 1992–1993*, Stockton Press, New York.

Crutzen, P.J., Mosier, A.R., Smith, K.A., and Winiwarter, W. (2007) $N_2O$ release from agro-diesel production negates global warming reduction by replacing fossil fuels. *Atmos. Chem. Phys. Discuss.*, **7**, 11191–11205.

Dameris, M., Peter, T., Schmidt, U., and Zellner, R. (2007) Das Ozonloch und seine Ursachen. *Chem. Z.*, **41** (3), 152–168.

Darnall, K.R., Lloyd, A.C., Winer, A.M., and Pitts, J.N. Jr., (1976) Reactivity scale for atmospheric hydrocarbons based on reaction with hydroxyl radical. *Environ. Sci. Technol.*, **10**, 692–696.

De Meester, B., Dewulf, J., Janssens, A., and Van Langenhove, H. (2006) An improved calculation of the exergy of natural resources for Exergetic Life Cycle Assessment (ELCA). *Environ. Sci. Technol.*, **40**, 6844–6851.

Derwent, R.G., Jenkin, M.E., and Saunders, S.M. (1996) Photochemical ozone creation potentials for a large number of reactive hydrocarbons under European conditions. *Atmos. Environ.*, **30**, 181–199.

Derwent, R.G., Jenkin, M.E., Saunders, S.M., and Pilling, M.J. (1998) Photochemical ozone creation potentials for organic compounds in northwest Europe calculated with a master chemical mechanism. *Atmos. Environ.*, **32**, 2429–2441.

Deutsche Forschungsgemeinschaft (Hrsg.) (2007) *MAK- und BAT-Werte-Liste 2007. Maximale Arbeitsplatzkonzentrationen und Biologische Arbeitsstofftoleranzwerte*. Senatskommission zur Prüfung gesundheitsschädlicher Arbeitsstoffe, vol. 43, CD-Rom, Wiley-VCH Verlag GmbH, Weinheim.

Deutscher Bundestag, Referat Öffentlichkeitsarbeit (Hrsg.) (1988) *Schutz der Erdatmosphäre: Eine internationale Herausforderung; Zwischenbericht der Enquete-Kommission des 11. Deutschen Bundestages "Vorsorge zum Schutz der Erdatmosphäre"*, Deutscher Bundestag, Referat Öffentlichkeitsarbeit (Hrsg.), Bonn. ISBN: 3-924521-27-1.

Deutscher Bundestag (1991) *Enquête-Kommission "Schutz der Erdatmosphäre" des*

12. *Deutschen Bundestages: Schutz der Erde. Eine Bestandsaufnahme mit Vorschlägen zu einer neuen Energiepolitik*. 3. Bericht, Teilband I, Economica Verlag, Bonn, and Verlag C.F. Müller, Karlruhe.

Deutscher Bundestag (1992) *Enquête-Kommission "Schutz der Erdatmosphäre" des 12. Deutschen Bundestages: Klimaänderung gefährdet globale Entwicklung. Zukunft ichern – Jetzt handeln*, Economica Verlag, Bonn, and Verlag C.F. Müller, Karlruhe.

Dewulf, J. and Van Langenhove, H. (2002) Assessment of the sustainability of technology by means of a thermodynamically based life cycle analysis. *Environ. Sci. Pollut. Res.*, **9** (4), 267–273.

DIN (1978a) DIN 1301/1. *Deutsche Normen: Einheiten – Einheitennamen, Einheitenzeichen*, DIN.

DIN (1978b) DIN 1301/2. *Deutsche Normen: Einheiten – Allgemein angewendete Teile und Vielfache*, DIN.

DIN/NAGUS (1996) DIN/NAGUS AA3/UA2. *Nationales Papier zu DIN ISO 14042*, Draft February 1996, DIN.

EC (2006) Regulation (EC) No 1907/2006 of the European Parliament and of the Council of 18 December 2006 concerning the Registration, Evaluation, Authorisation and Restriction of Chemicals (REACH).

EC (2013) Commission Recommendation of 9 April 2013 on the use of common methods to measure and communicate the life cycle environmental performance of products and organisations. *Off. J. Eur. Union*, **L 124**, PEF ergänzen.

Eganhouse, R.P. and Pontolillo, J. (2002) Assessing the reliability of physico-chemical property data (Kow, Sw) for hydrophobic organic compounds: DDT and DDE as a case study. *SETAC Globe*, **3** (4), 34–35.

Erlandsson, M. and Lindfors, L.-G. (2003) On the possibilities to apply the result from an LCA disclosed to public. *Int. J. Life Cycle Assess.*, **8** (2), 65–73.

EU (1999) *Externalities of Fuel Cycles – ExternE Project*, Methodology, 2nd edn, vol. 7, European Commission DGXII, Science, Research and Development, JOULE, Brussels-Luxembourg.

EU (2005) *ExternE – Externalities of Energy: Methodology 2005 Update*, Office for Official Publication of the European Communities, Luxembourg.

European Commission (2010) *ILCD Handbook: Analysing of existing Environmental Impact Assessment Methodologies for Use in Life Cycle Assessment*, Joint Research Center, Institute for Environment and Sustainability, Ispra, http://lct.jrc.ec.europa.eu/pdf-directory/ILCD-Handbook (accessed 18 December 2013).

European Environment Agency (2001) Late Lessons from Early Warning: The Precautionary Principle 1896–2000. Environmental Issue Report No 22, European Environment Agency, Copenhagen. ISBN: 92-9167-323-4.

Fabian, P. (1992) *Atmosphäre und Umwelt*, 4. Auflage, Springer-Verlag, Berlin.

Farman, J.C., Gardiner, B.G., and Shanklin, J.D. (1985) Large losses of total ozone in antarctica reveal seasonal ClOx/NOx interaction. *Nature*, **315** (1985), 207.

Faulstich, M. and Greiff, K.B. (2008) KLimaschutz durch Biomasse. *Umweltwiss. Schadst. Forsch.*, **20** (3), 171–179.

Fava, J., Consoli, F.J., Denison, R., Dickson, K., Mohin, T., and Vigon, B. (eds) (1993) *Conceptual Framework for Life-Cycle Impact Analysis*. Workshop Report. SETAC and SETAC Foundation for Environmental Education, Sandestin, FL, February 1–7, 1992, SETAC, March 1993.

Fenner, K., Scheringer, M., Stroebe, M., Macleod, M., McKone, T., Matthies, M., Klasmeier, J., Beyer, A., Bonnell, M., Le Gall, A.C., Mackay, D., van de Meent, D., Pennington, D., Scharenberg, B., Suzuki, N., and Wania, F. (2005) Comparing estimates of persistence and long-range transport potential among multimedia models. *Environ. Sci. Technol.*, **39**, 1932.

Finkbeiner, M. (2009) Carbon footprinting – opportunities and threats. *Int. J. Life Cycle Assess.*, **14** (2), 91–94.

Finlayson-Pitts, B.J. and Pitts, J.N. Jr., (1986) *Atmospheric Chemistry. Fundamentals and Experimental Techniques*, John Wiley & Sons, Inc., New York. ISBN: 0-471-88227-5.

Finnveden, G., Hauschild, M.Z., Ekvall, T., Guinée, J., Heijungs, R., Hellweg, S., Koehler, A., Pennington, D., and Suh, S.

(2009) Recent developments in life cycle assessment. *J. Environ. Manage.*, **91**, 1–21.

Finnveden, G. and Lindfors, L.-G. (1998) Data quality of life cycle inventory data – rules of thumb. *Int. J. Life Cycle Assess.*, **3** (2), 65–66.

Finnveden, G. and Östlund, P. (1997) Exergies of natural resources in life cycle assessment and other applications. *Energy*, **22** (9), 923–931.

Finnveden, G. and Potting, J. (1999) Eutrophication as an impact category. State of the art and research needs. *Int. J. Life Cycle Assess.*, **4** (6), 311–314.

Finnveden, G. and Potting, J. (2001) in *Best Available Practice in Life Cycle Assessment of Climate Change, Stratospheric Ozone Depletion, Photo-oxidant Formation, Acidification, and Eutrophication. Backgrounds on Specific Impact Categories* (eds W. Klöpffer, J. Potting, J. Seppälä, J. Risbey, S. Meilinger, G. Norris, L.-G. Lindfors, and M. Goedkoop), Chapter 5, RIVM Report 550015003/2001, RIVM, Bilthoven, NL, S.57-64, http://www.rivm.nl/bibliotheek/rapporten/550015003.html (accessed 17 October 2013).

Frische, R., Klöpffer, W., Esser, G., and Schönborn, W. (1982) Criteria for assessing the environmental behavior of chemicals: selection and preliminary quantification. *Ecotox. Environ. Safety*, **6**, 283–293.

Frischknecht, R., Steiner, R., and Jungbluth, N. (2009) *Methode der ökologischen Knappheit – Ökofaktoren 2006 Methode für die Wirkungsabschätzung in Ökobilanzen*, Umwelt-Wissen Nr. 0906, Bundesamt für Umwelt (BAFU), Bern.

Gabathuler, H. (1998) The CML story. How environmental sciences entered the debate on LCA. *Int. J. Life Cycle Assess.*, **2** (3), 187–194.

Giegrich, J., Mampel, U., Duscha, M., Zazcyk, R., Osorio-Peters, S., and Schmidt, T. (1995) *Bilanzbewertung in produktbezogenen Ökobilanzen. Evaluation von Bewertungsmethoden, Perspektiven*, UBA Texte 23/95, Endbericht des Instituts für Energie- und Umweltforschung Heidelberg GmbH (IFEU) an das Umweltbundesamt, Berlin, Heidelberg, März, 1995. ISSN: 0722-186X.

Giegrich, J. and Sturm, K. (2000) in *Naturraumbeanspruchung waldbaulicher Aktivitäten als Wirkungskategorie für Ökobilanzen, Teilbericht* (ed. A. Tiedemann (Hrsg.)), UBA Texte 22/2000, Ökobilanzen für graphische Papiere, Berlin, März 2000.

Goedkoop, M. (1995) The Eco-Indicator 95. Final Report (No 9523). National Reuse of Waste Research Programme (NOH), National Institute of Public Health and Environment Protection (RIVM) and Netherlands Agency for Energy and the Environment (Novem). Amersfoort.

Goedkoop, M., Hofstetter, P., Müller-Wenk, R., and Spriensma, R. (1998) The Eco-indicator 98 explained. *Int. J. Life Cycle Assess.*, **3** (6), 352–360.

Goedkoop, M. and Spriensma, R. (2001) *The Eco-indicator 99 – A Damage Oriented Method for Life Cycle Impact Assessment, Methodology Report*, Product Ecology Consultants (PRE), Amersfoort.

Gorree, M., Guinée, J., Huppes, G. and van Oers, L. (2000) Environmental Life Cycle Assessment of Linoleum. CML Report 151. Centre for Environmental Sciences, Leiden, 56 Seiten.

Güereca, L.P., Agell, N., Gassó, S., and Baldasano, J.M. (2007) Fuzzy approach to life cycle impact assessment. *Int. J. Life Cycle Assess.*, **12** (7), 488–496.

Guinée, J.B., Gorée, M., Heijungs, R., Huppes, G., Kleijn, R., de Koning, A., van Oers, L., Wegener Sleeswijk, A., Suh, S., Udo de Haes, H.A., de Bruijn, H., van Duin, R., and Huijbregts, M.A.J. (2002) *Handbook on Life Cycle Assessment – Operational Guide to the ISO Standards*, Kluwer Academic Publishers, Dordrecht. ISBN: 1-4020-0228-9.

Guinée, J. and Heijungs, R. (1993) A proposal for the classification of toxic substances within the framework of life cycle assessment of products. *Chemosphere*, **26**, 1925–1944.

Guinée, J., Heijungs, R., van Oers, L.F.C.M., Wegener Sleeswijk, A., van de Meent, D., Vermeire, T., and Rikken, M. (1996a) USES: Uniform System for the Evaluation of Substances. Inclusion of fate in LCA characterisation of toxic releases applying USES 1.0. *Int. J. Life Cycle Assess.*, **1** (3), 133–138.

Guinée, J., Heijungs, R., van Oers, L., van de Meent, D., Vermeire, T., and Rikken, M. (1996b) LCA impact assessment of toxic releases, in *Generic Modelling of Fate, Exposure and Effect for Ecosystems and Human Beings with Data for About 100 Chemicals*. Report by, CML, Leiden and RIVM, Bilthoven to the Dutch Ministry of Housing, Spatial Planning and Environment.

Guinée, J.B., de Koning, A., Pennington, D.W., Rosenbaum, R., Hauschild, M., Olsen, S.I., Molander, S., Bachmann, T.M., and Pant, R. (2004) Bringing science and pragmatism together. A tiered approach for modelling toxicological impacts in LCA. *Int. J. Life Cycle Assess.*, **9** (5), 320–326.

Guinée, J.B., Van Oers, L., De Koning, A. and Tamis, W. (2006) Life Cycle Approaches for Conservation Agriculture. CML Report 171, CML, Leiden, http://www.leidenuniv.nl/cml/ssp/index.html (accessed 16 October 2013).

Hauschild, M.Z., Huijbregts, M.A.J., Jolliet, O., MacLeod, M., Margni, M., van de Meent, D., Rosenbaum, R.K., Russel, A., and McKone, T.E. (2008) Building a consensus model for life cycle impact assessment of chemicals: the search for harmony and parsimony. *Environ. Sci. Technol.*, **42** (19), 7032–7036.

Hauschild, M.Z., Jolliet, O., and Huijbregts, M.A.J. (2011) A bright future for addressing chemical emissions in life cycle assessment. Editorial special issue USEtox. *Int. J. Life Cycle Assess.*, **16** (8), 697–700.

Hauschild, M. and Potting, J. (2001) *Spatial Differentiation in Life Cycle Impact Assessment; Guidance Document*, Danish Environmental Protection Agency, Copenhagen.

Hauschild, M.Z., Potting, J., Hertel, O., Schöpp, W., and Bastrup-Birk, A. (2006) Spatial differentiation in the characterisation of photochemical ozone formation – the EDIP2003 methodology. *Int. J. Life Cycle Assess.*, **11** (1) Special Issue, 72–80.

Hauschild, M. and Wenzel, H. (1998) *Environmental Assessment of Products: Scientific Background*, vol. 2, Chapman & Hall, London. ISBN: 0-412-80810-2.

Heijungs, R. (1995) Harmonisation of methods for impact assessment. *Environ. Sci. Pollut. Res.*, **2** (4), 217–224.

Heijungs, R. and Guinée, J. (1993) CML on actual versus potential risks. *LCA News- A SETAC- Eur. Publ.*, **3** (4), 4.

Heijungs, R., Guinée, J.B., Huppes, G., Lamkreijer, R.M., Udo de Haes, H.A., Wegener Sleeswijk, A., Ansems, A.M.M., Eggels, P.G., van Duin, R., and de Goede, H.P. (1992) *Environmental Life Cycle Assessment of Products. Guide (Part 1) and Backgrounds (Part 2) October 1992*, Prepared by , CML, TNO and B&G, Leiden. English Version 1993.

Heijungs, R., Guinée, J., Kleijn, R., and Rovers, V. (2007) Bias in normalization: causes, consequences, detection and Remedies. *Int. J. Life Cycle Assess.*, **12** (4), 211–216.

Held, M. and Geißler, K.A. (eds) (1993) *Ökologie der Zeit*, Edition Universitas, S. Hirzel, Stuttgart. ISBN: 3-8047-1264-9.

Held, M. and Geißler, K.A. (1995) *Von Rhythmen und Eigenzeiten. Perspektiven einer Ökologie der Zeit*, Edition Universitas, S. Hirzel, Stuttgart. ISBN: 3-8047-1414-5.

Held, M. and Klöpffer, W. (2000) Life cycle assessment without time? Time matters in life cycle assessment. *Gaia*, **9**, 101–108.

Henderson, A.D., Hauschild, M.Z., van de Meent, D., Huijbregts, M.A.J., Larsen, H.F., Margni, M., McKone, T.E., Payet, J., Rosenbaum, R.K., and Jolliet, O. (2011) USEtox fate and ecotoxicity factors for comparative assessment of toxic emissions in life cycle analysis: sensitivity to key chemical properties. Special issue USEtox. *Int. J. Life Cycle Assess.*, **16** (8), 701–709.

Hertwich, E.G. (1997) *Int. J. Life Cycle Assess.*, **2** (2), 62, Comment to Hogan, L.M., Beal, R.T. and Hunt, R.G. (1996) Threshold inventory interpretation methodology. A case study of three juice container system. *Int. J. LCA*, **1** (3), 159–167, Reply by Hunt, R.G. S. 63.

Hertwich, E., Matales, S.F., Pease, W.S., and McKones, T.E. (2001) Human toxicity potentials for life-cycle assessment and toxic release inventory risk screening. *Environ. Toxicol. Chem.*, **20**, 928–939.

Hertwich, E.G., Pease, W.S., and McKone, T.E. (1998) Evaluating toxic impact assessment methods: what works best? *Environ. Sci. Technol.*, **32**, 138A–145A.

Hettelingh, J.-P., Posch, M., and De Smet, P.A.M. (2001) Multi-effect critical loads used in multi-pollutant reduction agreements in Europe. *Water Air Soil Pollut.*, **130**, 1133–1138.

Hettelingh, J.-P., Posch, M., and Potting, J. (2005) Country-dependent characterization factors for acidification in Europe – a critical evaluation. *Int. J. Life Cycle Assess.*, **10** (3), 177–183.

Hofstetter, P. (1998) *Perspectives in Live Cycle Assessment: A Structured Approach to Combine Models of the Technosphere, Ecosphere and Valuesphere*, Kluver Academic Publishers, Boston, MA. ISBN: 0-7923-8377-X.

Hogan, L.M., Beal, R.T., and Hunt, R.G. (1996) Threshold inventory interpretation methodology. A case study of three juice container system. *Int. J. Life Cycle Assess.*, **1** (3), 159–167.

Huijbregts, M.A.J., Rombouts, L.J.A., Ragas, A.M.J., and van de Meent, D. (2005a) Human-toxicological effect and damage factors of carcinogenic and noncarcinogenic chemicals for life cycle impact assessment. *Integr. Environ. Assess. Manage.*, **1** (3), 181–244.

Huijbregts, M.A.J., Struijs, J., Goedkoop, M., Heijungs, R., Hendriks, A.J., and van de Meent, D. (2005b) Human population intake fractions and environmental fate factors of toxic pollutants in life cycle impact assessment. *Chemosphere*, **61**, 1495–1504.

Huijbregts, M.A.J., Schöpp, W., Verkuilen, E., Heijungs, R., and Reinders, L. (2000a) Spatially explicit characterisation of acidifying and eutrophying air pollution in life cycle assessment. *J. Ind. Ecol.*, **4** (3), 125–142.

Huijbregts, M.A.J., Thissen, U., Guinée, J.B., Jager, T., Van de Meent, D., Ragas, A.M.J., Wegener Sleeswijk, A., and Reijnders, L. (2000b) Priority assessment of toxic substances in life cycle assessment, I: calculation of toxicity potentials for 181 substances with the nested multi-media fate, exposure and effects model USES-LCA. *Chemosphere*, **41**, 541–573.

Huijbregts, M.A.J. and Seppälä, J. (2001) Life cycle impact assessment of pollutants causing aquatic eutrophication. *Int. J. Life Cycle Assess.*, **6** (6), 339–343.

IFEU (2006) Detzel, A. and Böß, A. *Ökobilanzieller Vergleich von Getränkekartons und PET-Einwegflaschen. Endbericht*, Institut für Energie und Umweltforschung (IFEU) Heidelberg an den Fachverband Kartonverpackungen (FKN) Wiesbaden, Wiesbaden, August 2006.

IPCC (1990) Houghton, J.T., Jenkins, G.J. and Ephraums, J.J. (eds) *Climate Change: The IPCC Scientific Assessment*, Cambridge University Press, Cambridge. 1990 (reprinted 1991 and 1993). ISBN: 0-521-40720-6 (paperback).

IPCC (1992) Houghton, J.T., Callander, B.A. and Varney, S.K. (eds) *Climate Change: The Supplimentary Report to the IPCC Scientific Assessment*, Cambridge University Press, Cambridge (reprinted 1992 and 1993).

IPCC (1995a) *Climate Change 1994: Radiative Forcing of Climate Change and an Evaluation of the IPCC IS92 Emission Scenarios*, Cambridge University Press.

IPCC (1995b) *Climate Change 1995: The IPCC Synthesis*, Cambridge University Press.

IPCC (1995c) *Climate Change 1995: Scientific-Technical Analyses of Impacts, Adaptations, and Mitigation of Climate Change*, Cambridge University Press.

IPCC (1996a) Houghton, J.T., Meira Filho, L.G., Callander, B.A., Harris, N., Kattenberg, A., Maskell, K. (eds) *Climate Change 1995: The Science of Climate Change*, Cambridge University Press.

IPCC (1996b) Bruce, J.P., Lee, H. and Haites, E.F. (eds) *Climate Change 1995: The Economic and Social Dimension of Climate Change*, Cambridge University Press.

IPCC (2003) *Good Practice Guidance for Land Use, Lande-Use Change and Forestry*, Institute for Global Environmental Strategies (IGES) for the Intergovernmental Panel on Climate Change, Kanagawa.

IPCC (2001) Houghton, J.T., Ding, Y., Griggs, D.J., Noguer, M., van der Linden, P.J., Dai, X., Maskell, K. and Johnson,

C.A. (eds) *Climate Change 2001: The Scientific Basis*. Published for IPCC, Cambridge University Press, Cambridge.

IPCC (2006) *2006 IPCC Guidelines for National Greenhouse Gas Inventories*, Institute for Global Environmental Strategies (IGES) for the Intergovernmental Panel on Climate Change, Kanagawa.

IPCC (2007) *Climate Change 2007: The Physical Science Basis. Contribution of Working Group 1 to the Fourth Assessment Report of the Intergovernmental Panel on Climate Change* (eds S. Solomon, D. Qin, M. Manning, Z. Chen, M. Marquis, K.B. Averyt, M. Tignor, and H.L. Miller), Cambridge University Press, Cambridge, New York.

IPCC/TEAP (2007) de Jager, D., Manning, M. and Kuijpers, L. (coord. Lead authors) Special Report. Safeguarding the Ozone Layer and the Global Climate System: Issues Related to Hydrofluorocarbons and Perfluorocarbons. Technical Summary.

ISO (1981) ISO 1000. *SI Units and Recommendations for the Use of their Multiples and of Certain Other Units*, (*Unités SI et recommandations pour l'emploi de leurs multiples et de certaines autres unités*) 2nd edn – 1981-02-15, International Standard Organization.

ISO (2000a) ISO 14042. *Environmental Management – Life cycle Assessment: Life Cycle Impact Assessment*, International Standard Organization.

ISO (2002) ISO/TR 14047. *Environmental Management – Life Cycle Impact Assessment – Examples of Application of ISO 14042*, International Standard Organization.

ISO (2006a) ISO 14040. *International Standard Organization TC 207/SC 5: Environmental Management – Life Cycle Assessment – Principles and Framework*, International Standard Organization.

ISO (2006b) ISO 14044. *International Standard Organization TC 207/SC 5: Environmental Management – Life Cycle Assessment – Requirements and Guidelines*, International Standard Organization.

ISO (2013) ISO/TS 14067. *International Standard Organization TC207/SC 7: Environmental Management – Greenhouse Gas Management and Related Activities: Carbon Footprint of Products – Requirements and Guidelines for Quantification and Communication*, International Standard Organization.

Jolliet, O. et al. (1996) Impact assessment of human and eco-toxicity in life cycle assessment, in *Towards a Methodology for Life Cycle Impact Assessment* (ed. H.A. Udo de Haes), SETAC-Europe, Brussels, pp. 49–61.

Jolliet, O., Margni, M., Humbert, C.R., Payet, J., Rebitzer, G., and Rosenbaum, R. (2003) Impact 2002+: a new life cycle impact assessment methodology. *Int. J. Life Cycle Assess.*, **8** (6), 324–330.

Jolliet, O., Müller-Wenk, R., Bare, J., Brent, A., Goedkoop, M., Heijungs, R., Itsubo, N., Peña, C., Pennington, D., Potting, J., Rebitzer, G., Steward, M., Udo de Haes, H., and Weidema, B. (2004) The LCA midpoint-damage framework of the UNEP/SETAC life cycle initiative. *Int. J. Life Cycle Assess.*, **9** (6), 394–404.

Jørgensen, A. and Hauschild, M. (eds) (2011) LCIA of impacts on human health and ecosystems (USEtox) Special issue of . *Int. J. Life Cycle Assess.*, **16** (8), 697–847.

Klaschka, U. (2008) The infochemical effect – a new chapter in ecotoxicology. *Environ. Sci. Pollut. Res.*, **15** (6), 452–462.

Klaschka, U. (2009) A new challenge – development of test systems for the infochemical effect. *Environ. Sci. Pollut. Res.*, **16** (4), 370–1388.

Klöpffer, W. (1989) Persistenz und Abbaubarkeit in der Beurteilung des Umweltverhaltens anthropogener Chemikalien. *UWSF-Z. Umweltchem. Ökotox.*, **1** (2), 43–51.

Klöpffer, W. (1993) Environmental hazard assessment of chemicals and products. Part I. General assessment principles. *Environ. Sci. Pollut. Res.*, **1** (1), 47–53.

Klöpffer, W. (1994a) Environmental hazard assessment of chemicals and products. Part IV. Life cycle assessment. *Environ. Sci. Pollut. Res.*, **1** (5), 272–279.

Klöpffer, W. (1994b) Review of life-cycle impact assessment, in *Integrating Impact Assessment into LCA* (eds H.A. Udo de Haes, A.A. Jensen, W. Klöpffer, and L.-G. Lindfors) Proceeding of the LCA symposium held at the Fourth SETAC-Europe Annual Meeting, Brussels, April 11–14, 1994, Society of Environmental

Toxicology and Chemistry – Europe, Brussels, pp. 11–15.

Klöpffer, W. (1994c) Environmental hazard assessment of chemicals and products. Part II. Persistence and degradability. *Environ. Sci. Pollut. Res.*, **1** (2), 108–116.

Klöpffer, W. (1995) Exposure and hazard assessment within life cycle impact assessment. *Environ. Sci. Pollut. Res.*, **2** (1), 38–40.

Klöpffer, W. (1996a) Reductionism versus expansionism in LCA. Editorial. *Int. J. Life Cycle Assess.*, **1** (2), 61.

Klöpffer, W. (1996b) *Verhalten und Abbau von Umweltchemikalien: Physikalisch-chemische Grundlagen*, Reihe Angewandter Umweltschutz, ecomed Verlag, Landsberg, Lech. ISBN: 3-609-73210-5.

Klöpffer, W. (1996c) Environmental hazard assessment of chemicals and products. Part VII. A critical survey of exposure data requirements and testing methods. *Chemosphere*, **33**, 1101–1117.

Klöpffer, W. (1997a) Life cycle assessment – from the beginning to the current state. *Environ. Sci. Pollut. Res.*, **4**, 223–228.

Klöpffer, W. (1997b) In defense of the cumulative energy demand. Editorial. *Int. J. Life Cycle Assess.*, **2** (2), 61.

Klöpffer, W. (1998a) Subjective is not arbitrary. Editorial. *Int. J. Life Cycle Assess.*, **3** (2), 61.

Klöpffer, W. (1998b) Book review of Ciambrone 1997 (loc. cit.). *Int. J. Life Cycle Assess.*, **3** (5), 280.

Klöpffer, W. (2001) Kriterien für eine ökologisch nachhaltige Stoff- und Gentechnikpolitik. *UWSF- Z. Umweltchem. Ökotox.*, **13** (3), 159–164.

Klöpffer, W. (2002) *Fachgespräche über Persistenz und Ferntransport von POP-Stoffen*, UBA-Texte 16/02, UBA (Umweltbundesamt Berlin), Berlin, S. 58–61. ISSN: 0722-186X.

Klöpffer, W. (2004) *Physikalisch-chemische Kenngrößen von Stoffen zur Bewertung ihres atmosphärisch-chemischen Verhaltens: Datenqualität und Datenverfügbarkeit. 10. BUA-Kolloquium: Stofftransport und Transformation in der Atmosphäre am 25. November 2003*, GDCh Monographie, Bd. **28**, Gesellschaft Deutscher Chemiker, Frankfurt am Main, S. 133–136. ISBN: 3-936028-22-2.

Klöpffer, W. (2005) The critical review process according to ISO 14040–43: an analysis of the standards and experiences gained in their application. *Int. J. Life Cycle Assess.*, **10**, 98–102.

Klöpffer, W. (2008) Life-cycle based sustainability assessment of products (with Comments by Helias, A., Udo de Haes, p. 95). *Int. J. Life Cycle Assess.*, **13** (2), 89–95.

Klöpffer, W. (2012a) The critical review of life cycle assessment studies according to ISO 14040 and 14044: origin, purpose and practical performance. *Int. J. Life Cycle Assess.*, **17** (9), 1087–1093.

Klöpffer, W. (2012b) *Verhalten und Abbau von Umweltchemikalien: Physikalisch-chemische Grundlagen*, 2., völlig überarbeitete Auflage, Wiley-VCH Verlag GmbH, Weinheim. ISBN: 978-3-527-32673-0.

Klöpffer, W. and Grahl, B. (2009) *Ökobilanz (LCA) – Ein Leitfaden für Ausbildung und Beruf*, Wiley-VCH Verlag GmbH, Weinheim. ISBN: 978-3-527-32043-1.

Klöpffer, W. and Meilinger, S. (2001a) in *Best Available Practice in Life Cycle Assessment of Climate Change, Stratospheric Ozone Depletion, Photo-oxidant Formation, Acidification, and Eutrophication. Backgrounds on Specific Impact Categories* (eds W. Klöpffer, J. Potting, J. Seppälä, J. Risbey, S. Meilinger, G. Norris, L.-G. Lindfors, and M. Goedkoop), Chapter 1, RIVM Report 550015003/2001, RIVM, Bilthoven, NLS. 13–24, http://www.rivm.nl/bibliotheek/rapporten/550015003.html (accessed 17 October 2013).

Klöpffer, W. and Meilinger, S. (2001b) in *Best Available Practice in Life Cycle Assessment of Climate Change, Stratospheric Ozone Depletion, Photo-oxidant Formation, Acidification, and Eutrophication. Backgrounds on Specific Impact Categories* (eds W. Klöpffer, J. Potting, J. Seppälä, J. Risbey, S. Meilinger, G. Norris, L.-G. Lindfors, and M. Goedkoop), Chapter 2, RIVM Report 550015003/2001, RIVM, Bilthoven, NLS. 25–36, http://www.rivm.nl/bibliotheek/rapporten/550015003.html (accessed 17 October 2013).

Klöpffer, W., Potting, J., Seppälä, J., Risbey, J., Meilinger, S., Norris, G., Lindfors, L.-G., and Goedkoop, M. (eds) (2001a) *Best*

Available Practice in Life Cycle Assessment of Climate Change, Stratospheric Ozone Depletion, Photo-oxidant Formation, Acidification, and Eutrophication. Backgrounds on Specific Impact Categories. RIVM Report 550015003/2001, Directorate-General of the National Institute of Public Health and the Environment (RIVM), Bilthoven, NL, http://www.rivm.nl/bibliotheek/rapporten/550015003.html (accessed 17 October 2013).

Klöpffer, W., Potting, J., and Meilinger, S. (2001b) in *Best Available Practice in Life Cycle Assessment of Climate Change, Stratospheric Ozone Depletion, Photo-oxidant Formation, Acidification, and Eutrophication. Backgrounds on Specific Impact Categories* (eds W. Klöpffer, J. Potting, J. Seppälä, J. Risbey, S. Meilinger, G. Norris, L.-G. Lindfors, and M. Goedkoop), Chapter 3, RIVM Report 550015003/2001, RIVM, Bilthoven, NLS. 37–50, http://www.rivm.nl/bibliotheek/rapporten/550015003.html (accessed 17 October 2013).

Klöpffer, W. and Renner, I. (1995) *Methodik der Wirkungsbilanz im Rahmen von Pro¬dukt-Ökobilanzen unter Berücksichtigung nicht oder nur schwer quantifizierbarer Umwelt-Kategorien*, UBA-Texte 23/95, Bericht der C.A.U. GmbH, Dreieich, an das Umweltbundesamt (UBA), Berlin. ISSN: 0722-186X.

Klöpffer, W. and Renner, I. (2007) Lebenszyklusbasierte Nachhaltigkeitsbewertung von Produkten. *Technikfolgenabschätzung – Theorie Prax.*, **16** (3), 32–38.

Klöpffer, W., Renner, I., Tappeser, B., Eckelkamp, C., and Dietrich, R. (1999) *Life Cycle Assessment gentechnisch veränderter Produkte als Basis für eine umfassende Beurteilung möglicher Umweltauswirkungen*, Monographien, Bd. **111**, Umweltbundesamt GmbH/ Federal Environment Agency Ltd., Wien. ISBN: 3-85457-475-4.

Klöpffer, W. and Schmidt, E. (2003) Comparative determination of the persistence of semivolatile organic compounds (SOC) using SimpleBox 2.0 and Chemrange 1.0/2.1. *Fresenius Environ. Bull. (FEB)*, **12** (6), 490–496.

Klöpffer, W. and Volkwein, S. (1995) in *Enzyklopädie der Kreislaufwirtschaft, Management der Kreislaufwirtschaft* (Kapitel 6.4 in K.J. Thomé-Kozmiensky (Hrsg.)), EF-Verlag für Energie- und Umwelttechnik, Berlin, pp. 336–340.

Klöpffer, W. and Wagner, B.O. (2007a) *Atmospheric Degradation of Organic Substances – Data for Persistence and Long-range Transport Potential*, Wiley-VCH Verlag GmbH, Weinheim.

Klöpffer, W. and Wagner, B.O. (2007b) Persistence revisited. Editorial. *Environ. Sci. Pollut. Res.*, **14** (3), 141–142.

Koellner, T. (2000) Species-pool effect potentials (SPEP) as a yardstick to evaluate land-use impacts on biodiversity. *J. Clean. Prod.*, **8**, 293–311.

Koellner, T., de Baan, L., Beck, T., Brandao, M., Civit, B., Margni, M., Milà i Canals, L., Saad, R., Maia de Souza, D., and Müller-Wenk, R. (2013a) UNEP-SETAC guideline on global land use impact assessment on biodiversity and ecosystem services in LCA. *Int. J. Life Cycle Assess.*, **18** (6), 1188–1202.

Koellner, T., de Baan, L., Beck, T., Brandao, M., Goedkoop, M., Margni, M., Milà i Canals, L., Müller-Wenk, R., Weidema, B., and Wittstock, B. (2013b) Principles for life cycle inventories of land use on a global scale. *Int. J. Life Cycle Assess.*, **18** (6), 1203–1215.

Koellner, T. and Geyer, R. (2013) Special Issue: Global land use impacts on biodiversity and ecosystem services in LCA. *Int. J. Life Cycle Assess.*, **18** (6), 1185–1277.

Koellner, T. and Scholz, R.W. (2007) Assessment of land use impacts on the natural environment. Part 1. An analytical framework for pure land accupation and land use change. *Int. J. Life Cycle Assess.*, **12** (1), 16–23.

Koellner, T. and Scholz, R.W. (2008) Assessment of land use impacts on the natural environment. Part 2. Generic characterization factors for local species diversity in central Europe. *Int. J. Life Cycle Assess.*, **13** (1), 32–48.

Koroneos, C.J., Rovas, D.D. and Dompros, A.Th. (eds) (2011) Proceedings of ELCAS 2011, 2nd International Exergy, Life Cycle Assessment and Sustainability Workshop and Symposium. Final Conference of Cost Action C24, Nisyros Island, Greece, June 19–21, 2011.

Korte, F., Bahadir, M., Klein, W., Lay, J.P., and Parlar, H. (1992) *Lehrbuch der Ökologischen Chemie*, 3rd edn, Thieme, Stuttgart.

Kowarik, I. (1999) Natürlichkeit, Naturnähe und Hemerobie als Bewertungskategorien, in *Handbuch Naturschutz und Landschaftspflege* (eds W. Konold, R. Böcker, and U. Hampicke), Ecomed, Landsberg.

Krewitt, W., Trukenmüller, A., Bachmann, T.M., and Heck, T. (2001) Country-specific damage factors for air pollutants: a step towards site dependent life cycle impact assessment. *Int. J. Life Cycle Assess.*, **6** (4), 199–210.

Krishnan, S.S., McLachlan, D.R., Dalton, A.J., Krishnan, B., Fenton, S.S.A., Harrison, J.E., and Kruck, T. (1988) in *Essential and Toxic Trace Elements in Human Health and Disease* (ed. R. Alan), Liss, Inc., pp. 645–659.

Krol, M., van Leeuwen, P.J., and Lelieveld, J. (1998) Global OH trend inferred from methylchloroform measurements. *J. Geophys. Res.*, **103** (D9), 10,697–10,711.

Kummert, R. and Stumm, W. (1989) *Gewässer als Ökosysteme*, B.G.Teubner, Stuttgart.

Kurth, S., Schüler, D., Renner, I., and Klöpffer, W. (2004) *Entwicklung eines Modells zur Berücksichtigung der Risiken durch nicht bestimmungsgemäße Betriebszustände von Industrieanlagen im Rahmen von Ökobilanzen (Vorstudie)*, Forschungsbericht 201 48 309, UBA-FB 000632, UBA Texte 34/04, UBA (Umweltbundesamt Berlin/Dessau), Berlin.

Labouze, E., Honoré, C., Moulay, L., Couffignal, B., and Beekmann, M. (2004) Photochemical ozone creation potentials. A new set of characterization factors for different gas species on the scale of Western Europe. *Int. J. Life Cycle Assess.*, **9** (3), 187–195.

Lafleche, V. and Sacchetto, F. (1997) Noise assessment in LCA – a methodology attempt: a case study with various means of transportation on a set trip. *Int. J. Life Cycle Assess.*, **2** (2), 111–115.

Landsiedel, R. and Saling, P. (2002) Assessment of toxicological risks for life cycle assessment and eco-efficiency analysis. *Int. J. Life Cycle Assess.*, **7** (5), 261–268.

Lane, J. and Lant, P. (2012) Including $N_2O$ in ozone depletion models for LCA. *Int. J. Life Cycle Assess.*, **17** (2), 252–257.

Larsen, H.F. and Hauschild, M. (2007a) Evaluation of ecotoxicity effect indicators for use in LCIA. *Int. J. Life Cycle Assess.*, **12** (1), 24–33.

Larsen, H.F. and Hauschild, M.Z. (2007b) GM-Troph. A low data demand ecotoxicity effect indicator for use in LCIA. *Int. J. Life Cycle Assess.*, **12** (2), 79–91.

Larsen, H.F., Birkved, M., Hauschild, M., Pennington, D.W., and Guinée, J.B. (2004) Evaluation of selection methods for toxicological impacts in LCA. Recommendations for OMNITOX. *Int. J. Life Cycle Assess.*, **9** (5), 307–319.

Lindeijer, E.W. (1998) A framework for LCIA of effects of land use changes and occupation applied in cases. Paper presented and Informal Workshop at the 8th Annual Meeting of SETAC-Europe, Bordeaux, France, April 14–18, 1998.

Lindeijer, E.; Müller-Wenk, R., and Steen, B. (2002) in *Life-Cycle Impact Assessment: Striving Towards Best Practice*, Chapter 2 (eds Udo de Haes, H.A., et al.) SETAC Press, Pensacola, FL, pp. 11–64.

Lindfors, L.-G., Christiansen, K., Hoffmann, L., Virtanen, Y., Juntilla, V., Hanssen, O.-J., Rønning, A., Ekvall, T., and Finnveden, G. (1995) *Nordic Guidelines on Life-Cycle Assessment*, Nord 1995:20, Nordic Council of Ministers, Copenhagen.

Lindfors, L.-G., Christiansen, K., Hoffmann, L., Virtanen, Y., Juntilla, V., Leskinen, A., Hansen, O.-J., Rønning, A., Ekvall, T., Finnveden, G., Weidema, B.P., Ersbøll, A.K., Bomann, B., and Ek, M. (1994) LCA-NORDIC. Technical Reports No 10 and Special Reports No 1–2, Tema Nord 1995:503, Nordic Council of Ministers, Copenhagen. ISBN: 92-9120-609-1.

Lovelock, J.E. (1975) Natural halocarbons in the air and in the sea. *Nature*, **256**, 193–194.

Lovelock, J. (1982) *Gaia: A New Look at Life on Earth*, Paperback edition, Oxford University Press, Oxford.

Lovelock, J. (1990) *The Ages of Gaia: A Biography of Our Living Earth*, Oxford University Press, Oxford, First published 1988, Paperback edition, ISBN: 0-19-286090-9.

Lovelock, J.E., Maggs, R.J., and Wade, R.J. (1973) Halogenated hydrocarbons in and over the Atlantic. *Nature*, 241, 194–196.

Mackay, D. (1991) *Multimedia Environmental Models: The Fugacity Approach*, Lewis Publishing, Boca Raton, FL.

MacLeod, M., Woodfine, D.G., Mackay, D., McKone, T.E., Bennet, D., and Maddalena, R. (2001) BETR North America: a regionally segmented multimedia contaminant fate model for North America. *Environ. Sci. Pollut. Res.*, 8 (3), 156–163.

Mauch, W. and Schäfer, H. (1996) in *Ganzheitliche Bilanzierung. Werkzeug zum Planen und Wirtschaften in Kreisläufen* (ed. P. Eyrer (Hrsg.)), Springer, Berlin, S. 152–180. ISBN 3-540-59356-X.

McCabe, L.C.(Chairman) (1952) *Air Pollution: Proceedings of the United States Technical Conference on Air Pollution*, McGraw-Hill Book Company, New York.

McIntyre, M.E. (1989) On the Antarctic ozone hole. *J. Atmos. Terrest. Phys.*, 51, 29–43.

McKone, T., Bennett, D., and Maddalena, R. (2001) *CalTOX 4.0 Technical Support Document*, vol. 1 LBNL −47254, Lawrence Berkely National Laboratory, Berkely, CA.

Meadows, D.H., Meadows, D.L., Randers, J., and Behrens, W.W. III, (1972) *The Limits to Growth: A Report for the Club of Rome's Project on the Predicament of Mankind*, Universe Books, New York. ISBN: 0-87663-165-0.

Michelsen, O. (2008) Assessment of land use impact on biodiversity. *Int. J. Life Cycle Assess.*, 13 (1), 22–31.

Milà i Canals, L., Bauer, C., Depestele, J., Dubreuil, A., Freiermuth Knuchel, R., Gaillard, G., Michelsen, O., Müller-Wenk, R., and Rydgren, B. (2007a) Key elements in a framework for land use impact assessment within LCA. *Int. J. Life Cycle Assess.*, 12 (1), 5–15.

Milà i Canals, L., Bauer, C., Depestele, J., Dubreuil, A., Freiermuth Knuchel, R., Gaillard, G., Michelsen, O., Müller-Wenk, R., and Rydgren, B. (2007b) Response to the comment by Helias Udo de Haes. *Int. J. Life Cycle Assess.*, 12 (1), 2–4.

Milà i Canals, L., Romanya, J., and Cowell, S. (2007) Method for assessing impacts on life support functions (LSF) related to the use of "fertile land" in life cycle assessment (LCA). *J. Clean. Prod.*, 15, 1426–1440.

Milá i Canals, L., Chenoweth, J., Chapagain, A., Orr, S., Antón, A., and Clift, R. (2009) Assessing freshwater use impacts in LCA: Part I − inventory modelling and characterisation factors for the main impact pathways. *Int. J. Life Cycle Assess.*, 14 (1), 28–42.

Milá i Canals, L., Orr, S., Chenoweth, J., Anton, A., and Clift, R. (2010) Assessing freshwater use impacts in LCA, Part 2: case study of broccoli production in the UK and Spain. *Int. J. Life Cycle Assess.*, 15 (6), 598–607.

Molander, S., Lidholm, P., Schowanek, D., Recasens, M., Fullana, I., Palmer, P., Christensen, F.M., Guinée, J.B., Hauschild, M., Jolliet, O., Carlson, R., Pennington, D.W., and Bachmann, T.M. (2004) OMNITOX − operational life-cycle impact assessment models and information tools for practitioners. *Int. J. Life Cycle Assess.*, 9 (5), 282–288.

Molina, M.J. and Rowland, F.S. (1974) Stratospheric sink for chlorofluoro-methanes: chlorine atom catalyzed destruction of ozone. *Nature*, 249, 810–814.

Möller, D. (2010) *Chemistry of the Climate System*, De Gruyter, Berlin.

Monod, J. (1970) *Le hazard et la nécessité. Essai sur la philosophie naturelle de la biologie moderne*, Édition du Seuil, Paris.

Motoshita, M., Itsubo, N., and Inaba, A. (2008) Development of impact assessment method on health damages of undernourishment related to agricultural water scarcity. Proceedings of the 8th International Conference on EcoBalance, Tokyo, Japan, December 10–12, 2008.

Motoshita, M., Itsubo, N., Inaba, A. and Aoustin, E. (2009) Development of a damage assessment model for infectious deseases arising from domestic water consumption. Proceedings of the SETAC Europe: 19th Annual Meeting, Göteborg, Sweden, 31 May − 4 June, 2009.

Müller-Herold, U. (1996) A simple general limiting law for the overall decay of organic compounds with global pollution potential. *Environ. Sci. Technol.*, 30, 586–591.

Müller-Wenk, R. (1999a) *Life-Cycle Impact Assessment of Road Transport Noise*.

IWOE-Diskussionsbeitrag Nr. 77, Institute for Economy and the Environment, http://www.iwoe.unisg.ch (accessed 17 October 2013).

Müller-Wenk, R. (2002a) in *Life-Cycle Impact Assessment: Striving Towards Best Practice* (eds H.A. Udo de Haes et al), Chapter 2, SETAC Press, Pensacola, FL, pp. 26–39. ISBN: 1-880611-54-6.

Müller-Wenk, R. (2002b) *Attribution to Road Traffic of the Impact of Noise on Health*, Environmental Series, vol. 339, Swiss Agency for Environment, Forest and Landscape, Bern 70 pp.

Müller-Wenk, R. (2004) A method to include in LCA road traffic noise and its health effects. *Int. J. Life Cycle Assess.*, **9** (2), 76–85.

Norris, G. (2001) in *Best Available Practice in Life Cycle Assessment of Climate Change, Stratospheric Ozone Depletion, Photo-oxidant Formation, Acidification, and Eutrophication. Backgrounds on Specific Impact Categories* (eds W. Klöpffer, J. Potting, J. Seppälä, J. Risbey, S. Meilinger, G. Norris, L.-G. Lindfors, and M. Goedkoop), Chapter 4, RIVM Report 550015003/2001, RIVM, Bilthoven, NLS. 51–64, http://www.rivm.nl/bibliotheek/rapporten/550015003.html (accessed 17 October 2013).

Norris, G.A. (2002) Impact characterization in the tool for the reduction and assessment of chemical and other environmental impacts. Methods for acidification, eutrophication, and ozone formation. *J. Ind. Ecol.*, **6** (3/4), 79–78.

Owens, J.W. (1996) LCA impact assessment categories – technical feasibility and accuracy. *Int. J. Life Cycle Assess.*, **1** (3), 151–158.

Owens, J.W. (1997) Life cycle assessment – constraints from moving from inventory to impact assessment. *J. Ind. Ecol.*, **1** (1), 37–49.

Owens, J.W. (1998) Life cycle impact assessment: the use of subjective judgements in classification and characterization. *Int. J. Life Cycle Assess.*, **3** (1), 43–46.

Pant, R., Christensen, F.M., and Pennington, D.W. (2004) Increasing the acceptance and practicality of toxicological effects assessment in LCA. A 5th European research framework programme project. OMNITOX Editorial. *Int. J. Life Cycle Assess.*, **9** (5), 281.

Papke, H.E., Krahl-Urban, B., Peters, K., and Chimansky, C. (1987) *Waldschäden. Ursachenforschung in der Bundesrepublik Deutschland und in den Vereinigten Staaten von Amerika*, 2. Auflage, Projektträgerschaft Biologie, Ökologie und Energie der KFA, Jülich.

Pennington, D.W., Margni, M., Amman, C., and Jolliet, O. (2005) Multimedia fate and human intake modelling: spatial versus nonspatial insights for chemical emissions in Western Europe. *Environ. Sci. Technol.*, **39**, 1119.

Pennington, D.W., Potting, J., Finnveden, G., Lindeijer, E., Jolliet, O., Rydberg, T., and Rebitzer, G. (2004) Life cycle assessment. Part 2: current impact assessment practice. *Environ. Int.*, **30** (5), 721–739.

Peper, H., Rohner, H.-S., and Winkelbrandt, A. (1985) Grundlagen zur Beurteilung der Bedarfsplanung für Bundesfernstraßen aus der Sicht von Naturschutz und Landschaftspflege am Beispiel des Raumes Wörth-Pirmasens. *Natur und Landschaft*, **60**, 397–401.

Pfister, S., Koehler, A., and Hellweg, S. (2009) Assessing the environmental impact of freshwater consumption in LCA. *Environ. Sci. Technol.*, **43**, 4098–4104.

Pfister, S., Saner, D., and Koehler, A. (2011) The environmental relevance of freshwater consumption in global power production. *Int. J. Life Cycle Assess.*, **16**, 580–591.

Popper, K.R. (1934) *Logik der Forschung*, Julius Springer, Wien; *The Logic of Scientific Discovery*, 1st English edn, Hutchison, London, 1959.

Posch, M., Hettelingh, J.-P., and De Smet, P.A.M. (2001) Characterization of critical load exceedances in Europe. *Water Air Soil Pollut.*, **130**, 1139–1144.

Potting, J. (2000) *Spatial Differentiation in Life Cycle Impact Assessment*. Proefschrift Universiteit Utrecht. Printed by, Mostert & Van Onderen, Leiden. ISBN: 90-393-2326-7.

Potting, J. and Blok, K. (1994) in *Integrating Impact Assessment into LCA. Proceeding of the LCA Symposium Held at the 4th SETAC-Europe Annual Meeting, April 11–14, 1994* (eds A.A. Jensen, W. Klöpffer, and L.-G. Lindfors), Published by

Society of Environmental Toxicology and Chemistry – Europe, Brussels, S. 11–15.

Potting, J. and Hauschild, M. (1997a) Predicted environmental impact and expected occurrence of actual environmental impact. Part I: the linear nature of environmental impact from emissions in life cycle assessment. *Int. J. Life Cycle Assess.*, **2** (3), 171–177.

Potting, J. and Hauschild, M. (1997b) Predicted environmental impact and expected occurrence of actual environmental impact. Part II: spatial differentiation in life cycle assessment via the site-dependent characterisation of environmental impact from emissions. *Int. J. Life Cycle Assess.*, **2** (4), 209–216.

Potting, J., Klöpffer, W., Seppälä, J., Norris, G., and Goedkoop, M. (2002) in *Life-Cycle Impact Assessment: Striving Towards Best Practice* (eds H.A. Udo de Haes et al), Chapter 3, SETAC Press, Pensacola, FL, pp. 65–100.

Potting, J., Klöpffer, W., Seppälä, J., Risbey, J., Meilinger, S., Norris, G., Lindfors, L.-G., and Goedkoop, M. (eds) (2001) *Best Available Practice in Life Cycle Assessment of Climate Change, Stratospheric Ozone Depletion, Photo-Oxidant Formation, Acidification, and Eutropiphication, Backgrounds on General Issues* RIVM Report 550015002/2001,, RIVM, Bilthoven, NL, http://www.rivm.nl/bibliotheek/rapporten/550015002.html (accessed 17 October 2013).

Potting, J., Schoepp, W., Blok, K., and Hauschild, M. (1998) Site-dependent life-cycle assessment of acidification. *J. Ind. Ecol.*, **2** (2), 61–85.

Poulsen, P.B., Jensen, A.A., Antonsson, A.-B., Bengtsson, G., Karling, M., Schmidt, A., Brekke, O., Becker, J., and Verschoor, A.H. (2004) *The Working Environment in LCA*, SETAC Press, Pensacola, FL. ISBN: 1-880611-68-6.

Ramanathan, V. (1975) Greenhouse effect due to chlorofluorocarbons: climatic implications. *Science*, **190** (1975), 50–52.

Ravishankara, A.R., Daniel, J.S., and Portmann, R.W. (2009) Nitrous oxide ($N_2O$): the dominant ozone-depleting substance emitted in the 21st century. *Science*, **326**, 123–125.

Reap, J., Roman, F., Duncan, S., and Bras, B. (2008a) A survey of unresolved problems in life cycle assessment. Part 1: goal and scope and inventory analysis. *Int. J. Life Cycle Assess.*, **13** (4), 290–300.

Reap, J., Roman, F., Duncan, S., and Bras, B. (2008b) A survey of unresolved problems in life cycle assessment. Part 2: impact assessment and interpretation. *Int. J. Life Cycle Assess.*, **13** (5), 374–388.

Redfield, A.C. (1934) in *James Johanson Memorial Volume* (ed. R.J. Daniel), University Press of Liverpool, S. 177–192.

Redfield, A.C., Ketchum, B.H., and Richards, F.A. (1993) *Proceedings of the 2nd International Water Pollution Conference, Tokyo, Japan*, Pergamon, S. 215–143.

Reichl, F.-X., Schwenk, M. (Hrsg.) (2004) *Regulatorische Toxikologie*, Springer-Verlag, Berlin. ISBN: 3-540-00985.

Renner, I., Schleyer, R., and Mühlhausen, D. (1990) Gefährdung der Grundwasserqualität durch anthropogene organische Luftverunreinigungen. *VDI Bericht Nr.*, **837** (1990), 705–727.

Ridoutt, B.G. and Pfister, S. (2010) Reducing humanity's water footprint. *Environ. Sci. Technol.*, **44**, 6019–6021.

Ridoutt, B.G. and Pfister, S. (2013) A new water footprint calculation method integrating consumptive and degradative water use into a single stand-alone weighted indicator. *Int. J. Life Cycle Assess.*, **18** (1), 204–207.

Rippen, G., Zietz, E., Frank, R., Knacker, T., and Klöpffer, W. (1987) Do airborne nitrophenols contribute to forest decline? *Environ. Technol. Lett.*, **8**, 475–482.

Rosenbaum, R.K., Bachmann, T.M., Hauschild, M.Z., Huijbregts, M.A.J., Jolliet, O., Juraske, R., Köhler, A., Larsen, H.F., MacLeod, M., Margni, M., McKone, T.E., Payet, J., Schuhmacher, M., and van de Ment, D. (2008) USEtox – The UNEP-SETAC toxicity model: recommended characterisation factors for human toxicity and freshwater ecotoxicity in life cycle impact assessment. *Int. J. Life Cycle Assess.*, **13** (7), 532–546.

Rosenbaum, R.K., Huijbregts, M.A.J., Henderson, A.D., Margni, M., McKone, T.E., van de Meent, D., Hauschild, M.Z., Shaked, S., Sheng Li, D., Gold, L.S., and

Jolliet, O. (2011) USEtox human exposure and toxicity factors for comparative assessment of toxic emissions in life cycle analysis: sensitivity to key chemical properties. *Int. J. Life Cycle Assess.*, **16**, 710–727.

Rowland, F.S. (1994) The scientific basis for policy decisions: a 20 year retrospective on the CFC-stratospheric ozone problem. Preprint of Papers Presented at the 208th ACS National Meeting, Washington, DC, August 21–25, 1994, Vol. 34 (2), p. 731.

Rowland, F.S. and Molina, M.J. (1975) Chlorofluoromethanes in the environment. *Rev. Geophys. Space Phys.*, **13**, 1–35.

RSU (1983) *Der Rat von Sachverständigen für Umweltfragen: Waldschäden und Luftverunreinigungen*. Sondergutachten, W. Kohlhammer, Stuttgart.

Saad, R., Margni, M., Koellner, T., Wittstock, B., and Deschenes, L. (2011) Assessment of land use impacts on soil ecological functions: development of spatially differentiated characterization factors within Canadian context. *Int. J. Life Cycle Assess.*, **16**, 198–211.

Saling, P., Kicherer, A., Dittrich-Krämer, B., Wittlinger, R., Zombik, W., Schmidt, I., Schrott, W., and Schmidt, S. (2002) Eco-efficiency analysis by BASF: the method. *Int. J. Life Cycle Assess.*, **7** (4), 203–218.

Samuelsson, M.-O. (1993) Life Cycle Assessment and Eutrophication: A Concept for Calculation of the Potential Effects of Nitrogen and Phosphorous. IVL Report B1119, Swedish Environmental Research Institute (IVL), Stockholm.

Schenk, R. (2001) Land use and biodiversity indicators for life cycle impact assessment. *Int. J. Life Cycle Assess.*, **6** (2), 114–117.

Scheringer, M. (1999) *Persistenz und Reichweite von Umweltchemikalien*, Wiley-VCH Verlag GmbH, Weinheim.

Scheringer, M. and Hungerbühler, K. (2008) Datenprobleme und Datenbedarf in der Umweltrisikobewertung von Chemikalien. *Mitt. Umweltchem. Ökotox. (GDCh)*, **14** (1), 3–10.

Schmid, P.E. and Schmid-Araya, J.M. (2001) Eine kritische Betrachtung zum Stand der ökologischen Forschung. *UWSF-Z-Umweltchem. Ökotox.*, **13** (5), 255–257.

Schmidt-Bleek, F. (1993) MIPS Re-visited. *Fresenius Envir. Bull.*, **2**, 407–412.

Schmidt-Bleek, F. (1994) Wieviel Umwelt braucht der Mensch?, in *MIPS – Das Maß für ökologisches Wirtschaften*, Birkhäuser Verlag, Berlin.

Schmidt, A., Jensen, A.A., Clausen, A., Kamstrup, O., and Postlethwaite, D. (2004) A comparative life cycle assessment of building insulation products made of stone wool, paper wool and flax. Part 1: background, goal and scope, life cycle inventory, impact assessment and interpretation. *Int. J. LCA*, **9** (1), 53–66.

Schmitz, S., Oels, H.-J., and Tiedemann, A. (1995) *Ökobilanz f. Getränkeverpackungen. Teil A: Methode zur Berechnung und Bewertung von Ökobilanzen für Verpackungen. Teil B: Vergleichende Untersuchung der durch Verpackungssysteme für Frischmilch und Bier hervorgerufenen Umweltbeeinflussungen*, UBA Texte 52/95, Umweltbundesamt, Berlin.

Schmitz, S. and Paulini, I. (1999) *Bewertung in Ökobilanzen. Methode des Umweltbundesamtes zur Normierung von Wirkungsindikatoren, Ordnung (Rangbildung) von Wirkungskategorien und zur Auswertung nach ISO 14042 und 14043*. Version '99, UBA Texte 92/99', Umweltbundesamt, Berlin.

Schwarzenbach, R.P., Gschwend, P.M., and Imboden, D.M. (2002) *Environmental Organic Chemistry*, 2nd edn, John Wiley & Sons, Inc., Hoboken, NJ.

Sedlbauer, K., Braune, A., Humbert, S., Margni, M., Schuller, O., and Fischer, M. (2007) Spatial differentialisation in LCA. Moving forward to more operational sustainability. *Technikfolgenabschätzung – Theorie Prax.*, **16** (3), 24–31.

Seppälä, J. and Hämäläinen, R.P. (2001) On the meaning of the distance-to-target weighting method and normalisation in life cycle impact assessment. *Int. J. Life Cycle Assess.*, **6** (4), 211–218.

Seppälä, J., Knuutila, S., and Silvo, K. (2004) Eutrophication of aquatic ecosystems. A new method for calculating the potential contribution of nitrogen and phosphorous. *Int. J. Life Cycle Assess.*, **9** (2), 90–100.

Seppälä, J., Posch, M., Johansson, M., and Hettelingh, J.-P. (2006) Country-dependent characterization factors for acidification and terrestrial eutrophication based on accumulated exceedance as an impact category indicator. *Int. J. Life Cycle Assess.*, **11** (6), 403–416.

SETAC (1993) *Society of Environmental Toxicology and Chemistry (SETAC): Guidelines for Life-Cycle Assessment: A "Code of Practice"*, 1st edn. From the SETAC Workshop held at Sesimbra, Portugal, 31 March – 3 April 1993, SETAC, Brussels and Pensacola, FL, August 1993.

SETAC (1996) Clift, R. (ed.) Inventory Analysis. Report of the SETAC-Europe Working Group, Brussels (unpublished).

Society of Environmental Toxicology and Chemistry – Europe (Ed.) (1992a) *Life-Cycle Assessment*. Workshop Report, December 2–3, 1991, Leiden, SETAC-Europe, Brussels.

SETAC Europe (ed.) (1992b) Leyden workshop progresses in life cycle assessment. *LCA Newslett.*, **2** (1), 3–4.

Sinden, G. (2009) The contribution of PAS 2050 to the evolution of international greenhouse gas emission standards. *Int. J. Life Cycle Assess.*, **14**, 195–203.

Solberg-Johansen, B., Clift, R., and Jeapes, A. (1997) Irradiating the environment. Radiological impacts in life cycle assessment. *Int. J. Life Cycle Assess.*, **2** (1), 16–19.

Solomon, S. and Albritton, D.L. (1992) Time dependent ozone depletion potentials for short and long-term forecasts. *Nature*, **357**, 33–37.

Sorbe, G. (1998) *Internationale MAK-Werte*, 4. Auflage, ecomed-Verlagsgesellschaft, Landsberg/Leech.

Steen, B. and Ryding, S.-O. (1992) The EPS Enviro-Accounting Method. IVL Report, Swedish Environmental Research Institute, Göteborg

Stephenson, M.E. (1977) An approach to the identification of organic compounds hazardous to the environment and human health. *Ecotox. Environ. Safety*, **1**, 39–48.

Steward, M. and Weidema, B. (2004) A consistent framework for assessing the impacts from resource use. A focus on resource functionality. *Int. J. Life Cycle Assess.*, **10** (4), 240–247.

Stoklasa, J. (1923) *Die Beschädigung der Vegetation durch Rauchgase und Fabriksexhalationen*, Urban & Schwarzenberg, Berlin/Wien.

Struijs, J., van Dijk, A., Slaper, H., van Wijnen, H.J., Velders, G.J.M., Chaplin, G., and Huijbregts, M.A.J. (2010) Spatial- and time-explicit human damage modeling of ozone depleting substances in life cycle impact assessment. *Environ. Sci. Technol.*, **44** (1), 204–209.

Stumm, W. (1977) Die Beeinträchtigung aquatischer Ökosysteme durch die Zivilisation. *Naturwissenschaften*, **64**, 157–165.

Suter, G.W. II, (1993) A critique of ecosystem health concept and indexes. *Environ. Toxicol. Chem.*, **12**, 1533–1539.

Suter, P., Walder, E. (Projektleitung), Frischknecht, R., Hofstetter, P., Knoepfel, I., Dones, R., Zollinger, E. (Ausarbeitung), Attinger, N., Baumann, Th., Doka, G., Dones, R., Frischknecht, R., Gränicher, H.-P., Grasser, Ch., Hofstetter, P., Knoepfel, I., Ménard, M., Müller, H., Vollmer, M. Walder, E., and Zollinger, E. (AutorInnen) (1995) Ökoinventare für Energiesysteme, *Villingen im Auftrag des Bundesamtes für Energiewirtschaft (BEW) und des Nationalen Energie-Forschungs-Fonds NEFF*, 2. Auflage, ETH Zürich und Paul Scherrer Institut.

Szargut, J. (2005) *Exergy Method: Technical and Ecological Applications*, WIT Press, Southampton.

Szargut, J., Morris, D.R., and Steward, F.R. (1988) *Exergy Analysis of Thermal, Chemical, and Metallurgical Processes*, Hemisphere Publishing Corporation, New York.

Tetra Pak (2005) *Getränkemarkt in Zahlen 2005*, Tetra Pak GmbH, Hochheim.

Tetra Pak (2006) *Getränkemarkt in Zahlen 2006*, Tetra Pak GmbH, Hochheim.

TGD (1996) Technical Guidance Document (TGD), basierend auf den Richtlinien und Verordnungen über neue Stoffe und Altstoffe 92/32/EEC, 93/67/EEC, 93/793/EEC und 94/1488/EEC, Brüssel.

TGD (2003) *Technical Guidance Document (TGD) on Risk Assessment of Chemical Substances following European Regulations and Directives*, 2nd edn, European Chemicals Bureau (ECB) JRC-Ispra (VA), Italy, http://echa.europa.eu/documents/10162/16960216/tgd (accessed 18 December 2013).

Thiel, C., Seppelt, R., Müller-Pietralla, W., and Richter, O. (1999) An integrated approach for environmental assessments.

Linking and integrating LCI, environmental fate models and ecological impact assessment using fuzzy expert systems. *Int. J. Life Cycle Assess.*, **4** (3), 151–160.

Toffoletto, L., Bulle, C., Godin, J., Reid, C., and Deschênes, L. (2007) LUCAS – a new LCIA method used for a canadian-specific context. *Int. J. Life Cycle Assess.*, **12** (2), 93–102.

Töpfer, K. (2002) The launch of the UNEP-SETAC life cycle initiative (Prague, April 28 2002). *Int. J. Life Cycle Assess.*, **7** (4), 191.

Trapp, S. and Matthies, M. (1996) *Dynamik von Schadstoffen – Umweltmodellierung mit CemoS. Eine Einführung*, Springer, Berlin.

Trapp, S. and Matthies, M. (1998) *Chemodynamics and Environmental Modeling. An Introduction*, Springer, Berlin.

Tyndall, J. (1873) *Heat Considered as a Mode of Motion*, Appleton & Co., New York, p. 415.

UBA (1998) *Ökobilanz Graphischer Papiere*, Umweltbundesamt Berlin (Hrsg.), Berlin.

UBA (1999) Braunschweig, A.) *Bewertung in Ökobilanzen. Projektbericht und Projektdokumentation (auf Basis der Ergebnisse einer Projektgruppe im Auftrag des Umweltbundesamtes Berlin)*, UFO-Plan Nr. 101 02 165, UBA, Berlin.

UBA (2000) Plinke, E., Schonert, M., Meckel, H., Detzel, A., Giegrich, J., Fehrenbach, H., Ostermayer, A., Schorb, A., Heinisch, J., Luxenhofer, K. and Schmitz, S. *Ökobilanz für Getränkeverpackungen II, Zwischenbericht (Phase 1) zum Forschungsvorhaben FKZ 296 92 504 des Umweltbundesamtes Berlin – Hauptteil*, UBA Texte 37/00, Umweltbundesamt, Berlin. ISSN: 0722-186X.

Udo de Haes, H.A. (ed) (1996) *Towards a Methodology for Life Cycle Impact Assessment*, SETAC-Europe, Brussels. ISBN: 90-5607-005-3.

Udo de Haes, H.A. (2006) How to approach land use in LCIA or, how to avoid the Cinderella effect? Comments on "Key Elements in a Framework for Land Use Impact Assessment Within LCA". *Int. J. Life Cycle Assess.*, **11** (4), 219–222.

Udo de Haes, H.A., Finnveden, G., Goedkoop, M., Hauschild, M., Hertwich, E.G., Hofstetter, P., Jolliet, O., Klöpffer, W., Krewitt, W., Lindeijer, E., Müller-Wenk, R., Olsen, S.I., Pennington, D.W., Potting, J., and Steen, B. (eds) (2002) *Life-Cycle Impact Assessment: Striving Towards Best Practice*, SETAC Press, Pensacola, FL. ISBN: 1-880611-54-6.

Udo de Haes, H.A., Jolliet, O., Finnveden, G., Hauschild, M., Krewitt, W., and Müller Wenk, R. (1999a) Best available practice regarding impact categories and category indicators in life cycle impact assessment. Part 1. *Int. J. Life Cycle Assess.*, **4** (2), 66–74.

Udo de Haes, H.A., Jolliet, O., Finnveden, G., Hauschild, M., Krewitt, W., and Müller-Wenk, R. (1999b) Best available practice regarding impact categories and category indicators in life cycle impact assessment. Part 2. *Int. J. Life Cycle Assess.*, **4** (3), 167–174.

Udo de Haes, H.A. and Wrisberg, N. (eds) (1997) LCANET European network for strategic life-cycle assessment research and development, in *Life Cycle Assessment: State-of-the Art and Research Priorities. LCA Documents*, vol. 1 (eds W. Klöpffer and O. Hutzinger), Ecoinforma Press, Bayreuth. ISBN: 3-928379-53-4.

Ulrich, B. (1984) in *Metalle in der Umwelt – Verteilung, Analytik und biologische Relevanz* (ed. E. Merian (Hrsg.)), Verlag Chemie, Weinheim, S. 163–170.

Umweltbundesamt Berlin (Hrsg.) (1979) *Proceedings der 2. Internationalen Konferenz über Fluorchlorkohlenwasserstoffe in München, Dezember 6–8, 1978*, Mercedes-Druck, Berlin-West.

Umweltundesamt (Hrsg.) (1995) *Ökobilanzen für Getränkeverpackungen*, UBA Texte 52/95, Umweltbundesamt, Berlin.

United Nations – Economic Commission for Europe (1991) *Protocol to the Convention on Long-Range Transboundary Air Pollution Concerning the Control of Emissions of Volatile Organic Compounds or their Transboundary Fluxes*, UNO, Geneva.

USA (1976) The Toxic Substances Control Act (TSCA), PL 94-469 (USA).

Valdivia, S., Ugaya, C.M.L., Sonnemann, G., and Hildenbrand, J. (eds), UNEP/SETAC (2011) *Towards a Life Cycle Sustainability Assessment – Making Informed Choices On Products*, UNEP/SETAC, Paris. ISBN: 978-92-807-3175-0, www.unep.org/publications/contents/

pub'details'search.asp?ID=6236 (accessed 18 December 2013).

Velders, G.J.M. and Madronich, S. (coord. Lead authors) (2007) *IPCC/TEAP Special Report. Safeguarding the Ozone Layer and the Global Climate System: Issues Related to Hydrofluorocarbons and Perfluorocarbons*, Chapter 2, Cambridge University Press, Cambridge, pp. 133–181.

VDI (1997) *Kumulierter Energieaufwand (Cumulative Energy Demand). Begriffe, Definitionen, Berechnungsmethoden*, VDI-Richtlinie VDI 4600, Deutsch und Englisch, Verein Deutscher Ingenieure, VDI-Gesellschaft Energietechnik Richtlinienausschuß Kumulierter Energieaufwand, Düsseldorf.

Volkwein, S., Gihr, R., and Klöpffer, W. (1996) The valuation step within LCA. Part II: A formalized method of prioritization by expert panels. *Int. J. Life Cycle Assess.*, **1** (4), 182–192.

Volkwein, S. and Klöpffer, W. (1996) The valuation step within LCA. Part I: general principles. *Int. J. Life Cycle Assess.*, **1** (1), 36–39.

WBCSD and WRI (2010) *Product Accounting and Reporting Standard. Draft for Stakeholders*, World Business Council for Sustainable Development, World Resources Institute.

WBCSD and WRI (2011) *Guidance for Calculating Scope 3 Emissions. Calculation Guidance for Implementing the GHG Protocol Corporate Value Chain (Scope 3). Accounting and Reporting Standard. DRAFT FOR PUBLIC COMMENT*, World Business Council for Sustainable Development, World Resources Institute, August 2011.

WBGU (2006) *Sondergutachten "Die Zukunft der Meere – zu warm, zu hoch, zu sauer"*, Wissenschaftlicher Beirat der Bundesregierung Globale Umweltveränderungen. ISBN: 3-936191-13-1.

Weckenmann, A. and Schwan, A. (2001) Environmental life cycle assessment with support of fuzzy sets. *Int. J. Life Cycle Assess.*, **6** (1), 13–18.

Wegener Sleeswijk, A. and Heijungs, R. (1996) Modelling fate for LCA. *Int. J. Life Cycle Assess.*, **1** (4), 237–240.

White, P., De Smet, B., Udo de Haes, H.A., and Heijungs, R. (1995) LCA back on track, but is it one track or two? *LCA News- A SETAC-Europe Publ.*, **5** (3), 2–4.

WHO (1996) Murray, C. The Global Burden of Disease, The World Health Organisation, Geneva.

WMO (1994) Scientific Assessment of Ozone Depletion: 1994. WMO Report No 37, World Meteorological Organization (WMO)/United Nations Environment Programme (UNEP). ISBN: 928071449X.

WMO (1999) Scientific Assessment of Ozone Depletion: 1998. Global Ozone Research and Monitoring Project – Report No. 44, World Meteorologic Organization, Geneva.

Wright, M., Allen, D., Clift, R., and Sas, H. (1997) Measuring corporate environmental performance. The ICI environmental burden system. *J. Ind. Ecol.*, **1**, 117–127.

Wuebbles, D.J. (2009) Nitrous oxide: no laughing matter. *Science*, **326**, 56–57.

Würdinger, E., Roth, U., Wegener, A., Peche, R., Rommel, W., Kreibe, S., Nikolakis, A., Rüdenauer, I., Pürschel, C., Ballarin, P., Knebel, T., Borken, J., Detzel, A., Fehrenbach, H., Giegrich, J., Möhler, S., Patyk, A., Reinhardt, G.A., Vogt, R., Mühlberger, D., Wante, J., and BIFA (2002) *Kunststoffe aus nachwachsenden Rohstoffen: Vergleichende Ökobilanz für Loose-fill-Packmittel aus Stärke bzw. Polystyrol*, Projektgemeinschaft BIfA/IFEU/Flo-Pack, Bayerisches Institut für Angewandte Umweltforschung und Technik GmbH, Augsburg.

# 5
# Life Cycle Interpretation, Reporting and Critical Review

Interpretation is the phase of a life cycle assessment (LCA) where conclusions are drawn from the results of the inventory and the impact assessment and recommendations are made according to the objective of the study. Hence it refers to the reasons for the accomplishment of the study.

Care should be taken for a thorough elaboration of issues that are substantially relevant to the study. Boundary conditions and comprehensibility of all conclusions should again be critically assessed. In matters of depth, detail and legibility of the report, compromises have to be found. Purely formal processing of objectives according to ISO 14044 may produce documents of poor legibility that do not have the desired benefit for the reader.

## 5.1
### Development and Rank of the Interpretation Phase

In the 20 years prior to the harmonisation efforts by Society of Environmental Toxicology and Chemistry (SETAC), LCAs or 'proto LCAs'[1] were inventories, sometimes supplemented by rudimentary impact assessment. The first SETAC workshop on LCA (held 18–23 August 1990, at Smugglers Notch, Vermont) had already proposed a mandatory impact assessment and the investigation of possible improvements on the basis of a product-related environmental analysis[2] now called *life cycle assessment* (see Section 1.2). Although LCAs have been used, in one form or another, even before the term was coined, the report of this workshop is in fact the first document that made use of this term for defining the method. At that time, the third component of the 'SETAC triangle'[3] was 'improvement analysis,' which was regarded as a breakthrough. It became the fourth component with the introduction of a first phase (definition of goal and scope) at the SETAC workshop in Sandestin, Florida, 1992. This structure for LCA was preserved in the guideline 'Code of Practice' developed at the SETAC workshop in Sesimbra, Portugal, 1993.[4] Shortly after the workshop of Sesimbra, the ISO standardisation process of LCA

---

1) Klöpffer (2006), and SETAC (2008).
2) SETAC (1991).
3) Model of the structure of an LCA.
4) SETAC (1993).

*Life Cycle Assessment (LCA): A Guide to Best Practice*, First Edition.
Walter Klöpffer and Birgit Grahl.
© 2014 Wiley-VCH Verlag GmbH & Co. KGaA. Published 2014 by Wiley-VCH Verlag GmbH & Co. KGaA.

(1993–2000) was initiated. ISO adapted the four-phase LCA structure with one major exception: the ambitious 'improvement analysis' phase was discarded.[5] As a replacement, a new phase, 'interpretation', was developed and introduced as international standard ISO 14043.[6] This inclusion into the standard turned out to be a meaningful and beneficial phase of LCA for the following reasons:

- Interpretation (fourth phase) is the counterpart of the scientifically similar 'soft' phase 'definition of goal and scope' (first phase). Here, an examination of consistency between results and goal has to be made.
- It has to be examined and documented whether the quality of the data and applied methods is sufficient to support the results.
- Strong regulations apply for the interpretation phase if the LCA is to be used for comparative assertions made publicly available.

Reporting and critical review are, strictly speaking, not part of the interpretation phase, as they report on all four phases and evaluate them. They, however, follow the interpretation phase, so a joint consideration of both makes sense.

- Rules for a critical review are prescribed, which are strictly mandatory for comparative assertions, but otherwise optional.
- Reporting is also regulated as regards confidential data. On the one hand, the reports comply with expectations of the economic operators regarding the confidentiality of data, and on the other hand, the critical reviews, if commissioned, are integrated as a fixed part into the reports.

The issues addressed above will be explained on the basis of the standards ISO 14040:2006 and 14044:2006 and on some scientific papers. Interpretation has, however, been less inspiring to scientists than, for example, life cycle impact assessment (LCIA). This may be due to the fact that substantial aspects of weighting of results on the basis of value choices, defined as transparently as possible, are still part of the impact assessment instead of being included in the interpretation phase (see Sections 4.3.3.2 and 4.3.3.3).

Since implicitly evaluated results are obtained via the retained value choices, which belong to the human and social sciences, an improved delimitation to the phases of 'inventory analysis' and 'impact assessment' primarily based on natural sciences and technology (the 'hard-core' of LCA[7]) would be possible.

This opportunity for a more logical structuring was missed out in the former standard of ISO 14043:2000 and, has not been made use of even following the revision of the LCA standards completed in 2006.[8]

---

5) This is in as much justified as an LCA besides a product optimization can have different applications. A small list was integrated into ISO 14040 (1997).
6) Saur (1997), Lecouls (1999), ISO (2000) and Marsmann (2000).
7) Klöpffer (1998).
8) ISO (2000, 2006a) and Finkbeiner et al. (2006).

## 5.2
## The Phase Interpretation According to ISO

### 5.2.1
### Interpretation in ISO 14040

In Section 5.5 of the framework of ISO 14040,[9] interpretation is described as follows:
> Interpretation is the phase of LCA in which the findings from the inventory analysis and the impact assessment are considered together or, in the case of LCI studies, the findings of the inventory analysis only. The interpretation phase should deliver results that are consistent with the defined goal and scope and which reach conclusions, explain limitations and provide recommendations.

It is to be noted that life cycle inventory studies as well require a final phase, that is, 'interpretation'. In the second paragraph of Section 5.5, the following is recalled:
> ... the LCIA results are based on a relative approach, they indicate potential environmental effects, and they do not predict actual impacts on category endpoints, the exceeding of thresholds or safety margins, or risks.

This includes a warning of over-interpretation of results of the impact assessment. This 'wagging finger' stretches through the entire 14040 framework.

### 5.2.2
### Interpretation in ISO 14044

The steps of interpretation in ISO 14044[10] in Section 4.5 are composed of

1. *identification of the significant issues based on the results of the LCI and LCIA phases of LCA;*
2. *an evaluation that considers completeness, sensitivity and consistency checks;*
3. *conclusions, limitations and recommendations.*

Interrelations of these steps and to LCA phases 1–3 (see Chapters 2–4) are schematically illustrated in Figure 5.1. Applications are out of scope of an LCA and not referenced (Figure 1.4).

According to Figure 5.1, the identification of significant parameters is the direct result of the three preceding phases of an LCA and is interrelated with the evaluation. Besides, information from phase 1 is necessary for the adaptation to the objective of the study. This reflects the insight that no LCA matches another: each depends on the objectives of phase 1 (definition of goal and scope). Therefore, the evaluation can only take place within the coverage of the goal and scope definition, which can, however, be adapted, owing to the iterative approach of the method, if during the study it is realised that the original prerequisites do not meet the objectives (e.g. if important data cannot be procured).

---
9) ISO (2006a).
10) ISO (2006b).

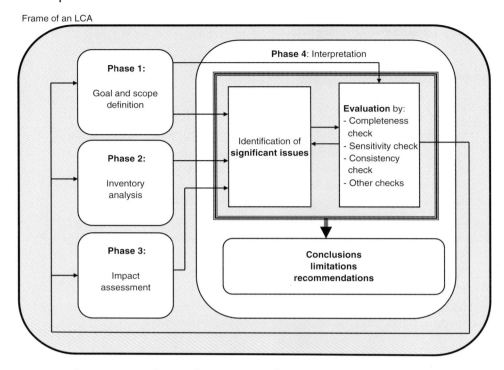

**Figure 5.1** Interrelations of constituents in the interpretation and other phases of an LCA (according to ISO 14040).

Owing to the iterative approach of LCAs, the evaluation requires some experience. This is taken into account in Supplement B of ISO 14044 'Examples for an Interpretation'. It is meant as a support for practitioners to understand how to conduct an interpretation.

## 5.2.3
### Identification of Significant Issues

The standard does not provide prerequisites for materiality thresholds for the identification of significant parameters. Every study therefore needs to articulate, depending on data quality, the significance criteria that are valid. The identification of significant issues is aimed at identifying result parameters for which a significant quantitative difference, involving data uncertainties, really exists. A careful identification of significant parameters is aimed at preventing over- and misinterpretations.

Significant issues can have a varying background. ISO 14044:2006 provides examples of the following in Section 4.5.2.2:

- Inventory data
- Impact categories

- Contributions of life cycle sections, for example, single unit processes or groups of unit processes (e.g. transports or energy production).

The deduction of significance according to the same standard (Section 4.5.2.3) is carried out with the help of the following four types of information:

- The results of the already finished phases of life cycle inventory and LCIA;
- Elements of the methodological approach, for example, allocation rules and system boundary in the inventory, impact categories and models (characterisation factors) of the impact assessment;
- Value choices applied in the study on the basis of the goal of the study and applied in the optional steps of grouping/weighting and normalisation in the phase impact assessment;
- Role and responsibility of interested parties and results of the critical review.[11]

Now follows a determination of consistency of results not explicitly described by the standards, but noted in ISO 14044, Section 4.5.2.3:

> When the results from preceding phases (LCI, LCIA) have been found to meet the demands of the goal and scope of the study, the significance of these results shall be determined.

### 5.2.4
**Evaluation**

Evaluation[12] according to the standards aims to enforce trust into the reliability of results of an LCA and into significant parameters. A further intent is to provide a clear and comprehensible overview of the results of the study.

The use of the following three techniques shall be considered:

- *Completeness check*
- *Sensitivity check*
- *Consistency check*.

The *completeness check* relates to all relevant information especially to the provision of 'significant parameters' (Section 5.2.3). In case of gaps, the inventory or the impact assessment should be repeated with optimised data in the sense of an iterative approach. As an alternative, the goal and scope can be adapted to the information provided. This implies that in most cases expectations need to be lowered in order to provide consistency.

The *sensitivity check* is probably the most frequently applied quantitative technique for evaluation. According to ISO 14044, it is mandatory if a choice between several allocation rules is possible.

---

11) As the critical review includes the interpretation phase, only intermediate results of an interactive review are possible.
12) Evaluation should not be confused with valuation. The former is orientated towards critical scientific quality standards.

The goal of the sensitivity check is to estimate the uncertainties in the results of an LCA due to data quality, cut-off criteria, choice of allocation rules and selection of impact categories. Mostly the scenarios that are investigated are those that differ in the modelling of the product system, which is the main scenario, with respect to one single parameter of investigation. The allocation rule may, for instance, be altered. It would then be examined whether this would imply profound changes in the results.

A sensitivity check allows in a descriptive way to determine and document the influence of the altered parameter on the final result. The following are the possible results of sensitivity analyses:

- The altered parameter does not modify or insignificantly modifies the results.
- Further detailed sensitivity analyses are required.
- The results are only valid within margins, which needs to be considered within the conclusions.

The *consistency check* provides reference to the first phase of LCA (goal and scope definition). ISO 14044 states (see Section 4.5.3.4),

> *The objective of the consistency check is to determine whether the assumptions, methods and data are consistent with the goal and scope.*

Besides the already addressed consistency within a product system, an examination is particularly necessary as to whether for comparative LCAs on different product systems, the following issues are identical or at least similar:

- Data quality
- Regional and time-related validity of data
- Allocation rules and system boundaries
- Constituents of the impact assessment.

In the following section, the techniques necessary for an evaluation are discussed on the basis of the scientific literature.

## 5.3
## Techniques for Result Analysis

### 5.3.1
### Scientific Background

LCA is often regarded as a support for decision-making within product comparison and optimisation[13] applied in the context of the 'environmental pillar' of sustainability (see Chapter 6). ISO 14040 lists the following direct applications of an LCA, which are, however, not part of the standard:

---

13) Grahl and Schmincke (1996), Hofstetter (1998), Seppälä (1999), Tukker (2000), Heijungs (2001), Hertwich and Hammit (2001), Werner and Scholz (2002), Hertwich (2005) and Heijungs *et al.* (2005).

- *Product development and improvement*
- *Strategic planning*
- *Public policy making*
- *Marketing.*

All of these applications may steer decisions even if those exclusively depend on results of an LCA only in rare cases. It is therefore advantageous to formulate results as quantitatively as possible and to indicate uncertainties of data and techniques. This problem was already observed by SETAC and several authors at the time when the LCA standardisation was still in process.[14]

Following the first round of standardisations (ISO 14040 to ISO 14043: 1997–2000), the quality-oriented method development became, under the slogan 'uncertainty', an established working field.[15] It was realised that quality management goes beyond the obvious problem of data quality and also concerns methodological objectives such as

- selection of system boundaries;
- allocation rules, impact categories and indicators, including weighting factors if present;
- the underlying framework of values.

Data as such are best suited for a mathematical examination, whereas the influence of assumptions is best evaluated by alternative scenarios in the form of sensitivity analyses.

### 5.3.2
**Mathematical Methods**

In view of the specific difficulties in the procurement of highly suited inventory data or of their adaptation to a specific case, classical error calculations after Gauss (± standard deviation) are seldom applied.

Heijungs and co-workers[16] discern five numerical types of analysis suited for data analysis in the interpretation phase (partly already in the inventory analysis!):

- *Contribution analysis*
- *Perturbation analysis*
- *Uncertainty analysis*

---

14) Fava et al. (1994), Chevalier and Le Téno (1996), Kennedy, Montgomery and Quay (1996), Kennedy et al. (1997), Coulon et al. (1997), Le Téno (1999) and Hildenbrand (1999).
15) Braam et al. (2001), Huijbregts et al. (2001, 2003), Huijbregts, Heijungs and Hellweg (2004), Heijungs and Kleijn (2001), (Marsmann, 1997), Ross et al. (2002), Ciroth (2001, 2006), Ciroth, Fleischer and Steinbach (2004), Heijungs et al. (2005), Heijungs and Frischknecht (2005) and Lloyd and Ries (2007).
16) Heijungs and Kleijn (2001) and Heijungs et al. (2005).

- Comparative analysis
- Discernibility analysis.

*Contribution analysis* generally serves the quantitative determination of parts to the total result. For LCAs this may imply determining the contribution of one or some life cycle sections to an impact category (indicator result). This approach is often called the *sectoral analysis* and can be adequately plotted as a coloured bar diagram (see Figure 4.4). Dominance analysis can be regarded as specific form of contribution analysis in such that it determines the dominant contributions to the result and therefore is suited for the identification of significant parameters. For sectoral analysis, however, it has to be considered that, in addition to the selected sectors, upstream processes (partial inventories related to raw material production) are often determined as well. A dominance of the 'production' sector must therefore by no means indicate that most emissions occur at one single production site!

Heijungs and Kleijn (2001, loc. cit.) point out that contribution analysis is suitable for a variety of objectives, both on the level of inventory and of impact assessment (with or without normalisation also followed by weighting, if required). It is possible to determine the impact of emissions or of individual unit processes, and so on, on intermediate or final results. Questions of utmost importance can be clarified during this process, for example, where improvements of product systems can be accomplished with highest efficiency. These objectives are particularly important for non-comparative LCAs that predominantly serve product optimisation. Graphical representations should be handled with prudence: pie charts, especially, are not suitable for negative entries resulting from credit entries; for bar diagrams, the latter can be represented by a zero baseline.

*Perturbation analysis* is related to sensitivity analysis, but mathematically more exactly defined. The impact of marginal[17] (very small) alterations of LCA input parameters is quantitatively examined (marginal analysis). It can thus be determined on how sensitively the calculated results react to uncertainties of input parameters. Hence it can be deduced which input parameters require great precision and which are suitable for estimated values, because their influences on the final result are small. The advantages of perturbation analysis are that existing uncertainties need not be known and that further data material is not necessary; it can be conducted with every LCA. The disadvantages can be the time demand, especially for complex systems and correspondingly high computation times (Heijungs and Kleijn, 2001, loc. cit.).

The results of perturbation analysis are extremely valuable for system improvements as those inputs can be identified whose reduction would imply large reductions of environmental loads.

*Uncertainty analysis* is a systematic analysis of the propagation of input uncertainties to output uncertainties, at which preferably Monte Carlo simulations are applied.[18] If values of input parameters obey a certain probability distribution (e.g. Gauss) a large number of arbitrarily chosen simulations (hence the name

---

17) Heijungs (1994).
18) Heijungs and Kleijn (2001), Morgan and Henrion (1990) and Huijbregts (1998).

'Monte Carlo') result in a just as large number of results (mostly per aggregation as impact indicator result). For a sufficiently high number of simulations (e.g. 1000) a result can be represented as a probability distribution that in turn, in simple cases, can be characterised by average values and standard deviations. Then a numerical result like '120 kg $CO_2$ equivalents' can be provided, for example, as '120 ± 10 kg $CO_2$ equivalents'.

For reasons of applicability, only inputs as normal distributions (Gauss) with average value and standard deviation are possible along with even distributions with highest or lowest value: the triangular distribution and logarithmic normal distribution for individual parameters have been proposed (Huijbregts, 1998, loc. cit.). Sometimes discrete single values with defined probabilities are also used as input. A graphical representation is best suited for an immediate illustration of the output. Tables of all average values and deviations tend to be confusing. Limitations to the methodology concern the availability or non-availability of statistical figures for the input data (which are often mere estimations) and computing times for up to 10 000 iterations.

*Comparative analysis* covers the systematic simultaneous listing of product systems alternatives. Since the most common application of LCA is targeted towards a comparison of product systems, this method is particularly important. Inventory and indicator value results often only differ within small margins and one has to resist the temptation to stress minor advantages of one system in relation to the other. This would lead to an over-interpretation of results. A comparative analysis can be made at all levels of results: inventory, characterisation (indicator result), normalisation and weighting, if required. For this purpose, either absolute values or presentations on a percentage basis with the highest value corresponding to 100% can be used. By routine, both approaches are already being used in the impact assessment (see Figure 4.5). In the interpretation phase, a critical analysis should be made as to whether the first impression complies with the data analysis.

*Discernibility analysis* is of special relevance in comparative LCAs as well. Especially in the case where several product systems need to be compared, a ranking is targeted. Unfortunately, however, the results are, in the rarest cases, unambiguously in favour of one of the analysed systems as proponents of diverse single-point methods (aggregation of all results into one figure) would like them to be. A qualitative statement like 'product A is significantly better concerning the consumption of fossil resources compared to product B', statistically characterised with the (frequently used) significance benchmark of 0.05 should read, 'With 95% probability product A in this concern is superior to B' (Heijungs and Kleijn, 2001, loc. cit.). Discernibility analysis tends to combine comparative and uncertainty analyses. The most important tool is therefore the Monte Carlo simulation for as many results of compared LCAs as possible. This implies the product system to be ecologically 'favourable' with more parameters above a significance threshold or below the threshold for damage indicators. A quantitative method has been provided by Huijbregts.[19] For real LCAs, this method, which also neglects figure margins and only refers to 'bigger or smaller', is rarely applied.

The methods described here were supplemented by two others (*key issue analysis* and *structural analysis*) and were tested on the ecoinvent'96 data record.[20] They have also been implemented in the educational software 'Chain Management by Life Cycle Assessment (CMLCA)'.[21]

The issue of data format should rather be part of the inventory (see Section 3.4) but must be considered in the interpretation. Data formats are often provided by data bases and software. For newer applications, they include specifications on statistical distributions of input data.[22] The quality of statistical specifications must, however, always be critically reflected; standard deviations, for instance, can be approximated by semi-quantitative procedures.

### 5.3.3
### Non-numerical Methods

Mathematical methods cannot solve problems that result from value choices: however, increasing the significance of the greenhouse effect or of land use can have a higher impact on conclusions and recommendations than an increase of the significance threshold from 0.05 to 0.2. Therefore non-numerical methods are a fixed part of the interpretation.

Verbal-argumentative interpretation of quantitative and semi-quantitative results are mainly part of non-numerical methods. Thus, in spite of numerical auxiliary tools, it is often very difficult to provide numerical limits which, if exceeded, imply an insure distinction of results for statements like 'A is better or just as good as B'. Here it can be verbally referred to the context of earlier experiences or to issues within the same study. It is the responsibility of the critical review to question these statements, which are particularly suited for a 'whitewashing' of results. It should, however, not be ignored that seemingly objective figures can be used for swindling as well; this is just not as evident in times such as ours, bent on figures and numbers.

### 5.4
### Reporting

The LCA framework ISO 14040[23] states as follows in Chapter 6:

> *A reporting strategy is an integral part of an LCA. An effective report should address the different phases of the study. ... Results and conclusions of the LCA*

---

19) Huijbregts (1998) and Heijungs and Kleijn (2001).
20) Heijung and Suh (2002) and Heijungs et al. (2005).
21) Chain Management by Life Cycle Assessment (CMLCA); http://www.leidenuniv.nl/cml/ssp/software/cmlca.
22) Heijungs and Frischknecht (2005).
23) ISO (2006a).

## 5.4 Reporting

*in an adequate form to the intended audience, addressing the data, methods and assumptions applied in the study, and the limitations thereof.*

The Standard ISO 14044[24] provides detailed regulations concerning reporting. Sections 5.2 (*Additional requirements and guidance for third-party reports*) and 5.3 (*Further reporting requirements for comparative assertion intended to be disclosed to the public*) should especially be considered. 'Third parties' are interested parties '*besides the commissioner and practitioner of the study*'.[25] The report to third parties can be based on a documentation of the study, which may contain confidential data and as such cannot be made accessible. This is an important task of the critical review (see Section 5.5), which must state that non-published data that are accessible to the reviewers are appropriate for the study.

The requirement on reports according to the above Sections 5.2 and 5.3 of ISO 14044 are detailed, four and a half pages within the bilingual version, and taken literally would expand the report of even small LCAs to hundreds of pages. There is thus a conflict between an intention to avoid fraud in consequence of badly elaborated LCAs and legibility. We recommend to practitioners and reviewers of a third-party report (Section 5.2), even more if comparative assertions are included, to carefully read the respective sections of the standard and to adapt them as best as possible. However, the report needs to be legible and inspiring for the target group, 'exciting' to some extent, as for every comparative LCA there is always a concern on 'who wins' even though restrictions have to be addressed. In case of over-simplifications, the standard provides a means for reviewers and, later, targeted parties to demand a higher transparency.

If the report is not regarded as document[26] but as a scientific-technical publication, the principles of scientific perception of knowledge as well as publication conventions apply in addition to the rules defined by the standard.[27] The most important rule by Karl Popper[28] states that hypotheses and theories have to be formulated in such a way as to be eligible for falsification (thus refuted). It is only by multiple futile falsification efforts that a hypothesis or a theory can mature into a natural law, as long as it is not replaced by a more comprehensive one.[29] Even if the requirements of natural science cannot be met in all detail in LCAs, we should always thrive for the best possible adaptation to the ideal (Klöpffer, 2007, loc. cit.).

For an illustration of the report, an attractive graphical representation is adequate, for example, coloured bar diagrams for sectoral analysis, but an obtrusive marketing-orientated version should be avoided. To make the study accessible to further interested parties, a publication at the website of the commissioner or the practitioner is a good solution. The consent of critical reviewers to a legible

---

24) ISO (2006b).
25) Commissioner, practitioner or interested party generally are not individuals but enterprises, associations, social groups and so on.
26) A negative example concerning legibility is the report on good laboratory practice (GLP).
27) Klöpffer (2007).
28) Popper (1934).
29) ISO (2006a,b). Popper's theories have been strongly influenced by Einstein's 'falsification' of Newton's theory, which had been regarded as axiomatic.

short version is important. Within the grey zone to marketing, the temptation of whitewashing is eminently high!

The publication of a short version of the report in a scientific journal is always recommended if new insights into the methodology or concerning the applicability of the LCA under consideration to less investigated product systems have been obtained. New inventory data are also of high interest, but they are often not communicated for reasons of confidentiality.[30]

## 5.5
## Critical Review

A critical review, originally called a *peer review*[31], had already been proposed prior to ISO standardisation by SETAC in the 'Code of Practice' (1993). By an 'interactive' accompanying review, two objectives should be achieved:

- Improvement in technical and scientific quality;
- Increase in reliability.

This requirement has been taken up by ISO, refined and alleviated as follows: the review can now also be made a posteriori.[32] This modification accounts for the fact that an LCA may originally be meant for internal use, at which the critical review is optional, but later on, following revision if necessary, a publication may be intended. In this case, a renewed review can be made interactively during the update and improvement process (if such work is done); for the original study, however, it occurred a posteriori.

In the current version of the ISO standard,[33] two types of critical review are provided:

1. CR by internal and external experts (ISO 14044 6.2 and ISO 14040 7.3.2);
2. CR by a panel of interested parties (ISO 14044 6.3 and ISO 14040 7.3.3).

Variant 1 is suited for internal studies but not approved for studies with comparative assertions to be made available to the public. In this case, a critical review according to variant 2 has to be accomplished.

It is imperative in both cases that the reviewers are *independent*, which is not self-evident for internal experts. In the frequent case of large companies performing LCAs without external help, for example, expert colleagues of quality management, of work safety, environmental departments or other areas of the enterprise not involved in the LCA to be reviewed can be assigned as critical reviewer(s). The internal critical reviewers have to meet the same requirements as external ones.

---

30) Frischknecht (2004).
31) SETAC (1993).
32) Klöpffer (1997, 2000, 2005, 2012).
33) ISO 14040:1997 provides three types of critical review: by an internal expert, an external expert and by 'interested parties' (panel method comprising two surveyors at least).

The task of the reviewers is unambiguously described in ISO 14044/6.1:

*The critical review process shall ensure that:*

- *the methods used to carry out the LCA are consistent with this International Standard,*
- *the methods used to carry out the LCA are scientifically and technically valid,*
- *the data used are appropriate and reasonable in relation to the goal of the study,*
- *the interpretations reflect the limitations identified and the goal of the study,*
- *the study report is transparent and consistent.*

These criteria apply equally to a critical review by either internal or external experts or by a panel of interested parties. This requires a precise inspection of the LCA by the reviewers and emerging queries shall be discussed with the practitioner of the study. At least one face-to-face meeting is desirable and additional telephone conferences are recommended. The reviewers should also stay in contact with the commissioner. The 'triangle' formed by the representative of the commissioner, the practitioner (leader of the project team) and the expert (chair of the panel in case of a critical review by interested parties) should work properly.[34]

If a critical review is accomplished by a panel of interested parties (variant 2), further contacts between the panel members and the practitioner of the study are not excluded but even desired. Thus experts for a central technology (or application, etc. depending on G&S of the study) dealt within the LCA study, who are not necessarily LCA experts, can be members of the panel and may have questions for the practitioner of the study. Mutual communication and the open exchange of information between the members of the review panel is of course a prerequisite.

A critical review by a panel of interested parties is mandatory if comparative assertions (notably for competing product systems on the market) are included in the study (ISO 14044, 4.2.3.7):

*If the study is intended to be used for a comparative assertion intended to be disclosed to the public, interested parties shall conduct this evaluation as a critical review.*

This paragraph could give the impression that every critical review according to ISO 14044 6.3 and ISO 14040 7.3.3 shall involve interested parties beyond scientists and experts. According to ISO 14040, 7.3.3, these interested parties are, however, optional, probably for financial reasons or out of the awareness that competitors in such a panel would hardly get confidential data from the commissioner. Here competent representatives of professional associations could fill in and they often do this skilfully. ISO 14040 Section 7.3.3 reads as follows:

*An external independent expert should be selected by the original study commissioner to act as chairperson of a review panel of at least three members. Based on*

---

34) Klöpffer (2005, 2012).

*the goal, scope and budget available for the review, the chairperson should select other independent qualified reviewers. This panel may also include other interested parties affected by the conclusions drawn from the LCA, such as government agencies, non-governmental groups, competitors and affected industries.*

Interested parties addressed in the heading of Section 7.3.3 of ISO 14040 ('Critical review by a panel of interested parties') are therefore only optionally invited, a contradiction that has been adopted from the former standard ISO 14040:1997. Even larger is the responsibility of the critical reviewers who have to co-advocate the interests of absent interested parties, for example, environmental organisations or competitors, ensuring a fair conduct and interpretation of the LCA study. Thereby they also act in the interest of the practitioner of the study who may frequently also work (in other studies) for the 'competitors' of the commissioner. Credibility is therefore an asset of great value for all parties involved that has to be treated most carefully. For projects that last for a couple of years, establishing an advisory board can be recommended where 'interested parties' are represented and can participate in the elaboration of the LCA. This allows the review panel to concentrate on professional aspects.

Occasionally, uncertainties may be related to the definition of comparative assertions. In the scientific literature, comparative LCAs that are not critically reviewed according to ISO 14040/44 are published now and then. This is tolerated if comparisons are only conducted for methodological development or similar reasons without a commercial background. The study to be published is then peer reviewed according to the rules of the journal (Klöpffer, 2007, loc. cit.). This peer review is accomplished voluntarily by specialists or peers in an honorary capacity. Unfortunately, the scientific literature on critical reviews is not very abundant.[35]

It should finally be noted that the report of the critical review is part of the final report of an LCA either as an annex or as a separate chapter. It should also be quoted in the executive summary. Both the practitioner of the study and the commissioner are entitled to comment on the critical review in written form, and these comments are also included into the final report.

### 5.5.1
**Outlook**

The critical review is an innovative and very useful instrument that improves the quality of LCA studies. The formulation of the quoted sections in ISO 14040:2006 and ISO14044:2006 is clear with regard to aim, but there some formulations and procedural aspects that need a better explanation. This should be achieved in a new 'technical specification (TS)' ISO 14071 to extend ISO 14044.[36] The TS will give advice on how critical reviews should be properly performed and the requirements

---

35) Klöpffer (1997, 2000, 2005, 2006, 2007, 2012), Klöpffer, Grießhammer and Sundström (1995, 1996) and Fava and Pomper (1997).
36) ISO (2013).

that the reviewers and panel chairs should have. Clearly, the scientific and technical background should be appropriate for the LCA study to be reviewed, but also the expertise needed by the LCA as a method has to be taken into account.

According to the last working draft available, self-declarations will be requested from the panel members. The spirit of the free critical reviewer groups (no accreditation, individual invitation), so successful in the past, will be preserved. It should be noted that a 'verification', which is often needed outside ISO 14040ff (e.g. in ISO 14025), does not form part of the duties of critical reviewers.[37] On the other hand, however, ISO 14071 will support the use of ISO 14040 + 44 in other standards proposing a quality assurance for the base LCAs that may be used in these other life cycle methods as a solid basis.

## 5.6
## Illustration of the Component Interpretation Using an Example of Practice

As has been explained in Sections 5.2 and 5.3, data from the inventory analysis and the impact assessment are analysed according to defined rules, restrictions are precisely described, conclusions and recommendations are made. Just as the example study[38] in the preceding chapters merely served as conceptual illustration, in this chapter, the interpretation is not entirely reproduced. Sample excerpts of the text should rather clarify on how elements of the interpretation described in Sections 5.2 and 5.3 may be applied in praxis. Chapters of the quoted text refer to the original study.

### 5.6.1
### Comparison Based on Impact Indicator Results

The results of mandatory parts of the impact assessment provide the data basis for this comparison. These data are therefore neither normalised nor weighted.

> **Comparison of beverage carton and PET bottle (juice, storage)**
>
> A system comparison of the beverage carton and Polyethylene terephthalate (PET) bottle with reference to net results is included in Table 5.1. Illustrations and table indicate significantly smaller indicator values within six of eight regarded categories for the investigated beverage carton compared to the PET bottle.

---

37) Grahl and Schmincke (2011).
38) IFEU (2006).

**Table 5.1** Comparison of beverage carton with closure and PET bottle for the market segment juice/nectar storage with 1000 ml filling volume.

| Indicator | Cardboard container (1 l) versus PET (1 l) (%) |
|---|---|
| Greenhouse effect | −167 |
| Fossil resource consumptions | −164 |
| Summer smog (POPC) | 42 |
| Acidification | −23 |
| Terrestrial eutrophication | −37 |
| Aquatic eutrophication | −26 |
| Land use – forest area | 999 |
| Cumulative energy demand (CED) total | −48 |

Relative system differences related to the respective smaller result (computational differences without definition of a significance threshold).
*Negative values*: indicator value of beverage carton is smaller than the one of the PET bottle.
*Positive values*: indicator value of beverage carton is larger than the one of the PET bottle.

### 5.6.2
### Comparison Based on Normalisation Results

The normalised results for the product system 1 l-beverage carton with closure and for the 1 l-PET bottle in the market segment juice/nectar have already illustrated the usefulness of normalisation (see Figure 4.5).

This comparison is also discussed in the interpretation of the example study.

### 5.6.3
### Sectoral Analysis

Sectoral analysis on the level of the impact assessment has already been provided in Section 4.6.3 by the example of the 'greenhouse effect' to illustrate its usefulness (see Figure 4.4). It is formally part of the interpretation. In the example study, a sectoral analysis has been conducted and thoroughly discussed for all considered impact categories. The following text of the example clarifies the importance of finding consistent explanations for the environmental loads within the individual sectors.

> The carton system (1 l with closure in the market segment fruit juice/nectar) is dominated in all indicators by the production of the packing materials aluminium foil, polyethylene and beverage raw carton. In sum, these sectors contribute to approximately 50–70% of the system burdens. Particularly high single contributions due to the plastics production occur for summer smog

and fossil resources consumption (>30%) as well as for aluminium foil with acidification (30%) and summer smog (34%). The large relevance of the aluminium sector is the more remarkable, as the mass contribution of this material to the primary packing amounts to only 5%.

As expected the categories of medium priority (aquatic eutrophication and land use - forest area) are dominated by the production of raw carton (~80%). However, this sector is also important for acidification and terrestrial eutrophication (>20%). This is particularly due to the production of the process energy required. While the raw carton portion amounts to only 12% of fossil resources consumption, this sector contributes 41% to the entire cumulative energy demand (CED). Such difference of relevance in these two categories cannot be observed for the other sectors. It results from the fact that the energy content of the wood used for the raw carton production is accounted for in the CED.

With a contribution of 4–7% within the priority categories, the actual compound production is of minor importance. With a maximum of 5% the relevance of the sector 'filling' is even less.

The production of the closure contributes 17% of fossil resources consumption; for other categories, contributions are between 7 and 8%. The supply of secondary and tertiary packaging shows contributions of around 10%, primarily due to the corrugated cardboard production. For aquatic eutrophication and land use - forest area, the contribution of this sector amounts to around 20%. The distribution is only relevant for the environmental category 'terrestrial eutrophication' (8%).

Contributions of the sector 'disposal and recycling' vary strongly. It is of particular importance for terrestrial eutrophication (14%) and greenhouse effect (23%), as this sector accounts for emissions from the incineration of the packing components.

The relative importance of credit entries (secondary material and energy) is between 8% (summer smog) and 35% (greenhouse effect), and there is a predominant portion of credit entries for substituted energy.

For the PET bottle system (11 in the market segment fruit juice/nectar), the sectoral analysis is likewise interpreted in detail.

In all categories, the results of the PET multilayer bottles regarded are clearly dominated by the production of bottle material (bottle consists of 5% PA, polyamide, 95% PET). For the priority indicators, contributions of the respective sector are between 47% (greenhouse effect) and 79% (fossil resources consumption). Remarkably is the high contribution of this sector in the category aquatic eutrophication (90%). By detailed analysis, it was determined that more than half of this environmental impact is caused by the production of PA granulates despite its minor mass fraction.

A significant environmental load of 3–14% in the priority categories is due to the energy-intensive production of pre-forms and PET bottles.

As for the beverage carton, the sectors filling and distribution are of minor importance (in each case, below 5%). The closure production has a larger contribution in the category summer smog only (18%). The same is valid for the production of secondary and tertiary packing with 14% for summer smog. The production of paper labels contributes very little (~2%).

The relevance of the sector 'disposal and recycling' varies strongly by category. The contribution is lowest (2%) for fossil resource consumption and highest for the greenhouse effect (29%).

Credit entries are quite important. In the priority categories, the amount of the credit entry corresponds to 19–26% of the system burdens. Secondary materials and substituted energy contribute more or less evenly.

## 5.6.4
### Completeness, Consistency and Data Quality

In the inventory analysis, data bases and data quality are thoroughly described for every data record. In the interpretation, all included inventory data, computations for the impact assessment as well as methodological aspects are again analysed and commented on.

Relevant information and data for the interpretation of the packaging systems examined in this study were present. According to estimates of the practitioners, result-relevant data gaps are not included.

A certain restriction of data representativeness is caused by specifications concerning the packing of the examined PET bottles. Since for this segment no data of the total market are available, market patterns were referred to. In order to allow a maximum reliability of results for a system comparison with the beverage carton, rather light PET bottles were analysed or weight-optimised variants were examined by sensitivity analyses.

Altogether, the *data quality* and *data symmetry* of this LCA can be classified as good to very good.

Allocation rules, system boundaries and calculations concerning impact assessment were uniformly and similarly applied for all examined packaging systems and scenarios based on those systems.

## 5.6.5
**Significance of Differences**

Since an over-interpretation of differences between compared product systems must be avoided, the examination of their significance is central to the reliability of an LCA study.

According to ISO 14042 (new ISO 14044), depending on the definition of goal and scope, information and procedures that allow a deduction of significant results may become necessary. This applies, if, as in the present case, LCA results may be part of market-strategic or political decision-making.

Since an examination of significance on the basis of error computation with error propagation in a strict mathematical sense is regarded critically owing to the data structure in LCAs, the following references are meant to provide an orientation as to when differences between systems are to be regarded as relevant.

The dominance analysis (sectoral analysis) represented in Section 4.1 may be regarded as an important support. It was stated that for PET bottles, substantial impacts are mainly caused by the supply of the bottle material. Indicator results of the beverage carton systems are affected by several sectors. Particularly relevant is the production of the individual composite materials, namely, aluminium, polyethylene and raw cardboard, as well as the production of closures.

During the project processing, there was particular emphasis on the quality of data and assumptions in these sectors (see also Chapters 2 and 3). The remaining uncertainties regarding PET bottle weights were examined by sensitivity analyses.

The practitioners express the opinion that the data and assumptions used are applicable for the result-relevant sectors of the examined packaging systems and that they are, to a large extent, symmetrical in actuality. On this basis, the results can be regarded as sound and robust. Uncertainties concerning the accuracy and representativeness of the data such as those discussed are nevertheless inevitable to a certain extent. Consequently, small differences of indicator values for a comparison of packaging systems are less significant than larger ones.

Even if discrete declarations for significance thresholds in LCA studies cannot be reliably deduced, as a result of basic objections, a system comparison of the beverage carton and PET packagings was nevertheless made by using a significance threshold in order to avoid an over-interpretation of small differences.

The Institut für Energie und Umweltforschung (IFEU) usually applies a significance threshold value of 10% for the analysis of packaging systems. This is a pragmatic and common approach in LCA praxis and regarded as admissible by the authors of the study for assessment comparisons of scenarios where system boundaries only comprise a single product system with a comparatively small complexity.

In the case of the present study, possible developments of PET systems for the period from 2006 to 2010 are included. In view of the prospective character of these estimations, the significance threshold value was increased to 20% for the examination of the results for all compared scenarios.

This approach is by no means transferable as standard to all LCAs. It should be noted that the beverage packaging LCAs of the Federal Environmental Agency[39] did not apply discrete significance thresholds at all.

For a critical analysis of the comparison on the level of impact categories, the results in Table 5.1 are discussed on the basis of a fixed significance threshold.

For a comparison of relative differences in computation (Table 5.1) regarding a significance threshold of 20%, the following are the advantages and disadvantages: advantages of the beverage carton under the categories fossil resource demand, greenhouse effect, acidification, terrestrial eutrophication and aquatic eutrophication, and disadvantages under summer smog and land use - forest area. The differences in terms of fossil resource demand, greenhouse effect and land use - forest area are particularly large.

No significant changes occur if sensitivity analyses targeting an increased sorting depth of used PET bottles as well as the use of recycled PET amounting to 25% of the bottle production are considered.

## 5.6.6
### Sensitivity Analyses

In the example study, the key parameters of the assessment were varied in order to examine their relevance on the result. The following two types of sensitivity analyses that have been accomplished are presented as examples:

- Result relevance of technical improvements of PET bottles
- Result relevance of system allocation.

In Table 5.2, the results of two sensitivity scenarios are presented, assuming *technical improvements of the PET bottle* considered in the system modelling: for one scenario an input of recycling material of 25% PET (R-PET; open-loop material), for the other scenario, an assortment optimisation of the collected light-packaging stream was assumed.

Compared with Table 5.1, the result of the computation of the PET-bottle-based scenario, the study states as follows:

---

39) Umweltbundesamt (UBA) Germany.

Both the input of 25% R-PET for bottle production as well as an improved PET assortment have no major influence on the comparison of the packaging systems. The specific differences between carton and PET remain; there is only a limited change of the relative positioning of the systems.

**Table 5.2** Comparison of 1 l-Beverage Carton with closure (reference 2005) and PET single-use systems assuming certain technical optimisations.

| Indicator | Carton 2005 (base scenario 1 l) versus: | |
|---|---|---|
| | 25% R-PET input (%) | Improved assortment (%) |
| Greenhouse effect | −160 | −165 |
| Fossil resources consumption | −131 | −154 |
| Summer smog | +53 | +47 |
| Acidification | −16 | −20 |
| Terrestrial eutrophication | −31 | −35 |
| Aquatic eutrophication | −42 | −24 |
| Land use – forest area | +997 | +1000 |
| Cumulative energy demand (CED) total | −34 | −43 |

Relative system differences in each case related to the smaller result (computational differences without specification of a significance threshold).
*Negative values*: indicator value of the beverage carton is smaller than the one of the PET bottle.
*Positive values*: indicator value of the beverage carton is larger than the one of the PET bottle.
Filling material juice/nectar; market segment storage.

In a further scenario, the 100:0 allocation[40] was used in the system modelling instead of the 50:50 allocation as in the base scenario. Credit entries for secondary materials are completely assigned to the delivering system (see Section 3.3.4.2). Thus the relevance of the definition of system allocation is examined. Table 5.3 shows the results. Here column 1 is identical to Table 5.1, the simple comparison in the basis scenario with 50:50 allocation.

For the influence of the system allocation method on the result, the study states as follows:

Table 5.3 documents the relative system differences of PET and carton in dependence of the allocation rule. The results show the derived findings to be permanently independent of specifications concerning system allocation.

---

40) This allocation rule unloads the system delivering secondary material (usually designated 'A') and therefore works opposite to the *cut-off rule*, which charges A with raw material extraction and only the avoided burdens for end-of-life procedures has an exculpatory impact on A.

**Table 5.3** Comparison beverage carton with closure and PET single-use bottle in the market segment juice/nectar – storage (1 l filling) with application of different allocation rules.

| Indicator | 50:50-allocation | 100:0-allocation |
|---|---|---|
| | Carton system (1 l) versus PET system (1 l) corresponds to Table 5.1 (%) | Carton system (1 l) versus PET system (1 l) (%) |
| Greenhouse effect | −167 | −204 |
| Fossil resources consumption | −164 | −123 |
| Summer smog (POCP) | +42 | +81 |
| Acidification | −23 | −16 |
| Terrestrial eutrophication | −37 | −38 |
| Aquatic eutrophication | −26 | −51 |
| Land use – forest area | +999 | +622 |
| Cumulative energy demand (CED) total | −48 | −40 |

Relative system differences, in each case related to the smaller result (computational differences without specification of a significance threshold).
*Negative values*: indicator value of the beverage carton is smaller than the one of the PET bottle.
*Positive values*: indicator value of the beverage carton is larger than the one of the PET bottle.
POCP, photo oxidant creation potential.

For individual categories, the relative differences decrease for an application of the 100:0 allocation, while they increase for others. In no case, however, does the tendency or significance of the results change.

### 5.6.7
**Restrictions**

In principle, it is assumed in the study that the results are sound and robust.

In the opinion of the practitioners, the results of base scenarios of the examined packaging systems and of system comparisons based on those are sound and robust within the defined boundary conditions. In case of a deviation from these boundary conditions, the following restrictions should be considered for an application of the results of the study.

The study addresses a set of restrictions, of which only some are stated below as an illustration. In the study, the restrictions are, however, described in further detail. The following list is only meant to indicate the possible restrictions:

- Restrictions by selection of market segments:
- The results of this study for a comparison of beverage carton and PET bottles are only valid for the examined market segments. A transfer of results to other filling materials or packaging sizes cannot be easily made, owing to the complexity of the context.
- The evaluation method used in the present study (normalisation and grouping in the phase impact assessment) mainly considers the approach as applied in the Beverage LCA II of the Federal Environmental Agency.[41]
- The presented results are valid using the data records described in Chapter 3. If for individual processes other data bases are consulted, this could influence comparative results of the examined packaging systems.
- The elaboration of packaging is constantly being developed. The packaging specifications used in this study are valid for the average beverage carton of the year 2005 as well as for typical PET bottles of this year.
- Restrictions concerning future developments: The statements of the present LCA study are valid for the reference time only. Questions related to future assessments of the examined packagings were not subject of the study.
- Restrictions concerning packaging specifications for PET bottles: The mass of PET bottles examined in this study was adapted to market patterns regarded as representative in sense of a median. Besides, the ecological profile of light bottle types is determined in sensitivity scenarios. Bottles above an average weight are, however, not examined.

## 5.6.8
### Conclusions and Recommendations

In Section 2.3, the goal definition of the study was summarised. The results following the interpretation phase must now allow redemption of these goals. All the issues that had been specified are discussed, and a series of proposals for optimisation were deduced from the results. However, these have not been discussed here, as this would exceed the purely didactic purpose of the practical example.

## 5.6.9
### Critical Review

Since in the example study, comparative assertions are defined to be made available to the public, a critical review by interested parties was necessary. At the time of the study, ISO 14040:1997 was still valid, asking for a minimum number of two experts. Further 'interested parties' were not included, but an independent advisory board was present to articulate their points of view in the study.

The reviewers were appointed by name and therefore provided a personal liability concerning the quality of the study.

---

41) UBA (1999).

The study is subject to a Critical Review according to ISO 14040:1997 Section 7.3.3.

The reviewers are

- Dipl. Univ. – Chemist (M.Sc.) Paul W. Gilgen (Chairman), Department Manager; c/o Eidgenössische Materialprufungs- und Versuchsanstalt (EMPA), Unterland Straße 129, 8600 Dubendorf, Switzerland
- Hans Jürgen Garvens (employee of the Federal Environmental Agency) Wolfgang Heinz Straße 54, 13125 Berlin, FRG.

Within this accompanying (interactive) critical review, the reviewer communicated with the practitioner of the study and with the advisory board and discussed their results in view of a concluding report in the course of several meetings.

In the final critical review report, which is part of the final study, the study was evaluated by the experts as of excellent quality. Since the critical reviewers themselves are LCA experts, methodological objectives were also discussed in addition to the test criteria required according to ISO 14044 and formulated as recommendations for subsequent studies.

## References

Braam, J., Tanner, T.M., Askham, C., Hendriks, N., Maurice, B., Mälkki, H., Vold, M., Wessman, H., and de Beaufort, A.S.H. (2001) Energy, transport and waste models. Availability and quality of energy, transport and waste models and data. *Int. J. Life Cycle Assess.*, **6** (3), 135–139.

Chevalier, J.-L. and Le Téno, J.-F. (1996) Life cycle analysis with Ill-defined data and its application to building products. *Int. J. Life Cycle Assess.*, **1** (2), 90–96.

Ciroth, A. (2001) Fehlerrechnung in Ökobilanzen. Dissertation. TU Berlin, Fakultät III – Prozesswissenschaften Buchpublikation im Dr. Müller Verlag, 2008 (im Druck).

Ciroth, A. (2004) Uncertainty in life cycle assessments. *Int. J. Life Cycle Assess.*, **9** (3), 141–142.

Ciroth, A. (2006) Arbeitspaket Fehlerrechnung, Datenqualität, Unsicherheit. Netzwerk Lebenszyklusdaten, Arbeitskreis Methodik. Karlsruhe.

Ciroth, A., Fleischer, G., and Steinbach, J. (2004) Uncertainty calculation in life cycle assessments. A combined model of simulation and approximation. *Int. J. Life Cycle Assess.*, **9** (4), 216–226.

Coulon, R., Camobreco, V., Teulon, H., and Besnaimou, J. (1997) Data quality and uncertainty in LCI. *Int. J. Life Cycle Assess.*, **2** (3), 178–182.

SETAC (1994) Fava, J., Jensen, A.A., Lindfors, L., Pomper, S., De Smet, B., Warren, J., and Vigon, B. (eds) *Conceptual Framework for Life-Cycle Data Quality*. Workshop Report, SETAC and SETAC Foundation for Environmental Education, Inc., Wintergreen, VA, October 1992. Pensacola, FL, June.

Fava, J. and Pomper, S. (1997) Life-cycle critical review! Does it work? Implementing a critical review process as a key element of the aluminium beverage container LCA. *Int. J. Life Cycle Assess.*, **2** (3), 144–153.

Finkbeiner, M., Inaba, A., Tan, R.B.H., Christiansen, K., and Klüppel, H.-J. (2006) The new international standards for life cycle assessment: ISO 14040 and ISO 14044. *Int. J. Life Cycle Assess.*, **11** (2), 80–85.

Frischknecht, R. (2004) Transparency in LCA – a heretical request? *Int. J. Life Cycle Assess.*, **9** (3), 211–213.

Grahl, B. and Schmincke, E. (1996) Evaluation and decision-making processes in life cycle assessment. *Int. J. Life Cycle Assess.*, **1** (1), 32–35.

Grahl, B. and Schmincke, E. (2011) 'Critical review' and 'Verification' cannot be used synonymously. A plea for a differentiated and precise use of the terms. LCM Conference, Berlin, 2011, http://www.lcm2011.org/papers.html; Session: Critical Review and Verification of LCA (accessed 23 Ocotber 2013).

Heijungs, R. (1994) A generic method for the identification of options for cleaner products. *Ecol. Econ.*, **10** (1), 69–81.

Heijungs, R. (2001) *A Theory of the Environmental and Economic Systems – a Unified Framework for Ecological Economic Analysis and Decision Support*, Edward Elgar, Cheltenham. ISBN: 1-84064-643-8.

Heijungs, R. and Frischknecht, R. (2005) Representing statistical distributions for uncertain parameters in LCA. Relationships between mathematical forms, their representation in EcoSpold, and their representation in CMLCA. *Int. J. Life Cycle Assess.*, **10** (4), 248–254.

Heijungs, R. and Kleijn, R. (2001) Numerical approaches towards life cycle interpretation. Five examples. *Int. J. Life Cycle Assess.*, **6** (3), 141–148.

Heijungs, R. and Suh, S. (2002) *The Computational Structure of Life Cycle Assessment*, Kluwer Academic Publishers, Dordrecht. ISBN: 1-4020-0672-1.

Heijungs, R., Suh, S., and Kleijn, R. (2005) Numerical approaches to life cycle interpretation. The case of the ecoinvent '96 database. *Int. J. Life Cycle Assess.*, **10** (2), 103–112.

Hertwich, G. (2005) Life cycle approaches to sustainable consumption. A critical review. *Environ. Sci. Technol.*, **39** (13), 4673–4684.

Hertwich, G.H. and Hammit, J. (2001) A decision-analytic framework for impact assessment, part 1. *Int. J. Life Cycle Assess.*, **6** (1), 5–12.

Hildenbrand, J. (1999) Vergleichende Darstellung von Auswertungsverfahren in Ökobilanzen. Diplomarbeit TU-Berlin, FB 6, Technischer Umweltschutz.

Hofstetter, P. (1998) *Perspectives in Live Cycle Assessment. A Structured Approach to Combine Models of the Technosphere, Ecosphere and Valuesphere*, Kluwer Academic Publishers, Boston, MA. ISBN: 0-7923-8377-X.

Huijbregts, M.A.J. (1998) Application of uncertainty and variability in LCA. Part II. Dealing with parameter uncertainty due to choices in life cycle assessment. *Int. J. Life Cycle Assess.*, **3** (6), 343–351.

Huijbregts, M.A.J., Gilijamse, W., Ragas, A.M.J., and Reijnders, L. (2003) Evaluating uncertainty in environmental life-cycle assessment. A case study comparing two insulation options for a Dutch one-family dwelling. *Environ. Sci. Technol.*, **37** (11), 2600–2608.

Huijbregts, M.A.J., Heijungs, R., and Hellweg, S. (2004) Complexity and integrated resource management: uncertainty in LCA. *Int. J. Life Cycle Assess.*, **9** (5), 341–342.

Huijbregts, M.A.J., Norris, G., Bretz, R., Ciroth, A., Maurice, B., von Bahr, B., Weidema, B., and de Beaufort, A.S.H. (2001) Framework for modelling data uncertainty in life cycle inventories. *Int. J. Life Cycle Assess.*, **6** (3), 127–132.

IFEU (2006) Detzel, A. and Böß, A. *Ökobilanzieller Vergleich von Getränkekartons und PET-Einwegflaschen.* Endbericht, Institut für Energie und Umweltforschung (IFEU), Heidelberg an den Fachverband Kartonverpackungen (FKN) Wiesbaden, August.

International Standard (ISO); Norme Européenne (CEN) (2000) ISO EN 14043. Environmental Management – Life Cycle Assessment: Interpretation (Auswertung), International Standard Organization.

International Standard Organisation (ISO) (2006a) ISO EN 14040:2006–2010, ISO TC 207/SC 5. Environmental Management – Life Cycle Assessment – Principles and Framework, International Organization for Standardization, Genva.

International Standard Organisation (ISO) (2006b) ISO EN 14044:2006-2010, ISO TC 207/SC 5. Environmental Management – Life Cycle Assessment – Requirements and Guidelines,

International Organization for Standardization, Genva

International Standard Organisation (ISO) (2013) ISO/PDTS 14071. Environmental Management – Life Cycle Assessment – Critical Review Processes and Reviewer Competencies – Additional Requirements and Guidelines to ISO 14044:2006, WD3, International Organization for Standardization (ISO), Geneva.

International Standard Organisation (ISO) (1997) ISO 14040. Environmental Management – Life Cycle Assessment – Principles and Framework. International Organization for Standardization (ISO).

Kennedy, D.J., Montgomery, D.C., and Quay, B.H. (1996) Data quality. Stochastic environmental life cycle assessment modeling – a probabilistic approach to incorporating variable input data quality. Int. J. Life Cycle Assess., **1** (4), 199–207.

Kennedy, D.J., Montgomery, D.C., Rollier, D.A., and Keats, J.B. (1997) Data quality. Assessing input data uncertainty in life cycle assessment inventory models. Int. J. Life Cycle Assess., **2** (4), 229–239.

Klöpffer, W. (1997) Peer (Expert) review according to SETAC and ISO 14040. Theory and practice. Int. J. Life Cycle Assess., **2** (4), 183–184.

Klöpffer, W. (1998) Subjective is not arbitrary. Editorial. Int. J. Life Cycle Assess., **3** (2), 61.

Klöpffer, W. (2000) Ökobilanzen & Produktverantwortung. Dokumentation Stiftung Arbeit und Umwelt (Hrsg.), Buchwerkstätten Hannover GmbH, pp. 37–42. ISBN: 3-89384-041-9.

Klöpffer, W. (2005) The critical review process according to ISO 14040-43: an analysis of the standards and experiences gained in their application. Int. J. Life Cycle Assess., **10** (2), 98–102.

Klöpffer, W. (2006) The role of SETAC in the development of LCA. Int. J. Life Cycle Assess., **11** (1) Special Issue, 116–122.

Klöpffer, W. (2007) Publishing scientific articles with special reference to LCA and related topics. Int. J. Life Cycle Assess., **12** (2), 71–76.

Klöpffer, W. (2012) The critical review of life cycle assessment studies according to ISO 14040 and 14044: origin, purpose and practical performance. Int. J. Life Cycle Assess., **17** (9), 1087–1093.

Klöpffer, W., Grießhammer, R., and Sundström, G. (1995) Overview of the scientific peer review of the european life cycle inventory for surfactant production. Tenside Surfactants Deterg., **32**, 378–383.

Klöpffer, W., Sundström, G., and Grießhammer, R. (1996) The Peer reviewing process – a case study: european life cycle inventory for surfactant production. Int. J. Life Cycle Assess., **1** (2), 113–115.

Klüppel, H.-J. (1997) Goal and scope definition and life cycle inventory analysis. Int. J. Life Cycle Assess., **2** (1), 5–8.

Klüppel, H.-J. (2002) The ISO standardization process: Quo Vadis? Editorial. Int. J. Life Cycle Assess., **7** (1), 1.

Lecouls, H. (1999) ISO 14043: environmental management • life cycle assessment • life cycle interpretation. Editorial. Int. J. Life Cycle Assess., **4** (5), 245.

Le Téno, J.F. (1999) Visual data analysis and decision support methods for non-deterministic LCA. Int. J. Life Cycle Assess., **4** (1), 41–47.

Lloyd, S.M. and Ries, R. (2007) Characterizing, propagating, and analyzing uncertainty in life cycle assessment. A survey of quantitative approaches. J. Industrial Ecology, **11** (1), 161–179.

Marsmann, M. (1997) ISO 14040 – the first project. Editorial. Int. J. Life Cycle Assess., **2** (3), 122–123.

Marsmann, M. (2000) The ISO 14040 family. Int. J. Life Cycle Assess., **5** (6), 317–318.

Morgan, M.G. and Henrion, M. (1990) A Guide to Dealing with Uncertainty in Quantitative Risk and Policy Analysis, Cambridge University Press, New York.

Popper, K.R. (1934) Logik der Forschung, Julius Springer, Wien, 7. Auflage, Mohr, J.C.B. (Paul Siebeck), Tübingen, 1982.

Ross, S., Evans, D., and Webber, M. (2002) How LCA studies deal with uncertainty. Int. J. Life Cycle Assess., **7** (1), 47–52.

Saur, K. (1997) Life cycle interpretation – A brand new perspective? Int. J. Life Cycle Assess., **2** (1), 8–10.

Seppälä, J. (1999) in Decision Analysis as a Tool for Live Cycle Impact Assessment. (eds W. Klöpffer and O. Hutzinger), LCA Documents 4, Ecoinforma Press, Bayreuth. ISBN: 3-928379-56-9.

SETAC (1993) *Guidelines for Life-Cycle Assessment: A 'Code of Practice'*, 1st edn. From the SETAC Workshop held at Sesimbra, Portugal, 31 March – 3 April 1993, SETAC, Brussels and Pensacola, FL, August 1993.

SETAC (1991) Fava, J.A., Denison, R., Jones, B., Curran, M.A., Vigon, B., Selke, S., and Barnum, J., *SETAC Workshop Report: A Technical Framework for Life Cycle Assessments*, August 18–23, 1990, Smugglers Notch, Vermont, SETAC, Washington, DC January 1991.

SETAC (2008) Jensen, A.A. and Postlethwaite, D. Europe LCA steering committee – the early years. *Int. J. Life Cycle Assess.*, **13** (1), 1–6.

Tukker, A. (2000) Philosophy of science, policy sciences and the basis of decision support with LCA based on the toxicity controversy in Sweden and the Netherlands. *Int. J. Life Cycle Assess.*, **5** (3), 177–186.

UBA (1999) Schmitz, S. and Paulini, I. Bewertung in Ökobilanzen. Methode des Umweltbundesamtes zur Normierung von Wirkungsindikatoren, Ordnung (Rangbildung) von Wirkungskategorien und zur Auswertung nach ISO 14042 and 14043. Version '99. UBA Texte 92/99, Berlin.

Werner, F. and Scholz, R.W. (2002) Ambiguities in decision-oriented life cycle inventories. The role of mental models. *Int. J. Life Cycle Assess.*, **7** (6), 330–338.

# 6
# From LCA to Sustainability Assessment

## 6.1
## Sustainability

The discussions during the first Society of Environmental Toxicology and Chemistry (SETAC) Europe life cycle assessment (LCA) symposium 1991 in the Dutch city of Leiden[1] resulted – at least in Europe – in a restriction of the LCA to the impact of products on the environment. It had been clear from the outset that a complete sustainability analysis must include socio-economic dimensions. Around 10 years later these extensions to the product-related LCA shifted again to the foreground and developed into an objective of research, trial/testing and standardisation.

What exactly is meant by the notion of sustainability, which is often used with little precision?[2]

In Germany it was first used in forestry. A pioneer in this field was Hans Carl von Carlowitz with his book 'Sylvicultura Oeconomica', which was published 1713 in Leipzig.[3] Carlowitz was no forester, but in his position as Superintendent of the Saxon silver mines he needed large quantities of timber and found that the German forests were in a very bad condition. Forestry was his life time hobby, and he stated the principle of 'sustainable forestry', which proposed that only as much wood as would regenerate should be logged. He had already perceived interconnections between environmental factors and economic and social interests (as we would formulate today). Even if the book, because of its baroque language and gothic printing, is not easy to read, the message is nevertheless clear and quite relevant for today's sustainability discussion. The German word 'nachhaltig' was even translated into French ('soutenu')[4] and via this into the now familiar English term *sustainable*.[5]

To date this idea has been associated with the global development policy defined in the Brundtland report[6] from which the following lines are often quoted:

---

1) SETAC Europe (1992).
2) Kuhlman and Farrington (2010).
3) Carlowitz (2000; reprint 2000).
4) Now mainly: durable.
5) Grober (2010).
6) WCED (1987) and Hauff (1987).

*Life Cycle Assessment (LCA): A Guide to Best Practice*, First Edition.
Walter Klöpffer and Birgit Grahl.
© 2014 Wiley-VCH Verlag GmbH & Co. KGaA. Published 2014 by Wiley-VCH Verlag GmbH & Co. KGaA.

> *Sustainable development is development that meets the needs of the present without compromising the ability of future generations to meet their own needs.*

This statement addresses a worldwide responsibility of humans living today to future generations. This ambitious goal was rapidly included into a political discussion: 1992, the United Nations in Rio de Janeiro declared sustainability as the guiding principle for the twenty-first century, which was confirmed 10 years later at the succeeding conference in Johannesburg. The relation to the entire life cycle of products, *life cycle thinking*, was already recognised as an important principle. Beyond political declarations of intent the necessity for a quantification and operationalisation of sustainability remains though an abuse, for example, of product comparisons is to be avoided. This is addressed, for example, by a statement of the Advisory Board for Sustainability in Germany concerning indicators for the national sustainability strategy[7]: it is stressed that sustainability without quantified goals threatens to evolve into an empty phrase. In addition, the Advisory Board's definition of sustainability emphasises the global claim:

> *Sustainable development is the creation of economic and social development by means of preservation of natural fundamentals of life and an achievement of economic and social welfare for present and future generations – for us and globally.*

## 6.2
### The Three Dimensions of Sustainability

The definitions of sustainability given in Section 6.1 are not directly useful for the purpose of (mostly comparative) product assessments. The standard model, which is also accepted by industry, is called the *triple bottom line*,[8] an interpretation of sustainability based on *the three pillars of sustainability*. It basically states that for the achievement and, of course also, for the analysis of sustainability of anthropogenic activities ecological, economic and social aspects have to be considered.

This threefold interpretation is, however, not straightforward as it suggests that all three 'pillars' are evenly weighted within the framework of sustainability and that each 'pillar' can be developed independent of the others. Besides, their common basis is unclear. Figure 6.1 therefore assigns the micro- and macro-economic perspective as well as the demand for inter-cultural participation and justice, provided the natural resources of life are handled carefully, to the technosphere, which is embedded into the ecosphere.

There is no lack of effort in emphasising the role of the environment (or ecosphere, nature), which is the basis of human survival.

---

[7] Rat für Nachhaltige Entwicklung (2008).
[8] A similar popular formulation is 3P or PPP (People, Planet, Profit).

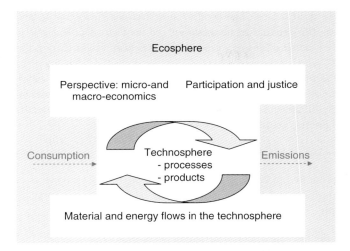

**Figure 6.1** Natural basis of life is the prerequisite for sustainable development.

An analysis made by the Austrian Ministry of Life[9] defined national sustainability by a twofold ('dualistic') model of spheres that has a certain resemblance to the functional environmental model[10] (technosphere + environment). The two spheres are defined as man/society and environment. Economy within this dualistic model is of course part of the society (or technosphere) and thus emphasises the integrating view of relations between economic and social phenomena. This Austrian model is also related to the sustainability model developed by the Institut für Energie- und Umweltforschung (IFEU) commissioned on behalf of the German Environmental Agency (Umweltbundesamt (UBA) Berlin).[11]

Since the current development of LCA adheres to the 3P (People, Planet, Profit) model,[12] it will be chosen for further discussion. There are no fundamental objections with regard to the combination of two dimensions (social and economy) to a dual system.

The three dimensions of sustainability were discussed – as has already been mentioned – on the occasion of the first SETAC Europe LCA workshop in Leiden, The Netherlands, 1991, and reflected the philosophy of the 'product line analysis' proposed by the Ökoinstitut in 1987.[13] This method – whose successor has recently been called *Product Sustainability Assessment (PROSA)* [14] – served as a precursor to LCAs, a 'proto LCA',[15] consisting of an inventory analysis, an impact assessment

---

9) Life Ministry (2006).
10) Frische *et al.* (1982) and Klöpffer (2001, 2012). This model, also called *the functional model of the environment*, defines the environment ex contrario: the technosphere is defined as everything that is controlled by humans, and the environment as everything that the technosphere is not.
11) Giegrich, Möhler and Borken (2003).
12) UNEP-DTIE (2011).
13) 'Produktlinienanalyse': Projektgruppe Ökologische Wirtschaft.
14) Grießhammer *et al.* (2007).
15) Klöpffer (2006a).

and comprising three dimensions. The product line analysis further proposed a demand analysis (is the product to be analysed required at all?). Today it is assumed that there is a demand for every product that has been established in the market.[16]

On the basis of the broad acceptance of the *triple bottom line* concept which has also been documented by the SETAC/UNEP (United Nations Environmental Programme) life cycle initiative,[17] the following scheme for a *life cycle sustainability assessment* – LCSA of products reads:

$$LCSA = LCA + LCC + SLCA \qquad (6.1)$$

This pattern was presented in 2003 as 'SustAss'[18] with the three components of LCA, LCC (life cycle costing) and SLCA (social life cycle assessment or societal life cycle assessment) with

- LCA, the (environmental) **Life Cycle Assessment** according to SETAC and ISO (International Standard Organization);
- LCC, **Life Cycle Costing** (compatible with LCA);
- SLCA, **Social (or societal) LCA**, the product-related social assessment.

For the use of the Equation 6.1 certain prerequisites must be fulfilled.[19]

The first and most important prerequisite is the use of consistent, ideally identical system boundaries for all three assessments. Because the involved technical disciplines have differing terminology, the terms must be consistently defined. An example: The relevant term regarded here of the physical life cycle of a product (from cradle to grave) differs fundamentally from the term *Product life cycle* used in marketing. This signifies the period from production development ('R&D') to product marketing. It ends with the product being taken from the market.

Ideally *one* inventory serves as the basis for all *three* dimensions. However, it must be assumed that the inventory for SLCA generally requires a stronger regional resolution than usually necessary for LCA and LCC. In Chapter 4 some efforts for an improved regional resolution of an ecological impact assessment were reported.[20] Finally, the '+' signs in Equation 6.1 are symbolic: they do not mean that the results of the three LCAs should be added.

The reason why the product-related sustainability assessment has to be life-cycle-based is obvious, and can be similarly applied in LCA: Only from a perspective of the entire life cycle, can problem shifting and apparent compensations (trade-offs) be observed and avoided. As required by the Brundtland Report (WCED, 1987, loc. cit.) for worldwide fairness across generations, a substantial issue in sustainability is to avoid shifting problems to the future or to other regions of the world.

---

16) There are actually products, which do have no perceivable use (or are even harmful) and which are enforced on consumers by marketing campaigns.
17) Remmen, Jensen and Frydendal (2007).
18) Klöpffer (2003a).
19) Klöpffer (2003, 2008), Klöpffer and Renner (2007, 2008) and UNEP-DTIE (2011).
20) José Potting is a pioneer in the area, Potting (2000).

## 6.3
## State of the Art of Methods

### 6.3.1
### Life Cycle Assessment – LCA

> *Current practices in ecological risk assessment generally do a poor job of considering biological and physical factors as most focus entirely or nearly so on chemical effects.*[21]

LCA, as described in the first five chapters of this book, is the sole ('one and only') internationally standardised method of environmental-oriented analysis of product systems. The now relevant international standards ISO 14040:2006 and 14044:2006 have been called *The constitution of LCA* by Finkbeiner.[22] Two key issues determine LCA: the analytic view on the entire life cycle 'from cradle to grave' and the functional unit, which allows the quantification of the benefit of goods or services ('reference flow'). The original series of international standards ISO 14040 to 14043:1997–2000 was replaced by the slightly modified standards ISO 14040 and 14044:2006.[23] The well-known structure – definition of goal and scope, inventory analysis, impact assessment and interpretation (see Chapter 1) – was developed by SETAC (1990–1993) and by ISO (1993–2000) in the course of harmonisation and standardisations. The standards provide strict requirements particularly for comparative assertions (see Chapter 5), which are to be publicly made available to prevent the abuse of LCA results. Thus LCA has reached a high level, and further progress will be adjusted more slowly. On the other hand, there are numerous weak points and corresponding improvement opportunities,[24] many of which have been discussed in the preceding chapters.

Some of the weaknesses attributed to LCA can, however, only be removed with a loss of simplicity and robustness of the method. These are mainly related to the restricted resolution of location and time in life cycle inventory (LCI) and life cycle impact assessment (LCIA) (see Chapters 3 and 4). The way out of the world of potential impacts of classical LCA into a world of life-cycle based quantitative risk assessment is costly and implies new uncertainties. Other problem fields like, for example, the choice of allocation rules and system expansions could in principle be solved by *conventions*.[25] Even the seemingly strict scientific metre convention and its modern successor 'Système International des Mesures et Quantités' (Système International d'unités, SI)[26] is by no means scientifically superior to the obsolete US unit-system but 'only' more consistent and practicable. As there is still no international LCA society,[27] SETAC would be the best suited forum to provide these conventions or at least activities for their preparation.

---

21) Anonymous in : SETAC Globe vol. 3(4) p. 59 (Ecological Risk Assessment section).
22) Finkbeiner (2013).
23) Finkbeiner et al. (2006).
24) Reap et al. (2008a,2008b) and Guinée et al. (2011).
25) Klöpffer (1998).
26) ISO (1981).
27) Klöpffer (1997).

An improvable phase of LCA, particularly with respect to sustainability assessment, is the impact assessment, the elaboration of which has deliberately been left open by ISO. In Chapter 4 numerous new developments, which are still being tested, have been discussed. A mismatch concerning the impacts of emissions of chemical origin and other emissions or stressors should be indicated here.[28] To illustrate this situation, impact categories that are arranged differently by from the usual mode (Section 4.5), and two categories are added[29]:

A: Consumption of resources
B: Impacts of chemical emissions
C: Impacts of physical emissions
D: Impacts of biological emissions
E: Other impacts.

*A: Consumption of resources*
    A1: Consumption of abiotic resources (including water)
    A2: Consumption of biotic resources
    A3: Land use.

These categories are entirely compatible to input related impact categories of the Centrum voor Milieukunde Leiden (CML) method.[30]

*B: Impact of chemical emissions*
    B1: Climate change
    B2: Stratospheric ozone depletion
    B3: Formation of photo-oxidants
    B4: Acidification
    B5: Eutrophication
    B6: Human toxicity
    B7: Eco-toxicity
    B8: Odour.

Comprising eight impact categories, often with additional subcategories,[31] this group is the strongest and practically dominates every impact assessment by the number of components. This 'chemical preponderance' is undoubtedly an unintended side-effect of the triumph of the CML method.[32]

*C: Impacts of physical emissions*
    C1: Noise
    C2: Ionising radiation (radioactivity)
    C3: Waste heat.

This group, practically not considered much, comprises physical emissions, which may have secondary physiological and psychological impacts.

---

28) This lead to the false impression that chemistry is the base of noxity, see Klöpffer (2003b).
29) Klöpffer and Renner (2003) and Klöpffer (2006a).
30) Heijungs *et al.* (1992) and Guinée *et al.* (2002).
31) Guinée *et al.* (2002).
32) Klöpffer (2006b).

*D: Impacts of biological emissions*
   D1: Impacts on ecosystems; modification of species and biodiversity
   D2: Impacts on humans (e.g. by pathogen organisms).

These 'new' impact categories, where indicator models are partly under research and not yet satisfactory for practical application in standard LCAs, were initially introduced because of their widespread neglect in LCIA (Renner and Klöpffer, 2005, loc. cit.). Undoubtedly ecosystems worldwide are threatened by invasive species (mostly neozoa and neophytes, though native species can also evolve into invasive species[33]) at least just as much as by the destruction of habitats and chemical and physical exposures.[34] Potential ecological impacts of genetically modified organisms (GMOs) are likewise part of this group c.[35] Regions that in the course of geologic history have been separated from the rest of the world, for example, Australia and New Zealand, are mostly threatened by these 'stressors'. Interestingly, neophytes and neozoa are spread predominantly via the technosphere, for example, ballast water of tankers, incrustation of hulls (ships), 'blind passengers' in the case of food transportation, tourists, and so on. The propagation of pathogenic germs can also take place via the ecosphere by wild animals, for example, migratory birds. A consideration of all these damaging impacts would even worsen the perception of worldwide trade completely neglecting the sphere 'environment', in addition to the already widespread bad perception for social reasons.

*E: Further categories*
   E1: Casualties;
   E2: Impacts on health at the working place (technosphere; exposition via the environment see B6);
   E3: Drainage, erosion and salting of soils (see also A3);
   E4: Destruction of landscapes (see also A3);
   E5: Disturbance of ecological systems and variety of species (biodiversity) (see also A3, B7 and D1);
   E6: (Solid) waste.

For reasons of completeness, further impact categories suggested in the scientific literature are listed in group E.[36] Some are more important for countries of the south (E3) and therefore of particular importance for the UNEP/SETAC life cycle initiative. Others represent serious problems for system boundaries: for instance, are parts of the technosphere (E1, E2, E6) to be included? E5 is in fact of central importance but overlaps with some other categories and is very difficult to quantify.[37] E6 is a relic of the time of the proto LCAs and surely not an impact category, at least not collected waste. Litter, however, should be

---

33) If for instance a predator is extinguished or hunting restrictions are enacted, climate changes favour a species, and so on.
34) *Around the world, invasive species are the second ranking cause of extinction of native species, after the destruction of habitats by human activity* Wilson (2006).
35) Klöpffer *et al.* (1999, 2001).
36) Renner and Klöpffer (2005).
37) Koellner and Geyer (2013).

considered as an impact category, especially plastic waste thrown into the sea from ships.

The inclusion of 'casualties' (E1) into impact assessment is difficult, but not impossible.[38] A routine inclusion of this category into impact assessment does not, however, seem to be adequate.

'Impacts on health at the working place' (E2) are more part of SLCA as in this part of the LCSA (see Section 6.3.3) the working place and its impact on the employees is at the centre of interest. However, efforts have been made, particularly in Scandinavia, to enforce the inclusion of the working place into the impact assessment.[39]

LCA remains an active area of research, where methodological developments can be expected in future. These, for instance, include the definition of difficult impact categories, the input/output and hybrid analysis,[40] the *consequential* LCA,[41] and the correct applications of LCA and related similar analysis tools within life cycle management (LCM).[42]

### 6.3.2
### Life Cycle Costing – LCC

LCC is modelled according to the LCA pattern and sums all costs, which occur in the life cycle of a product for one of the actors (e.g. supplier, producer, user, consumer, recycler); these costs must consist of *real money flows* to avoid overlaps between LCC and LCA.[43]

LCC is accomplished similar to the analogue LCA, both steady state models by nature. The definition of a functional unit and system boundaries similar to an LCA dealing with the same system are integrated. Ideally an inventory should already be available; LCC can, however, also be accomplished as a stand-alone analysis.

Although LCC calculations are actually older than LCA, it has not been standardised so far with some exceptions. Already during the years of LCA standardisation, practitioners of LCA were attracted by this method as a potential supplement to LCA.[44] Researchers, practitioners and scientists proposed a combination of simplified LCA and LCC under the name 'eco-efficiency analysis' for an easy and swift product valuation by use of valuing elements.[45] As discussed in Section 4.3.3, transparency of the latter is required for a comprehension of the results.

The topic was taken up by a SETAC Europe working group and published as a book.[46] A short version corresponding in purpose and length to LCAs 'Code

---

38) Kurth *et al.* (2004).
39) Poulsen *et al.* (2004).
40) Suh (2003).
41) Weidema (2001) and Steward and Weidema (2004).
42) Wrisberg *et al.* (2002), Rebitzer (2005) and Remmen, Jensen and Frydendal (2007).
43) Such an overlap may be caused by monetarisation of environmental damage in life cycle cost calculation.
44) White, Savage and Shapiro (1996), Norris (2001), Shapiro (2001), and Rebitzer (2002).
45) Saling *et al.* (2002), Landsettle and Saling (2002), Kicherer *et al.* (2007) and Huppes (2007).
46) SETAC (2008).

of Practice',[47] the result of the SETAC workshop in Sesimbra 1993, has been published with some additional elements.[48]

The suggested 'environmental' LCC method adheres to the form of LCA in accordance with ISO 14040 and covers the entire life cycle of a product including use and end-of-life phase. A monetarisation of possible external costs by environmental damage in the future is *not* included. Thus double counting is meant to be avoided, as the impact assessment of LCA accounts for (potential) environmental damage in physical units (LCIA).

In LCC calculation there is no stand-alone impact assessment. The aggregated result ideally corresponds to the actual costs, related to the selected functional unit, in a common currency.

Similar to LCA, no information should be lost by aggregation, and the exact analyses of the life cycle stages should be documented. Life cycle cost calculation, similar to LCA, also differs from common cost calculations by the fact that all included costs are assigned to the examined product and no 'overhead' exists. It is also to be distinguished from the environmental cost accounting.[49]

LCC provides a meaningful supplement to LCA and to product-related social assessments, because sustainable products should also be profitable (for the producer) and affordable (for the user/consumer) to be accepted by the market. The detailed LCC inventory indicates to persons responsible for the product system along the value chain where an improved cost-efficiency can be obtained.

The important actors in the life cycles are the consumers who not only decide upon the acquisition of a product, but also upon its use and, in many cases, upon disposal. In many cases an improved cost-efficiency is also an environmentally more compatible one, particularly regarding energy-intensive products. Unfortunately consumer decisions are often based exclusively on the price of purchase of a product. Thus information, like the one supplied by LCC calculation, is able to induce well-founded purchase decisions as, for example, the use phase is also included. The price of a product, frequently used as 'zeroth approximation' (cradle-to-point of sale) for LCCs, is generally not suited. It includes of course the costs 'further up' in the product tree including energy and raw material costs in a highly aggregated form, and not the use-phase and only in rare exceptional cases the costs of disposal (e.g. in Germany 'the green point'[50] for packaging). In addition, no reference concerning cost aggregations and profit margins is given.

LCC is even by itself a meaningful method to analyse a product, which can later be supplemented by an LCA or product-related SLCA and in this way can be promoted into a complete life-cycle based sustainability assessment (LCSA).[51]

---

47) SETAC (1993).
48) SETAC (2011) and Swarr *et al.* (2011).
49) Rikhardsson *et al.* (2005) and Schaltegger (2008).
50) German 'Der Grüne Punkt' of Duales System Deutschland (DSD).
51) Klöpffer (2008) and UNEP/SETAC (2011).

Usually it is assumed that first an LCA or at least an LCI study of the product exists that can then be accordingly extended.

### 6.3.3
**Product-Related Social Life Cycle Assessment – SLCA**

The third dimension of sustainability poses special difficulties for an operationalisation as humans are involved. Whereas humans, for ethical reasons in principle, are not assessed in LCA (unless targeted in the impact category 'human toxicity'), they are present in LCC as cost factors and consumers, and finally in SLCA, their well-being is the main content of the analysis. Thus the SLCA acts as corrective to the two previous 'dimensions': a product may be environmentally compatible and economically producible, but nevertheless not sustainable, if, for example, favourable LCCs are obtained by inhuman working conditions in certain countries and companies.

Even if the idea is not new,[52] the product-related social assessment is nevertheless still at its beginning. Currently the topic is a very active area of research related to numerous publications of which the most recent ones are outlined here. Approaches to a uniform methodology are slowly developing, but not yet generally observed.

Dreyer and co-worker[53] focus on the responsibility of the involved companies, even if the products are the points of reference. Thus it is inevitable that in the foreground processes and the involved persons are the focus of their emphasis. The responsibility of the management of an enterprise in social issues is beyond dispute and that can be more important than the technical processes assigned to the product system. On the other hand, responsibility is also required for the machinery (including safety measures) and for the environmental protection technology, if there is any.

Weidema[54] includes elements of cost benefit analysis (CBA) and proposes quality adjusted life years (QALYs) as a common measure for human health and human well-being.

Norris[55] is also concerned with social and socio-economic impacts, leading to health impairment. Norris is sceptical of an SLCA using LCA and LCC as model. He proposes an Internet-based instrument (life cycle of attribute assessment, LCAA) in addition to the classical life-cycle-based analysis methods. Recent advances in Internet-based social data collection on a global scale led to a much used social 'hot-spot analysis' and data bank.[56] Such generic data are very important in a social research field where individual plant owners and managers are reluctant to give information, especially on hot topics like child labour and bad or even criminal working conditions.

---

52) Projekt Gruppe Ökologische Wirtschaft (1987) and O'Brian, Doig and Clift (1996).
53) Dreyer et al. (2006).
54) Weidema (2006).
55) Norris (2006).
56) Benoît et al. (2010).

Labuschagne and Brent[57] strive at a completeness of the indicator set. However, their method does not primarily seem to be aimed at a product valuation.

Hunkeler[58] solves the problem of connecting social effects with the functional unit by using a proportionate *work time* for production (per functional unit) as a start for quantification. Work time can be included into the inventory which, however, has to show a high degree of regional resolution and differentiation. Here regionalisation is of greater importance than is usual in LCA. Work time is the relevant inventory parameter of SLCA. If foreground data are missing, the relation of wage per hour to goods and services[59] of vital importance, for example, food, medical supply, education, and so on, can be determined by national or supranational statistics. In combination with a proportionate work time per functional unit, this can provide a quantification of the social component. A kind of social impact assessment results depending on the purchase (food, etc.). By comparative social assessments it can be determined, in which of the products more low wages is involved, and thus which of the variants can be manufactured financially favourably only by exploitation of manpower.[60] In addition it is emphasised that the working place is the natural junction of the product life cycle and social dimensions. Thus not all, but some violations of human rights, for example, in low waged countries, can be included into the analysis.

Further SLCA methods developed in recent times are based on the eco-efficiency analysis, a combination of simplified LCAs and LCCs (see Section 6.3.2). Saling and co-authors[61] added a social component to the eco-efficiency analysis of BASF, which leads to SEE balance$^{®}$. Thus a two-dimensional eco-efficiency diagram transforms into a cube, which shows the position of the product in relation to the three dimensions of sustainability. Because the method depends on value-related weighting factors and is a result of multiple normalisations, transparency is hardly possible, and hence, according to ISO 14044 it should only be applied for internal purposes.

*Life cycle working time* (LCWT) implies the inclusion of working-place-related socio-economic aspects into the LCA-software GaBi, which thereby also considers a third life-cycle-based dimension of sustainability besides LCA and LCC. A feasibility study[62] for the UNEP/SETAC life cycle Initiative for an integration of social aspects into LCA was developed by Grießhammer and co-authors. Because many social indicators cannot be quantified, a qualitative valuation pattern in addition to quantitative results is used. The UNEP/SETAC guidelines for SLCA have been published and are now being tested in practice[63]

---

57) Labuschagne and Brent (2006).
58) Hunkeler (2006).
59) A descriptive, trivial version of this method is the so-called Big Mac-index: how long has a labourer to work (in different countries) in order to make 1 hamburger affordable?
60) An objection has been, in Europe as in the USA, that to a period of exploitation a period of relative prosperity for all can follow.
61) Saling *et al.* (2007).
62) Grießhammer *et al.* (2006).
63) Benoît and Mazijn (2009) and Benoît *et al.* (2010) Ciroth.

Pesonen[64] recently proposed *Sustainability SWOTs* (strengths, weaknesses, opportunities and threats)[65] as a simplified form of SLCA. Finally there is an overview by Jørgensen and co-authors[66] on publications as well as grey literature on SLCA.

It is surely too early for a standardisation of product-related social assessment; however, a certain measure of harmonisation could be achieved, if the different approaches were compared in case studies. As for finance, it could prove to be useful to have different indicators for the evaluation of the diverse aspects of SLCA. Thus experiences could be gained, and the most suitable method(s) would emerge. As to impacts and their indicators, it should be kept in mind that, for good reasons, for LCA impact assessments as well there is no absolutely valid list.

The main difficulties of product-related social assessments are as follows:

- How can existing indicators be linked to the functional unit of the examined system?
- How are specific data for the necessary regional resolution of SLCA procured?
- How can the choice between multiple qualitative indicators and a few quantifiable indicators be decided, for example, by an inventory of work time per functional unit (Hunkeler, 2006, loc. cit.)?
- How are impacts correctly quantified?

The last issue is probably the most difficult, and indeed the quantification of all impacts is not possible in LCA either. An example is the fact that there is still no suitable and generally accepted indicator for the important impact category of 'biodiversity'.

## 6.4
## One Life Cycle Assessment or Three?

There are several options (Equations 6.1–6.4) on how to integrate LCC and product-related social assessment into the sustainability analysis of products.

### 6.4.1
### Option 1

This option is based on one functional unit and three separated life-cycle-assessments with consistent, at best identical, system boundaries as already suggested in the introduction (Equation 6.1). The '+' signs are symbolic and do not suggest that the results should be added up. The two methods not standardised yet (LCC and SLCA), should be standardised in future based on existing guidelines.[67]

---

64) Pesonen (2007).
65) SWOT acronym for Strengths, Weaknesses, Opportunities and Threats (Wikipedia).
66) Jørgensen *et al.* (2008).
67) Swarr *et al.* (2011b) and Benoît and Mazijn (2009).

A weighting between the three dimensions should not take place. Thus transparency remains, which can surely be regarded as an advantage of this option. The assignment of pros and cons in comparative analyses is clear; there are no – and there should not be – compensational factors between the three dimensions ecology, economy and social aspects.

## 6.4.2
**Option 2**

$$LCSA = LCAnew \qquad (6.2)$$

(including LCC and SLCA as additional impact assessments in the impact assessment of the LCA ('LCA new').

This option means that on the basis of one (extended) inventory up to three impact assessments are accomplished (LCC has no formal impact assessment, the results are the costs in a common currency), which can refer, for example, to a common set of areas of protection. The advantage here is that only one inventory model must be defined in goal and scope. Also the results of an inventory of LCA (LCI) can be used as starting point for the product-related social assessment, as introduced in the method by Hunkeler.[68]

There are advocates for both options, and a possible future extension of ISO 14040 series is crucial for the discussion. Therefore the following important question arises: Is option 2 compatible with ISO 14040? In Section 4.1.3 it reads:

> LCA addresses the environmental aspects and impacts of a product system. Economic and social aspects and impacts are, typically, outside the scope of the LCA. Other tools may be combined with LCA for more extensive assessments.

Already the introduction quotes:

> LCA typically does not address the economic or social aspects of a product, but the life cycle approach and methodologies described in this International Standard can be applied to these other aspects.

These quotations from ISO 14040 clearly speak in favour of option 1 (Equation 6.1) and a separate standardisation of LCC and SLCA consistent with LCA would be a logical consequence. On the other hand, the standards ISO 14040 and 14044 could again be changed in the future in order to make option 2 (Equation 6.2) ISO-conformable ('LCA new'). This would, however, also have as a consequence that the already extensive standard ISO 14044 would have to be extended by detailed regulations for LCC and SLCA.

There are two further hypothetical possibilities to quantify LCSA (Equations 6.3 and 6.4).

---

68) Hunkeler, 2006.

$$\text{LCSA} = \text{Ecoefficiency} + \text{SLCA} \qquad (6.3)$$

$$\text{LCSA} = \text{LCA} + \text{Socio-economic assessment} \qquad (6.4)$$

In favour of the method in Equation 6.3 speaks a new international standard on ecoefficiency[69] consisting of an LCA + a 'value' assessment, which can be defined as a LCC assessment and also other quantifications of value over the life cycle. It should be noted that the enigmatic term *value* has been banned from LCA, at least (strictly!) for comparative assertions to be published in one way or the other.

Eco-efficiency has been used in LCM for about 10 years, for example, by BASF and also an extension by a SLCA as in Equation 6.3 is used in the SEEDbalance® sustainability assessment.[70] The method in Equation 6.3 seems to be preferred by industrial users of LCSA within the concept of LCM.

The fourth possibility (Equation 6.4) combines the economic pillar with the social one into a 'socio-economic' assessment with the argument that both the economy and the social effects produced by the economic activities belong together whereas the environment suffers – more or less passively – from a broad range of impacts caused by human activities in the form of toxic emissions and socially caused devastations (wars, deforestation, overfishing and loss of biodiversity, climate change, etc.). LCSA according to Equation 6.4 seems to be preferred by conservationist groups. The economic aspects are mixed into the socio-economic assessment and not clearly identified as such.

## 6.5
## Conclusions

It is often said that thinking in life cycles is already sufficient for an approach to implement the guideline of sustainability and that appropriate decisions do not always require quantified information. This may be true for the determination of hot spots but will not support a considered decision-making: if multiple proposals for a solution are made, quantitative methods are required to decide on one. One of the strengths of LCA is its capability to quantify, and this advantage should be preserved if supplemented by economic (LCC) and social (SLCA) aspects. This will be easy for LCC but difficult for a product-related SLCA. In view of the high goal, great efforts should be made to provide and to continuously improve the tools necessary.[71]

LCA itself needs to be improved and surely is capable of improvement. The development as hitherto should be balanced between the desired scientific accuracy and practical feasibility.[72]

---

69) ISO (2012).
70) Saling *et al.* (2002), Landsiedel and Saling (2002), Kicherer *et al.* (2007) and Saling *et al.* (2007).
71) Zamagni, Pesonen, and Swarr (2013).
72) Klöpffer (2013b) and Zamagni *et al.* (2009).

# References

Benoît, C. and Mazijn, B. (eds) (2009) *Guidelines for Social Life Cycle Assessment of Products*, Paris, p. 104 www.unep.fr/shared/publications/pdf/DTIx1164xPA-guidelines_sLCA.pdf (accessed 18 December 2013).

Benoît, C., Norris, G.A., Valdivia, S., Ciroth, A., Moberg, A., Bos, U., Prakash, S., Ugaya, C., and Beck, T. (2010) The guidelines for social life cycle assessment of products: just in time!. *Int. J. Life Cycle Assess.*, **15** (2), 156–163.

Carlowitz, D.-C. (1713) *Sylvicultura Oeconomica – Naturmäßige Anweisung zur Wilden Baum-Zucht*, Braun Verlag, LeipzigReprint TU: Bergakademie Freiberg, Akademische Buchhandlung Freiberg, 2000, ISBN: 3-86012-115-4.

Dreyer, L.C., Hauschild, M., and Schierbeck, J. (2006) A framework for social life cycle impact assessment. *Int. J. Life Cycle Assess.*, **11** (2), 88–97.

Finkbeiner, M. (2013) The international standards as constitution of life cycle assessment: the ISO 14040 series and its offspring, in *Encyclopedia Life Cycle Assess*, Chapter 6, vol. 1 (eds W. Klöpffer and M.A. Curran), Springer.

Finkbeiner, M., Inaba, A., Tan, R.B.H., Christiansen, K., and Klüppel, H.-J. (2006) The new international standards for life cycle assessment: ISO 14040 and ISO 14044. *Int. J. Life Cycle Assess.*, **11**, 80–85.

Frische, R., Klöpffer, W., Esser, G., and Schönborn, W. (1982) Criteria for assessing the environmental behavior of chemicals: selection and preliminary quantification. *Ecotox. Environ. Safety*, **6**, 283–293.

Giegrich, J., Möhler, S. and Borken, J. (2003) Entwicklung von Schlüsselindikatoren für eine nachhaltige Entwicklung, IFEU im Auftrag des UBA Dessau. FZK 200 12 119, Heidelberg.

Grießhammer, R., Benoît, C., Dreyer, L.C., Flysjö, A., Manhart, A., Mazijn, B., Méthot, A., and Weidema, BP. (2006) Feasibility study 2006: integration of social aspects into LCA. Discussion paper from UNEP-SETAC Task Force Integration of Social Aspects in LCA, Meeting in Bologna (January 2005), Lille (May 2005) and Brussels (November 2005), Freiburg, Germany.

Grießhammer, R., Buchert, M., Gensch, C.-O., Hochfeld, C., Manhart, A., Reisch, L., and Rüdenauer, I. (2007) *PROSA – Product Sustainability Assessment*, Öko-Institut e.V, Freiburg.

Grober, U. (2010) *Die Entdeckung der Nachhaltigkeit. Kulturgeschichte eines Begriffs*, Verlag Antje Kunstmann, München.

Guinée, J.B., Heijungs, R., Huppes, G., Zamagni, A., Masoni, P., Buonamici, R., Ekvall, T., and Rydberg, T. (2011) Life cycle assessment: past, present, and future. *Environ. Sci. Technol.*, **45**, 90–96.

Guinée, J.B., Gorée, M., Heijungs, R., Huppes, G., Kleijn, R., de Koning, A., van Oers, L., Wegener Sleeswijk, A., Suh, S., Udo de Haes, H.A., de Bruijn, H., van Duin, R., and Huijbregts, M.A.J. (2002) *Handbook on Life Cycle Assessment – Operational Guide to the ISO Standards*, Kluwer Academic Publisher, Dordrecht. ISBN: 1-4020-0228-9.

Hauff, V. (Hrsg.) (1987) *Unsere gemeinsame Zukunft. Der Brundtland-Bericht der Weltkommission für Umwelt und Entwicklung*, Eggenkamp Verlag, Greven.

Heijungs, R., Guinée, J.B., Huppes, G., Lamkreijer, R.M., Udo de Haes, H.A., Wegener Sleeswijk, A., Ansems, A.M.M., Eggels, P.G., van Duin, R., and de Goede, H.P. (1992) *Environmental Life Cycle Assessment of Products. Guide (Part 1) and Backgrounds (Part 2) October 1992*, prepared by, CML, TNO and B&G, Leiden, English Version 1993.

Hunkeler, D. (2006) Societal LCA methodology and case study. *Int. J. Life Cycle Assess.*, **11** (7), 371–382.

Huppes, G. (2007) Why we need better eco-efficiency analysis. From technological optimism to realism. *Technikfolgenabschätzung – Theorie und Praxis (TATuP)*, **16**, 38–45.

International Standard Organization (ISO) (1981) ISO 1000:1981. *SI Units and Recommendations for the Use of their Multiples and of Certain Other Units* (Unités SI et recommandations pour l'emploi de leurs multiples et de certaines autres unités).

International Standard Organization (ISO) Second edition – February.

International Standard Organization (ISO) (2012) ISO 14045. *Environmental Management – Eco Efficiency Assessment of Product Systems – Principles, Requirements and Guidelines*. International Standard Organization (ISO) Geneva.

Jørgensen, A., Le Bocq, A., Nazarkina, L., and Hauschild, M. (2008) Methodologies for social life cycle assessment – a review. *Int. J. Life Cycle Assess.*, **13** (2), 96–103.

Kicherer, A., Schaltegger, S., Tschochohei, H., and Ferreira Pozo, B. (2007) Eco-Efficiency. Combining life cycle assessment and life cycle costs via normalization. *Int. J. Life Cycle Assess.*, **12** (7), 537–543.

Klöpffer, W. (1997) Do we truly require an International Society for LCA practitioners? Editorial. *Int. J. Life Cycle Assess.*, **2** (1), 1.

Klöpffer, W. (1998) Subjective is not Arbitrary. Editorial. *Int. J. Life Cycle Assess.*, **3** (2), 61.

Klöpffer, W. (2001) Kriterien für eine ökologisch nachhaltige Stoff- und Gentechnikpolitik. *UWSF- Z. Umweltchem. Ökotox.*, **13**, 159–164.

Klöpffer, W. (2003a) Life-cycle based methods for sustainable product development. Editorial for the LCM section in. *Int. J. Life Cycle Assess.*, **8** (3), 157–159.

Klöpffer, W. (2003b) Gedanken zum Jahr der Chemie. Editorial. *UWSF-Z. Umweltchem. Ökotox.*, **15** (4), 214.

Klöpffer, W. (2006a) The role of SETAC in the development of LCA. *Int. J. Life Cycle Assess.*, **11** (Suppl. 1), 116–122.

Klöpffer, W. (2006b) *Sporen van een Gedreven Pionier. Verhalen bij het afscheid van Helias Udo de Haes ('Liber Amicorum')*, CML, Universiteit Leiden, Leiden, pp. 143–147 ISBN: 90-5191-149-1.

Klöpffer, W. (2008) Life cycle sustainability assessment of products. *Int. J. Life Cycle Assess.*, **13** (2), 89–94.

Klöpffer, W. (2012) *Verhalten und Abbau von Umweltchemikalien: Physikalisch-chemische Grundlagen*, 2nd edn, völlig überarbeitete Auflage, Wiley-VCH Verlag GmbH, Weinheim. ISBN: 978-3-527-32673-0.

Klöpffer, W. (2013a) Comment on the editorial note by Baitz et XXI aliis. Letter from the editor. *Int. J. Life Cycle Assess.*, **18** (1), 14–16.

Klöpffer, W. (2013b) Introducing life cycle assessment and its presentation in this encyclopedia, in *Encyclopedia Life Cycle Assess*, Chapter 3, vol. 1 (eds W. Klöpffer and M.A. Curran), Springer.

Klöpffer, W. and Renner, I. (2003) Life cycle impact categories – the problem of new categories & biological impacts – part I: systematic approach. SETAC Europe, 13th Annual Meeting Hamburg, Germany, 27 April – 1 May 2003.

Klöpffer, W. and Renner, I. (2007) Lebenszyklusbasierte Nachhaltigkeitsbewertung von Produkten. *Technikfolgenabschätzung – Theorie und Praxis*, **16** (3), 32–38.

Klöpffer, W. and Renner, I. (2008) Lifecycle based sustainability assessment of products, in *Environmental Management Accounting for Cleaner Production* (eds S. Schaltegger, M. Bennett, R. Burritt, and D. Jasch), Springer Publishers, Dordrecht.

Klöpffer, W., Renner, I., Schmidt, E., Tappeser, B., Gensch, C.-O., and Gaugitsch, H. (2001) *Methodische Weiterentwicklung der Wirkungsabschätzung in Ökobilanzen (LCA) gentechnisch veränderter Pflanzen*, Monographien Bd. **143**, Federal Environment Agency Ltd, Vienna. ISBN: 3-85457-597-1.

Klöpffer, W., Renner, I., Tappeser, B., Eckelkamp, C., and Dietrich, R. (1999) *Life Cycle Assessment gentechnisch veränderter Produkte als Basis für eine umfassende Beurteilung möglicher Umweltauswirkungen*, Monographien Bd. **111**, Federal Environment Agency Ltd, Vienna. ISBN: 3-85457-475-4.

Koellner, T. and Geyer, R. (2013) Global land use impacts on biodiversity and ecosystem services. *Int. J. Life Cycle Assess.*, **18** (Special Issue 6), 1185–1277.

Kuhlman, T. and Farrington, J. (2010) What is sustainability? *Sustainability*, **2010** (2), 3436–3448. doi: 10.3390/su2113436

Kurth, S., Schüler, D., Renner, I. and Klöpffer, W. (2004) Entwicklung eines Modells zur Berücksichtigung der Risiken durch nicht bestimmungsgemäße Betriebszustände von Industrieanlagen im Rahmen von Ökobilanzen (Vorstudie).

Forschungsbericht 201 48 309, UBA-FB 000632. UBA Texte 34/04, Berlin.

Life Ministry (2006) Monitoring nachhaltiger Entwicklung in Österreich. Indikatoren für nachhaltige Entwicklung. Vienna, Austria, October 2006.

Labuschagne, C. and Brent, A.C. (2006) Social indicators for sustainable project and technology life cycle management in the process industry. *Int. J. Life Cycle Assess.*, **11** (1), 3–15.

Landsiedel, R. and Saling, P. (2002) Assessment of toxicological risks for life cycle assessment and eco-efficiency analysis. *Int. J. Life Cycle Assess.*, **7** (5), 261–268.

Norris, G.A. (2001) Integrating life cycle cost analysis and LCA. *Int. J. Life Cycle Assess.*, **6** (2), 118–120.

Norris, G.A. (2006) Social impacts in product life cycles. Towards life cycle attribute assessment. *Int. J. Life Cycle Assess.*, **11** (Suppl. 1), 97–104.

O'Brian, M., Doig, A., and Clift, R. (1996) Social and Environmental Life Cycle Assessment (SELCA). *Int. J. Life Cycle Assess.*, **1** (4), 231–237.

Pesonen, H.-L. (2007) Sustainability SWOTs–new method for summarizing product sustainability information for business decision making. Platform presentation at the 3rd International Conference on Life Cycle Management, Zürich, Switzerland, 27-29 August.

Potting, J. (2000) *Spatial Differentiation in Life Cycle Impact Assessment*, Proefschrift Universiteit Utrecht, Leiden, Printed by Mostert & Van Onderen. ISBN: 90-393-2326-7.

Poulsen, P.B., Jensen, A.A., Antonsson, A.-B., Bengtsson, G., Karling, M., Schmidt, A., Brekke, O., Becker, J., and Verschoor, A.H. (2004) *The Working Environment in LCA*, SETAC Press, Pensacola, FL. ISBN 1-880611-68-6.

Projekt Gruppe Ökologische Wirtschaft (1987) *Produktlinienanalyse: Bedürfnisse, Produkte und ihre Folgen*, Kölner Volksblattverlag, Köln.

Rat für Nachhaltige Entwicklung (2008) Welche Ampel steht auf Rot? Stand der 21 Indikatoren der nationalen Nachhaltigkeitsstrategie – auf der Grundlage des Indikatorenberichts 2006 des statistischen Bundesamtes. Stellungnahme des Rates für Nachhaltige Entwicklung. Texte Nr. 22, April 2008. *www.nachhaltigkeitsrat.de* (accessed 15 October 2013).

Reap, J., Roman, F., Duncan, S., and Bras, B. (2008a) A survey of unresolved problems in life cycle assessment. Part 1: Goal and scope and inventory analysis. *Int. J. Life Cycle Assess*, **13** (4), 290–300.

Reap, J., Roman, F., Duncan, S., and Bras, B. (2008b) A survey of unresolved problems in life cycle assessment. Part 2: impact assessment and interpretation. *Int. J. Life Cycle Assess*, **13** (5), 374–388.

Rebitzer, G. (2002) in *Cost Management in Supply Chains* (eds S. Seuring and M. Goldbach), Physica-Verlag, HeidelbergS. 128-146.

Rebitzer, G. (2005) Enhancing the application efficiency of life cycle assessment for industrial uses. Thèse No 3307, École Polytechnique Féderale de Lausanne (EPFL).

Remmen, A., Jensen, A.A., and Frydendal, J. (2007) *Life Cycle Management. A Business Guide to Sustainability*, UNEP/SETAC Life Cycle Initiative, pp. 10–11 ISBN 978-92-807-2772-2.

Renner, I. and Klöpffer, W. (2005) Untersuchung der Anpassung von Ökobilanzen an spezifische Erfordernisse biotechnischer Prozesse und Produkte. Forschungsbericht 201 66 306 UBA-FB 000713. UBA Texte 02/05 Berlin, http://www.umweltbundesamt.de.

Rikhardsson, P., Bennett, M., Bouma, J., and Schaltegger, S. (eds) (2005) *Implementing Environmental Management Accounting: Status and Challenges*, Springer - Kluwer Academic Publishers.

Saling, P., Gensch, C.-O., Kreisel, G., Kralisch, D., Diehlmann, A., Preuße, D., Meurer, M., Kölsch, D., and Schmidt, I. (2007) Entwicklung der Nachhaltigkeitsbewertung SEEbalance®, in *BMBF-Projekt 'Nachhaltige Aromatenchemie'*, Karlsruher Schriften zur Geographie und Geoökologie, Karlsruhe.

Saling, P., Kicherer, A., Dittrich-Krämer, B., Wittlinger, R., Zombik, W., Schmidt, I., Schrott, W., and Schmidt, S. (2002) Eco-efficiency analysis by BASF: the method. *Int. J. Life Cycle Assess*, **7** (4), 203–218.

Schaltegger, S., Bennett, M., Burritt, R., and Jasch, D. (eds) (2008) *Environmental Management Accounting for Cleaner Production*, Springer, Dordrecht.

SETAC (2008) *Environmental Life Cycle Costing* (eds Hunkeler, D., Lichtenvort, K., and Rebitzer, G.). Pensacola, FL, in collaboration with CRC Press, Boca Raton, FL.

SETAC (2011) Swarr, T.E., Hunkeler, D., Klöpffer, W., Pesonen, H.-L., Ciroth, A., Brent, A.C., and Pagan, R.) *Environmental Life Cycle Costing: A Code of Practice*, SETAC Press, Pensacola, FL.

SETAC Globe (2002) **3** (4), 59.

Shapiro, K.G. (2001) Incorporating Costs in LCA. *Int. J. Life Cycle Assess*, **6** (2), 121–123.

Society of Environmental Toxicology and Chemistry – Europe (Ed.) (1992) *Life-Cycle Assessment*. Workshop Report, 2-3 December 1991, Leiden, SETAC-Europe, Brussels.

Society of Environmental Toxicology and Chemistry (SETAC) (1993) *Guidelines for Life-Cycle Assessment: A 'Code of Practice'*, 1st edn. From the SETAC Workshop held at Sesimbra, Portugal, 31 March – 3 April 1993., SETAC, Brussels and Pensacola, FL, August 1993.

Steward, M. and Weidema, B. (2004) A consistent framework for assessing the impacts from resource use. A focus on resource functionality. *Int. J. Life Cycle Assess*, **10** (4), 240–247.

Suh, S. (2003) Input-output and hybrid life cycle assessment. *Int. J. Life Cycle Assess*, **8** (5), 257.

Swarr, T.E., Hunkeler, D., Klöpffer, W., Pesonen, H.-L., Ciroth, A., Brent, A.C., and Pagan, R. (2011) Environmental life cycle costing: a code of practice. Editorial. *Int. J. Life Cycle Assess.*, **16** (5), 389–391.

UNEP-DTIE (2011) *Towards a Life Cycle Sustainability Assessment – Making Informed Choices on products* (eds Valdivia, S., Ugaya, C.M.L., Sonnemann, G. and Hildenbrand, J.). UNEP-DTIE, Paris, http://lcinitiative.unep.fr ISBN: 978-92-807-3175-0.

Weidema, B. (2001) Avoiding co-product allocation in life-cycle assessment. *J. Indust. Ecology*, **4** (3), 11–33.

Weidema, B.P. (2006) The integration of economic and social aspects in life cycle impact assessment. *Int. J. Life Cycle Assess*, **11** (Suppl. 1), 89–96.

White, A.L., Savage, D., and Shapiro, K. (1996) Life-cycle costing: concepts and application, in *Environmental Life-Cycle Assessment* (ed. M.A. Curran), Chapter 7, McGraw-Hill, New York, pp. 7.1–7.19 ISBN 0-07-015063-X.

Wilson, E.O. (2006) *The Creation. An Appeal to Save Life on Earth*, W.W. Norton, New York.

World Commission on Environment and Development (WCED) (1987) *Our Common Future*, Oxford University Press, Oxford.

Wrisberg, N., Udo de Haes, H.A., Triebswetter, U., Eder, P., and Clift, R. (eds) (2002) *Analytical Tools for Environmental Design and Management in a Systems Perspective – The Combined Use of Analytical Tools*, Kluwer Academic Publishers, Dordrecht, NL.

Zamagni, A., Buttol, P., Buonamici, R., Masoni, P.; Guinée, J.B., Huppes, G., Heijungs, R., van der Voet, E., Ekvall, T. and Rydberg, T. (2009) Co-ordination Action for innovation in Life-Cycle Analysis for Sustainability. D20 Blue Paper on Life Cycle Sustainability Analysis. Deliverable 20 of Work Package 7 of the CALCAS project. Revision 1 after the open consultation, August 2009, http://www.estis.net (accessed 15 October 2013).

Zamagni, A., Pesonen, H.-L., and Swarr, Th. (2013) Life cycle sustainability assessment: From LCA to LCSA. Special issue. *Int. J. Life Cycle Assess.*, **18** (9), 1637–1803.

# Appendix A
# Solution of Exercises

Solution of exercise: Provide an equal benefit of system variants (Section 2.2.5.3)

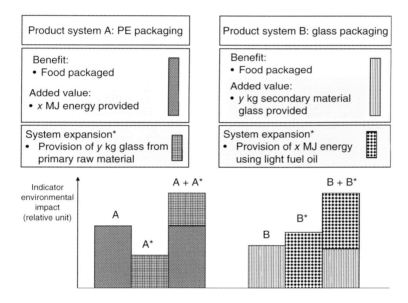

# Appendix A  Solution of Exercises

## Solution of exercise: Sample case for a calculation of $CO_2$-emissions (Section 3.1.3.2)

Calculation $CO_2$-emission/$m^3$ natural gas

| | | | | | Density | 821 g/$m^3$ | | | |
| | | | | | M ($CO_2$) | 44 g mol$^{-1}$ | | | |
| | | | | | LHV | 10.457 kWh/$m^3$ | | | |
| | | | | | 1 kWh | 3.6 MJ | | | |

### Composition

| | | A[a] | | B[a] | | | | | | |
|---|---|---|---|---|---|---|---|---|---|---|
| | | mol/100 mol | m(gas)/100 mol | m(gas)/$m^3$ | m(gas)/$m^3$ | Number C | n($CO_2$)/$m^3$ | m($CO_2$)/$m^3$ | m($CO_2$)/kWh | m($CO_2$)/MJ |
| Gas | M (g mol$^{-1}$) | n (mol) | m (g) | m (g) | n (mol) | — | n (mol) | m (g) | m (g) | m (g) |
| Methane | 16 | 87.54 | 1400.56 | 629.20 | 39.33 | 1 | 39.33 | 1730.30 | 165.47 | 45.96 |
| Ethane | 30 | 5.55 | 166.35 | 74.73 | 2.49 | 2 | 4.98 | 219.22 | 20.96 | 5.82 |
| Propane | 44 | 2.00 | 88.00 | 39.53 | 0.90 | 3 | 2.70 | 118.60 | 11.34 | 3.15 |
| i-Butane | 58 | 0.25 | 14.38 | 6.46 | 0.11 | 4 | 0.45 | 19.61 | 1.88 | 0.52 |
| n-Butane | 58 | 0.35 | 20.36 | 9.15 | 0.16 | 4 | 0.63 | 27.75 | 2.65 | 0.74 |
| i-Pentane | 72 | 0.06 | 4.03 | 1.81 | 0.03 | 5 | 0.13 | 5.53 | 0.53 | 0.15 |
| n-Pentane | 72 | 0.004 | 0.29 | 0.13 | 0.002 | 5 | 0.01 | 0.40 | 0.04 | 0.01 |
| Nitrogen[b] | 28 | 3.26 | 91.28 | 41.01 | 1.46 | 0 | 0.00 | 0.00 | 0.00 | 0.00 |
| $CO_2$ | 44 | 0.96 | 42.24 | 18.98 | 0.43 | 1 | 0.43 | 18.98 | 1.81 | 0.50 |
| Sum | | 99.96 | 1827.49 | 821.00 | 44.91 | | 48.65 | 2140.39 | 204.68 | 56.86 |

[a] In column A the mass ratio of gases is calculated based on the given values of mol% (calculated with rounded molar mass); 100 mol natural gas weighs 1827.5 g. Because the heat value is given in kilowatts hour per cubic metre a conversion of calculated masses per 100 mol (column A) into masses per cubic metre (column B) is necessary. As a simplification it is assumed that the density of all gases have the same density of 821 g m$^{-3}$. In this case the mass ratio of gases per 100 mol and per 1m$^3$ are the same. A more detailed calculation can be performed for a defined temperature and a defined pressure using the gas equation.
[b] Not relevant for $CO_2$ calculation.

## Solution of exercise: Calculation of emissions based on final energy (Section 3.2.3.2)

Low heat value (natural gas) = 46.1 MJ kg$^{-1}$
Low heat value (hard coal) = 29.65 MJ kg$^{-1}$.

With an efficiency of 35% the primary energy demand to generate 100 MJ final energy is 285.7 MJ.

| Energy carrier | m (energy carrier) (kg) | m (C) (kg) | m (CO$_2$) (kg) |
|---|---|---|---|
| Natural gas | 6.20 | 4.65 | 17.04 |
| Hard coal | 9.64 | 7.71 | 28.26 |

## Solution of exercise: Calculation of environmental loads by transport (without supply chain of the fuel) (Section 3.2.5)

1. Calculate the degree of utilisation of the truck.
   - Calculation of transported quantity: 24 loading positions are used. Per pallet 720 cartons are packed. This results in a total quantity of 17.280 pieces.
   - Calculation of weight if 24 pallets are fully loaded:

| Packaging (for 1 l beverage carton) | Weight | Calculation | Weight of load (kg) |
|---|---|---|---|
| Primary packaging (carton) | 31.5 g | 31.5 g × 17 280 | 544 |
| Secondary packaging (corrugated cardboard trays) | 128 g | 128 g × (12 × 5 × 24) | 184 |
| **Transportation packaging** | | | |
| Euro-pallets (wood) | 24 000 g | 24 kg × 24 | 576 |
| Pallet pattern | | | |
| Carton per tray | 12 pieces | | |
| Trays per layer | 12 pieces | | |
| Layers per pallet | 5 pieces | | |
| Cartons per pallet | 720 pieces | | |
| Sum packaging | | | 1 311 |
| Filling goods | | | 17 280 |
| Sum | | | 18 591 |

- degree of utilisation = $\dfrac{\text{actual load}}{\text{maximum payload}}$

  in this case: $\dfrac{18.6\ t}{25\ t} = 0.744$.

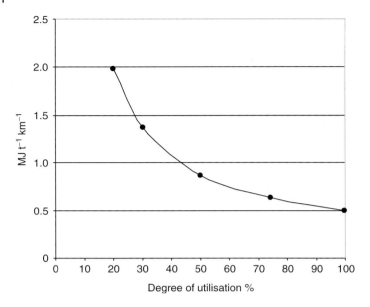

**Figure A.1** Specific energy consumption (MJ t$^{-1}$ km$^{-1}$) dependent on the degree of utilisation.

2. *Calculate the fuel consumption (in l Diesel) for a distance of 100 km for a truck loaded with 24 pallets. Assume a linear dependence between the fuel consumption and the utilisation rate. Finally, derive the specific energy consumption from the fuel consumption.*

| Degree of utilisation | Load | Energy consumption | Fuel consumption |
|---|---|---|---|
| 0% (empty) | 0 t | 9.29 MJ km$^{-1}$ | 26 l/100 km |
| 100% | 25 t | 0.50 MJ t$^{-1}$ km$^{-1}$ | 35 l/100 km |
| **Fuel consumption according to figure in exercise: y = 9 × degree of utilisation + 26** | | | |
| 74.4% | 18.6 t | 0.63 MJ t$^{-1}$ km$^{-1}$ | 32.7 l/100 km |

Contrary to the fuel consumption (see figure in exercise) there is no linear dependence between the specific energy consumption and the degree of utilisation (Figure A.1)!

3. *Calculate the fuel consumption (in l Diesel) for a distance of 100 km for a truck loaded with 24 pallets related to the functional unit.*
Diesel consumption (100 km) = 1.9 l fu$^{-1}$.

**Solution of exercise: Allocation per mass in a process chain (anonymised case example) (Section 3.3.2.2)**

Starting point for the consideration of allocation is the 33 t end product.

- Production end product: no allocation; 154 GJ are fully assigned to the 33 t end product.
- Production intermediate product: allocation according to mass:
    3106 GJ are assigned to 247 t co-product (88.2%) and
    415 GJ are assigned to 33 t intermediate product (11.8%).
    If no loads would be attributed to the co-product, it would not carry any loads of the process.
- Steam cracker: One product of the steam cracker is 280 t ethene (37.2% of products), that is processed in the plant 'production of intermediate product'. Of the loads that are allocated by mass in the steam cracker to the 280 t ethene (2590 GJ) only 11.8% (306 GJ) are to be assigned to the intermediate product. The remaining 88.2% have to be assigned to the co-product. Otherwise the co-product would not carry any load of the pre-production.
- Atmospheric distillation: One product of the atmospheric distillation is 935 t Naphtha (14.7%) that is processed in the steam cracker. Of the loads that are allocated by mass in the atmospheric distillation to the Naphtha (484 GJ) only 37.2% (180 GJ) are to be assigned to the Ethene. Otherwise the other products of the steam cracker would carry no loads of the atmospheric distillation. Of this 180 GJ only 11.8% (21 GJ) are to be assigned to the intermediate product. The remaining 159 GJ have to be assigned to the co-product.
- Consequently the loads for the production of 33 t end product sum up as follows:

| Process | GJ |
| --- | --- |
| Production end product | 154 |
| Production intermediate product | 415 |
| Steam cracker | 306 |
| Atmospheric distillation | 21 |
| Sum | 896 |

This corresponds to 27 MJ kg$^{-1}$ end product.

**Solution of exercise: Closed-loop recycling of production scraps (Section 3.3.3)**

1. *Calculation of raw material and energy consumption without recycling*:
    In this case the scrap is treated as waste and therefore no load is allocated.

| Process | MJ energy/80 kg product | MJ kg$^{-1}$ product | kg raw material/kg product |
| --- | --- | --- | --- |
| Punching and moulding | 200 | 2.5 | |
| Rolling | 300 | 3.75 | |
| Iron production | 1000 | 12.5 | 5 |
| Sum | | 18.75 | 5 |

2. *Calculation of raw material and energy consumption with recycling*:
   Punching and moulding: In this case the scrap is not waste but a co-product. Allocation of energy consumption according to mass (product: 160 MJ, scrap: 40 MJ) must not be conducted because in the case of closed-loop recycling the scrap remains inside the product system and therefore this applies also for the loads of the processes in which the scrap material was processed.
   Rolling: Remains unaffected, too, in this process.
   Iron production: Because 20 kg of pig iron is substituted by 20 kg of scrap only 80 kg pig iron must be produced from 320 kg iron ore. The resulting energy demand for the iron production is 800 MJ and 240 kg of slag is generated.

| Process | MJ energy/80 kg product | MJ kg$^{-1}$ product | kg raw material/kg product |
| --- | --- | --- | --- |
| Punching and moulding | 200 | 2.5 | |
| Rolling | 300 | 3.75 | |
| Iron production | 800 | 10.0 | 4 |
| Sum | | 16.25 | 4 |

3. *Savings*
   energy saving: 2.5 MJ kg$^{-1}$ product (13%)
   iron ore saving: 1 kg/kg product (20%).

# Appendix B

# Standard Report Sheet of Electricity Mix Germany (UBA 2000, Materials p. 179ff) Historic example, only for illustrative purposes

6. Energy supply
    6.1 Electricity grids
        6.1.1 Electricity grid Germany

| No. | Topic | Explanation | |
|-----|-------|-------------|---|
| A | *General* | | |
| A1 | Name of unit process | Electricity Germany | |
| A2 | Practitioner | Achim Schorb | |
| | | ifeu-Institut | |
| | | Wilckensstr. 3 | |
| | | 69120 Heidelberg | |
| A3 | Date | 12.11.1998 | |
| B | *Process specification* | | |
| B1 | Functional unit | 1 kJ energy, electric | |
| B2 | Synonyms | | |
| B3 | Technical description | Energy supply via public electricity grid, Germany. The power plant mix for 1996 was modelled after being simplified according to gross electricity generation 1996 (AGE, 1997) as follows: | |
| | | Nuclear energy | 30.0% |
| | | Lignite | 26.0% |
| | | Hard coal | 28.0% |
| | | Water power | 4.5% |
| | | Natural gas | 9.5% |
| | | Crude oil | 2.0% |

*Life Cycle Assessment (LCA): A Guide to Best Practice*, First Edition.
Walter Klöpffer and Birgit Grahl.
© 2014 Wiley-VCH Verlag GmbH & Co. KGaA. Published 2014 by Wiley-VCH Verlag GmbH & Co. KGaA.

# Appendix B  Standard Report Sheet of Electricity Mix Germany

| No. | Topic | Explanation |
|---|---|---|
| | | The percentage of electricity generation in the statistics listed as 'other' was distributed proportionately to water power (for proportional wind power) and crude oil (for proportional electricity from waste incineration plants). The respective power plant types were calculated according to (GEMIS 1997), but not using the energy mix reported there. Because (GEMIS, 1997) provides no information on carcinogenic pollutants, in a first approach the calculation was supplemented despite an asymmetric database, for example, with carcinogenic heavy metals according to ETH (1994) |
| B4 | Reference year | 1996 |
| B5 | Reference location | Power plant technology Germany/Switzerland |
| | | Electricity mix: Germany |
| B6 | System boundary | Electricity supply from German power plants including upstream processes for provision of energy carriers |
| B6a | Upstream processes | Mining of energy carriers |
| B6b | Energy | Power plant |
| B6c | Transport | Energy carrier from deposit |
| B7 | Allocation | None |
| B8 | Confidentiality | Public data |
| B9 | Data gaps | In GEMIS no information on carcinogenic pollutants |
| | | Data base for carcinogenic pollutants in ETH has only limited reliability. |
| B10 | Data quality (for processes) | Source  Generation  Type |
| | | ☐ Company ☐ Measured ☐ Single value |
| | | ☒ Literature ☐ Estimated ☐ Average |
| | | ☐ Other ☒ Calculated ☒ Other |
| B10a | Representativeness | Average data for plant in Germany (BRD) |
| B10b | Averaging | Weighted averages according to percentage of energy carriers (see above) |
| B11 | Data sources | AGE (1997) |
| | | GEMIS (1997) |
| | | ETH (1994) |

GEMIS, Gesamt-Emissions-Modell Integrierter Systeme.

| Data set: | Electricity grid Germany | | | 23.06.99 11:30:00 | |
|---|---|---|---|---|---|
| Input | | | Output | | |
| Material | Quantity | Unit | Material | Quantity | Unit |
| Cumulative energy demand (CED) | | | Waste | | |
| CED (nuclear erergy) | 1.08E+00 | kJ | Waste for disposal (wd) | | |
| CED (waterpower) | 6.07E−02 | kJ | Waste, other | | |
| CED fossil total | 2.28E+00 | kJ | Ash and slag | 7.24E−06 | kg |
| CED unspecified | 6.48E−07 | kJ | Sewage sludge | 1.10E−09 | kg |
| **Raw material in deposit (RiD)** | | | Hazardous waste | 9.40E−09 | kg |
| Energy carrier (RiD) | | | Waste for recovery (wr) | | |
| Natural gas | 5.57E−06 | kg | Waste, other (wr) | | |
| Crude oil | 1.45E−06 | kg | Ash and slag | 4.12E−06 | kg |
| Coal (RiD) | | | Waste, unspecified | 1.87E−08 | kg |
| Lignite | 1.08E−04 | kg | Emissions (air) | | |
| Hard coal | 3.70E−05 | kg | Dust | 1.25E−07 | kg |
| **Non energy carrier (RiD)** | | | Compounds inorganic (air) | | |
| Minerals (RiD) | | | Ammonia | 1.14E−09 | kg |
| Limestone | 1.46E−06 | kg | Hydrogen chloride | 3.37E−08 | kg |
| Water | | | Dinitrogen monoxide | 1.34E−09 | kg |
| Cooling water | 6.93E−03 | kg | Hydrogen fluoride | 4.65E−09 | kg |
| Water (process) | 1.49E−05 | kg | Carbon dioxide (air) | | |
| | | | Carbon dioxide, fossil | 2.11E−04 | kg |
| | | | Carbon monoxide | 2.48E−08 | kg |
| | | | Metals (air) | | |
| | | | Arsenic | 9.52E−13 | kg |
| | | | Cadmium | 2.76E−13 | kg |
| | | | Chromium | 1.67E−12 | kg |
| | | | Nickel | 1.63E−11 | kg |
| | | | $NO_x$ | 2.52E−07 | kg |
| | | | Sulphur dioxide | 8.98E−07 | kg |
| | | | VOC (air) | | |
| | | | Methane, fossil | 5.61E−07 | kg |
| | | | NMVOC (air) | | |
| | | | Benzene | 6.33E−11 | kg |
| | | | NMVOC, halog. (air) | | |
| | | | NMVOC, chlor. (air) | | |
| | | | NMVOC, chlor, aromat. (air) | | |
| | | | PCDD, PCDF | 8.90E−18 | kg |
| | | | NMVOC, unspecified | 6.40E−09 | kg |
| | | | PAH (air) | | |
| | | | Benzo(a)pyrene | 3.56E−16 | kg |
| | | | PAH without B(a)P | 1.78E−12 | kg |
| | | | PAH, unspec. | 9.72E−14 | kg |
| | | | VOC, unspecified | 3.18E−13 | kg |

| Data set: | Electricity grid Germany | | | 23.06.99 11:30:00 | |
|---|---|---|---|---|---|
| Input | | | Output | | |
| Material | Quantity | Unit | Material | Quantity | Unit |
| | | | Emissions (water) | | |
| | | | Emissions (w) | | |
| | | | Salts, inorganic | 1.37E−14 | kg |
| | | | Nitrogen compounds (w) | | |
| | | | Nitrogen compounds as N | 2.47E−15 | kg |
| | | | Indicator parameter | | |
| | | | AOX | 2.75E−18 | kg |
| | | | BOD-5 | 5.49E−17 | kg |
| | | | COD | 1.81E−15 | kg |
| | | | Energy carrier, secondary | | |
| | | | Energy, electric | 1.00E+00 | kJ |
| | | | Minerals | | |
| | | | Gypsum (FGD) | 2.64E−06 | kg |
| | | | Water | | |
| | | | Waste water (cooling water) | 6.63E−03 | kg |
| | | | Waste water (process) | 3.34E−06 | kg |
| Sum | Quantity | Unit | Sum | Quantity | Unit |
| kJ | 3.42E+00 | kJ | kJ | 1.00E+00 | kJ |
| kg | 7.10E−03 | kg | kg | 6.86E−03 | kg |

CED, cumulative energy demand; wd, waste for disposal; wr, waste for recovery; RiD, raw material in deposit; VOC, volatile organic compound; NMVOC, non methane volatile organic substances; FGD, flue gas desulphurisation.

### References

AGE (Arbeitsgemeinschaft Energiebilanzen) (1997) Energiebilanz 1996, VDEW, Frankfurt.

ETH (1994) Ökoinventar für Energiesysteme.

GEMIS (1997) GEMIS 3.0, Wiesbaden.

# Acronyms/Abbreviations

| | |
|---|---|
| a | Annum (year, SI abbr.) |
| ADI | Acceptable daily intake |
| AFNOR | Association Française de Normalisation |
| AoP | Area of protection |
| AP | Acidification potential |
| APME | Association of Plastics Manufacturers in Europe (later Plastics Europe) |
| BAFU | Bundesamt für Umwelt (Bern, "Swiss EPA") |
| BCF | Bioconcentration factor |
| BDI | Bundesverband der Deutschen Industrie e.V. |
| BEW | Bundesamtes für Energiewirtschaft (CH) |
| BOD | Biological oxygen demand |
| BTU | British thermal unit (obsolete) |
| BUS | Bundesamt für Umweltschutz, Bern (later BUWAL, BAFU), CH |
| BUWAL | Bundesamt für Umwelt, Wald und Landschaft, Bern, CH |
| CBA | Cost benefit analysis |
| CD | Committee draft |
| CED | Cumulative energy demand |
| CEFIC | Conseil Européen de l'Industrie Chimique |
| CEN | Comité Européen de Normalisation |
| CFC | Chlorofluorohydrocarbon |
| CLR | Closed loop recycling |
| CF | Carbon footprint |
| CH | Switzerland |
| CML | Centrum voor Milieukunde Leiden (Institute of Environmental Sciences Leiden) |
| COD | Chemical oxygen demand |
| CPM | Center for Environmental Assessment of Product and Material Systems (SE) |
| CR | Critical review |
| CSB | Chemischer sauerstoff-bedarf (COD) |
| CSA | Canadian Standards Association |

*Life Cycle Assessment (LCA): A Guide to Best Practice*, First Edition.
Walter Klöpffer and Birgit Grahl.
© 2014 Wiley-VCH Verlag GmbH & Co. KGaA. Published 2014 by Wiley-VCH Verlag GmbH & Co. KGaA.

| | |
|---|---|
| DALY | Disability-adjusted lost life years |
| DFG | Deutsche ForschungsGemeinschaft |
| DIN | Deutsches Institut für Normung |
| DIS | Draft international standard |
| DK | Denmark |
| DKR | Deutsche Gesellschaft für Kunststoff-Recycling mbH |
| DLCA | Dynamic life cycle assessment |
| DNOC | Dinitroorthocresol |
| DSD | Duales system Deutschland ("green dot") |
| EC | European Commission |
| ECOSOL | European Centre of Studies on LAB/LAS (a sector group of CEFIC) |
| EDIP | Environmental design of industrial products (DK) |
| EEA | European Environment Agency |
| EEV | Endenergieverbrauch (final energy consumption) |
| EI | Environmental increment |
| EINECS | European inventory of existing commercial chemical substances |
| ELCD | European union's European references life cycle data system |
| EMPA | Eidgenössische Materialprüfungs- und Versuchsanstalt |
| EN | European norm |
| EOL | End of life (waste management/recycling) |
| EP | Eutrophication potential |
| EPA | Environmental Protection Agency (USA) |
| EPS | Enviro-accounting (SE) |
| EUSES | European union system for the evaluation of substances |
| EVOH | Ethyl-vinyl-alcohol |
| FAL | Franklin Ass. Ltd. (Kansas) |
| FCHC | Fluoro-chloro-hydrocarbons |
| fu | Functional unit |
| FEFCO | Fédération Européenne des Fabricants de Carton Ondulé |
| FKN | Fachverband Getränkekarton |
| GEMIS | Gesamt-emissions-modell integrierter systeme |
| GHG | Green house gas |
| GMO | Genetically modified organism |
| GWP | Global warming potential |
| HC | Hydrocarbon |
| HDPE | High density polyethylene |
| HHV | Higher heating value ($H_o$) |
| HTP | Human toxicity potential |
| IAV | Inhabitant average value |
| IEA | International Energy Agency |
| IEAM | Integrated Environmental Assessment and Management (journal published by SETAC) |

| | |
|---|---|
| IFEU | Institut für Energie- und Umweltforschung Heidelberg |
| IIASA | International Institute for Applied Systems Analysis (Laxenburg, Austria) |
| ILCD | International reference life cycle data system (EC) |
| ILO | International Labour Organisation |
| Int. J. LCA | International Journal of Life Cycle Assessment, ecomed (1996-2007); since 2008: Int. J. Life Cycle Assess., Springer |
| IPCC | Intergovernmental Panel on Climate Change |
| ISO | International Standard Organization |
| ISO/TS | ISO/technical specification |
| IVV | Ingenieurgesellschaft für ver-kehrsplanung und verkehrssicherung |
| KEA | Kumulierter energieaufwand (CED) |
| KExA | Kumulierter exergieaufwand |
| kgce | Kilogram coal equivalent |
| KNA | Kumulierter nichtenergetischer aufwand (cumulative non-energy demand) |
| KPA | Kumulierter prozessenergie-aufwand |
| LAB | Linear alkylbenzene (mixture, mostly n-dodecylbenzene) |
| LAS | Linear alkyl benzene sulphonate ($Na^+$) |
| LCA | Life cycle assessment (acc. to SETAC, ISO, UNEP, EC); life cycle analysis (obsolete) |
| LCAA | Life cycle attribute assessment |
| LCC | Life cycle costing |
| LCI | Life cycle inventory analysis (ISO) |
| LCIA | Life cycle impact assessment (ISO) |
| LCSA | Life cycle sustainability assessment |
| LCWT | Life cycle working time |
| LD | Lethal dose |
| LDPE | Low density polyethylene |
| LHV | Lower heating value ($H_u$) |
| LOEL | Lowest observed effect level |
| LPB | Liquid packaging board |
| LSF | Life support function |
| MAK | Maximale arbeitsplatz konzentration (OEL) |
| MFA | Material-flow-analysis |
| MIC | Maximum (allowed) immission concentration |
| MIPS | Mass intensity per service unit (service unit = fu) |
| MIR | Maximum incremental reactivity |
| MSWI | Municipal solid waste incineration |
| MWIP | Municipal waste incineration plant |
| NAGUS | Normenausschuss grundlagen des umweltschutzes (DIN) |
| NCPOCP | Nitrogen corrected photochemical ozone creation potential |
| NDI | Naturalness degradation indicator |
| NDP | Naturalness degradation potential |

| | |
|---|---|
| NEC | No effect concentration |
| NEV | Nichtenergetischer verbrauch (non-energetic consumption) |
| NMVOC | Non methane volatile organic substances |
| NOEC | No observed effect concentration |
| NOEL | No observed effect level |
| NP | Nutrification potential |
| ODP | Ozone depletion potential |
| OECD | Organisation for Economic Co-operation and Development |
| OEL | Occupational exposure limit |
| OLR | Open loop recycling |
| ÖTP | Ökotoxizitäts-potential |
| ÖTPA | Ökotoxizitäts-potential aquatisch |
| ÖTPT | Ökotoxizitäts-potential terrestrisch |
| OTV | Odour threshold value |
| PA | Polyamide |
| PAS | Publicly available specification |
| PE | Polyethylen |
| PEC | Predicted environmental concentration |
| PET | Polyethylen terephtalat |
| PLA | Product line analysis (produktlinienanal.) |
| PNEC | Predicted no effect concentration |
| POCP | Photo oxidant creation potential |
| ProBas | Prozessorientierte basisdaten für umweltmanagement-instrumente (UBA/D) |
| PROSA | Product sustainability assessment |
| PS | Polystyrene |
| PU | Polyurethane |
| PVC | Polyvinylchloride |
| PWMI | The European Centre for Plastics in the Environment (later APME, Plastics Europe) |
| QALY | Quality adjusted life years |
| R11 | Refrigerant 11 (trichlorofluoromethan) |
| R12 | Refrigerant 12 (dichlorodifluoromethane) |
| RB | Re-fillable (-usable, -used) bottles |
| REACH | Registration, evaluation, authorisation and restriction of chemicals (EU) |
| REPA | Resource and environmental profile analysis (coined by FAL) |
| ROE | Rohöläquivalente (crude oil equivalent) |
| SBM | Streckblasverfahren |
| SC | Sub committee (ISO) |
| SE | Sweden |
| SEI | Stoffgebundener energieInhalt (inherent energy) |
| SETAC | Society of Environmental Toxicology and Chemistry |
| SI | Système International d'unités |

| | |
|---|---|
| SKE | Stein-kohle-einheit (average heating value of 1 kg hard coal); kilogram of coal equivalent (kgce) |
| SLCA | Social (or societal) life cycle assessment |
| SOM | Soil organic matter |
| SOC | Soil organic carbon |
| SPINE | Sustainable product information network for the environment (SE); data format for LCI data banks (CPM) |
| SPOLD | Society for the Promotion of LCA Development; Data format for LCI data transfer developed by SPOLD |
| TASi | Technische anleitung siedlungsabfall |
| TBS | Toxicity-based scoring |
| TC | Technical committee (ISO) |
| TN | Trip number (of re-fillable bottles) |
| TR | Technical report (ISO) |
| TRACI | Tool for the reduction and assessment of chemical and other environmental impacts |
| TREMOD | Transport emission model (IFEU) |
| TRK | Technische richt-konzentration |
| TRS | Total reduced sulphur |
| UBA | Umweltbundesamt (Dessau/Berlin: "German EPA"; UBA Vienna: "Austrian EPA") |
| UCPTE | Union pour la Coordination de la Production et du Transport de l'Électricité (until 1998) |
| UCTE | Union for the Co-ordination of Transmission of Electricity (since 1999) |
| UN | United Nations |
| UNEP | United Nations Environmental Programme |
| U.S. EPA (US EPA) | United States Environmental Protection Agency |
| USEtox | UNEP-SETAC toxicity model (in LCIA) |
| VDEW | Verband der Elektrizitätswirtschaft |
| VDI | Verein Deutscher Ingenieure |
| VOC | Volatile organic compounds |
| WBCSD | World Business Council for Sustainable Development |
| WCED | World Commission on Environment and Development |
| WHO | World Health Organization |
| WMO | World Meteorological Organization |
| WRI | World Resources Institute |

# Index

## a

abiotic resources consumption  214, 220
– impact indicators  214–215
– indicator model and characterisation factors  215–220
acceptable daily intake (ADI)  272
acidification  208, 254–256
– characterisation/quantification  257–258
– impact indicator and characterisation factors  256–257
– potential (AP)  297, 303
– regionalisation  259–261
AFNOR (Association Française de Normalisation, France)  14
APME (Association of Plastic Manufacturers in Europe)  130
aquatic eutrophication  261–262
Association of Carton Packaging for liquid foods (FKN)  47, 49, 50
Austrian Ministry of Life  359
avoided burden approach  104

## b

basket of benefit method  42
beverage cartons  138, 140, 141, 149–150
– comparison with PET bottle (juice, storage)  343–344
biochemical oxygen demand (BOD)  262, 265, 266
biotic resources consumption  222–224
black box method  29
Boustead Model (UK)  131
BUWAL (Swiss Federal Office for Environment, Forest and Landscape)  129, 136, 271

## c

carbon footprint  238–239
casualties  289–290
chemical oxygen demand (COD)  203, 262, 265, 266
Cleaner Production (Elsevier)  131
climate change  234–235
– carbon footprint  238–239
– characterisation  237–238
– greenhouse effect  235–236
– impact indicator and characterisation factors  236–237
closed loop recycling  34, 70
– allocation and recycling  105–107
Comité Européen de Normalisation (CEN)  14
comparative analysis  337
contribution analysis  335, 336
co-production and allocation
– approaches  101–102
– definition  92–93
– fair allocation  93–94
– – allocation per mass  94–96
– – system expansion  96–98
– proposed solutions  98–101
– system expansion  102–104, 105
co-products  33–34, 70
critical review  46–47, 57
critical volumes method  183–184
– criticism  185 186–187
– interpretation  184–185
crude oil equivalence factor (ROE)  298–300
CSA (Canadian Standards Association, Canada)  14
cumulative energy demand (CED)  8, 38, 39, 87, 182, 298, 301, 345
– balancing boundaries  79–80, 81
– definition  77
– exergy demand  220–222
– partial amounts  77–79

*Life Cycle Assessment (LCA): A Guide to Best Practice*, First Edition.
Walter Klöpffer and Birgit Grahl.
© 2014 Wiley-VCH Verlag GmbH & Co. KGaA. Published 2014 by Wiley-VCH Verlag GmbH & Co. KGaA.

## d

data availability and depth of study   43–44, 55–56
data collection template   66
diamond paradox   100
DIN-NAGUS   1, 12, 16
disability adjusted lost life years (DALYs)   277, 278, 288
discernability analysis   337–338
distance-to-target criterion   198
dominance analysis   336
down cycling   113
Dual System Germany (DSD)   118, 140, 141, 142

## e

ecoinvent (CH)   131, 132
ecological endangering   198
ECOSOL (European LCI Surfactant Study Group)   130
ecotoxicity   209
– chemicals and environment   280–282
– persistence and distribution inclusion to quantification   283–285
– protected objects   279, 280
– quantification without relation to exposure   282–283
electricity mix   85, 86
end-of-life (EOL) phase   113
enthalpy   82
environmental increments (EIs)   291
Eurostat   86
EUSES (European Union System for the Evaluation of Substances)   277
eutrophication   261
– aquatic eutrophication   261–262
– characterisation/quantification   267
– indicator and characterisation factor   263–267
– potential (EP)   296–297, 303
– regionalisation   267–268
– terrestrial eutrophication   267

## f

flow-pulse problem   276
fossil fuels scarcity   298–300
fresh water use   224–227
fuels and biomass   80
functional unit
– example   40
– impairment factors on comparison and negligible added value   40–41
– non-negligible added value procedure   41–42
– reference flow   37–39, 55

## g

GaBi ((University of Stuttgart and PE International, DE)   132
GEMIS (total release model of integrated systems)   130, 131
generic data sets   45
geographical system boundary   34, 35, 54
German Federal Environment Agency (Umweltbundesamt, UBA)   10, 46, 89, 197, 198, 199, 201, 203, 289, 298–299, 311
goal definition   27–28
greenhouse effect   208, 211, 235–236, 290, 294
– and global warming potential (GWP)   294–295, 302

## h

hemerobic level approach   227–229
higher heating value (HHV)   82, 83
human toxicity
– harmonised LCIA toxicity model   277, 278–279
– problem definition   269–270
– simple weighting using occupational exposure limit and indicative values   270–273
– supplementary exposure estimation characterisation   273–277
hydropower   80

## i

IMPACT2002+   269
impact assessment type   44–45, 56–57
inflammable materials energy content
– fossil fuels   81
– infrastructure   84–85
– quantification   81–84
input-related impact categories   212–214
– abiotic resources consumption   214, 220
– – impact indicators   214–215
– – indicator model and characterisation factors   215–220
– biotic resources consumption   222–224
– cumulative energy and exergy demand   220–222
– fresh water use   224–227
– land use   227
– – advanced concepts   231–233
– – characterisation using hemerobic level concept   229–231

## Index

– – hemerobic level approach   227–229
Institute for Energy and Environmental
  Research (IFEU Heidelberg)   47, 347
Integrated Environmental Assessment and
  Management (IEAM)   131
International Energy Agency (IEA)   85
International Institute for Applied Systems
  Analysis (IIASA)   8
International Journal of Life Cycle Assessment
  17, 18, 131, 135
ISO 14040   13, 27, 50, 65, 188, 191–192, 200,
  201, 207, 209–210, 292, 330, 334, 335,
  338–339, 341–342, 343, 351, 352, 361, 365,
  369
ISO 14043   330
ISO 14044   27, 30, 42, 45, 47, 50, 181, 182,
  197, 207, 220, 256, 292, 330, 333, 334, 339,
  341, 342, 352, 367, 369
– LCIA mandatory components   187
– – characterisation   191–192
– – classification   190–191
– – impact categories selection   187–190
– LCIA optional elements
– – data quality additional analysis   201
– – grouping   197–200
– – normalisation   192–197
– – weighting   200–201
ISO 14046   226
ISO 14048   134
ISO 14071   343

## j

Journal of Industrial Ecology (Wiley)   131

## l

land use   227
– advanced concepts   231–233
– characterisation using hemerobic level
  concept   229–231
– hemerobic level approach   227–229
life cycle assessment (LCA). *See also* individual
  entries
– component interpretation illustration using
  practice example   343
– – comparison based on impact indicator
    results   343–344
– – comparison based on normalisation
    results   344
– – completeness, consistency, and data
    quality   346
– – critical review   351–352
– – differences significance   347–348
– – recommendations   351
– – restrictions   350–351

– – sectoral analysis   344–346
– – sensitivity analyses   348–350
– critical review   340–342
– – outlook   342–343
– definition and limitations   1, 2
– early applications according to ISO 14040
    13
– functional unit   3
– history
– – 1980s   8, 9
– – early stages   6–7
– – energy analysis   8
– – environmental policy background   7–8
– – SETAC   9–10
– interpretation phase
– – development and rank   329–330
– – evaluation   333–334
– – ISO 14040   331
– – ISO14044   331–332
– – significant issues identification   332–333
– literature and information   17–18
– operational input–output analysis
    (gate-to-gate)   5–6
– product life cycle   2–3
– reporting   338–340
– result analysis techniques
– – mathematical methods   335, 336–338
– – non-numerical methods   338
– – scientific background   334–335
– standardisation
– – formation process   14–16
– – status quo   16–17
– structure
– – according to ISO   11–12
– – according to SETAC   10–11
– – valuation   12, 14
– as system analysis   4–5
life-cycle based sustainability assessment
    (LCSA)   365
life cycle impact assessment (LCIA)
– basic principles   181–183
– critical volumes method   183–184
– – criticism   185, 186–187
– – interpretation   184–185
– environmental problem fields   201–202
– – historical list   202–206
– – stressor-effect relationships and indicators
    206–212
– impact categories, impact indicators, and
    characterisation factors   212
– – accidents and radioactivity   289–291
– – input-related impact categories   212–233
– – nuisances by chemical and physical
    emissions   286–289

life cycle impact assessment (LCIA) (contd.)
– – output-based impact categories (global and regional impacts)   233–268
– – toxicity-related impact categories   268–286
– phase impact assessment illustration by practical example   291–293
– – characterisation   300–305
– – classification   300
– – grouping   310–311
– – impact categories selection   293–300
– – normalisation   305–310
– – weighting   311
– structure according to ISO 14040 and 14044
– – mandatory elements   187–192
– – optionalelements   187, 192–201
life cycle inventory (LCI)   10, 29, 30, 35, 41, 43, 44, 45, 46, 55
– allocation   117–118
– – by co-production example   92–104, 105
– – fundamentals   92
– – recycling for open-loop recycling   107–113
– – recycling in closed-loops and reuse   105–107
– – rules definition on process level   153, 156–157
– – rules definition on system level for open-loop recycling   157
– – within waste-LCAs   113–117
– calculation   158–159
– – input   159–161
– – output   162–170
– data aggregation and units   134, 135, 136
– data estimations   132, 133
– data procurement   119–127
– – preparation and system flow chart refining   118–119
– data quality and documentation   133–134
– energy analysis   74–77
– – cumulative energy demand (CED)   77–81
– – electricity supply   85–87
– – inflammable materials energy content   81–85
– – transports   88–91
– flow charts   69–72
– generic data   127–129
– – purchasable databases and software systems   131–132
– – reports, publications, and web sites   129–131
– literature on fundamentals   64–65
– phase illustration by example   137–138
– – allocation   153, 156–157

– – collection and sorting of used packaging   148–149
– – differentiated system flow chart with reference flows   153
– – distribution   148
– – electricity supply   152–153
– – examined product systems differentiated description   138–143
– – production by materials   146–148
– – production procedures of materials   143–146
– – recovery technologies (recycling)   149–151
– – system modelling   157–158
– – transportation by truck   152
– – transport packaging recycling   151
– reference values   72–74
– results presentation   136
– scientific principles   63–64
– studies   7
– unit process as smallest cell
– – balancing   67–68
– – integration into system flow chart   65–67
life cycle working time (LCWT)   367
life support function (LSF)   232
liquid packaging board (LPB)   54
lower heating value (LHV)   81, 82, 83

## m

Mackay model   276
Mass Intensity per Service Unit (MIPS)   136
material flow analysis (MFA)   73–74
maximum incremental reactivity (MIR)   249, 251–252
maximum working site concentration (MAK)   270, 271, 272
multi-input processes   156–157
multi-output processes   153, 156

## n

noise   287–289
non-negligible added value
– impairment factors on comparison   40–41
– procedure   41–42
nuclear energy   80
nuisances by chemical and physical emissions   286
– noise   287–289
– smell   286–287

## o

open loop recycling   34, 70
– allocation and recycling
– – allocation per equal parts   109–111

– – cut-off rule   111–113
– – overall load to system B   113
– – problem definition   107–109
– allocation rules definition on system level   157
Open Source Software   134
output-based impact categories (global and regional impacts)   233–234
– acidification   254–256
– – characterisation/quantification   257–258
– – impact indicator and characterisation factors   256–257
– – regionalisation   259–261
– climate change   234–235
– – carbon footprint   238–239
– – characterisation   237–238
– – greenhouse effect   235–236
– – impact indicator and characterisation factors   236–237
– eutrophication   261
– – aquatic eutrophication   261–262
– – characterisation/quantification   267
– – indicator and characterisation factor   263–267
– – regionalisation   267–268
– – terrestrial eutrophication   267
– photo oxidants formation (summer smog)   246–248
– – characterisation/quantification   252
– – impact indicator regionalisation   252–253
– – indicators and characterisation factors   249–252
– stratospheric ozone depletion   240–241
– – causing substances   241
– – characterisation   245–246
– – impact indicator and characterisation factors   242–245
– – ozone hole and legal measures   241–242
ozone depletion potential (ODP)   193, 243–245
ozone hole and legal measures   241–242

**p**

perturbation analysis   336
PET bottle   139–140, 141–142, 143, 147, 150–151
– comparison with beverage cartons   343–344
photochemical ozone creation potential (POCP)   249–251, 252, 253, 295–296, 302
photo oxidants formation   246–248, 295–296
– characterisation/quantification   252

– impact indicator regionalisation   252–253
– indicators and characterisation factors   249–252
photovoltaic energy   80
potential environmental impact   183
practice example and definition of goal and scope illustration   47–48
– goal definition   48–50
– scope   50–57
ProBas (process orientated base data for environmental management instruments)   130
Product Sustainability Assessment (PROSA)   359
product systems   28–29, 50–53
product tree   28
pseudo improvement, by outsourcing   5–6

**r**

radioactivity   290–291
RAINS (regional air pollution information and simulation) model   252, 253, 260
relative toxicity scale   271
resident equivalents (REQs)   195–197, 305, 308, 309, 310
Resource and Environmental Profile Analysis (REPA)   7
resource demand   298
– energy resources   298–300
– land use   300

**s**

sectoral analysis. *See* contribution analysis
SimaPro (Pré Consultants, NL)   132, 200
smell   286–287
Society of Environmental Toxicology and Chemistry (SETAC)   1, 9–11, 32, 45, 46, 202, 203, 207, 213, 223, 252, 262, 269, 274, 275, 329, 335, 357, 359, 361, 364, 365
– Code of Practice   12, 329
soil organic carbon (SOC)   232
SPINE (Swedish data format)   134
SPOLD (Society for the Promotion of LCA Development)   131, 133, 134
stratospheric ozone depletion   240–241
– causing substances   241
– characterisation   245–246
– impact indicator and characterisation factors   242–245
– ozone hole and legal measures   241–242
stressor-effect relationship and indicators   206
– impacts hierarchy   207–209
– potential versus actual impacts   209–212

sustainability 357–358
– dimensions 358–360
– life cycle assessment options 368–370
– state of the art of methods
– – life cycle assessment 361–364
– – life cycle costing (LCC) 364–366
– – product-related social life cycle assessment (SCLA) 366–368
system boundaries 4, 5, 6, 32

*t*
technical system boundary
– cut-off criteria 29, 30, 31, 32, 53
– demarcation towards system surrounding 32–33, 53–54
– – co-products 33–34
– – secondary raw material 34
temporal system boundary 55
– and time horizon 35–36
terrestrial eutrophication 267
toxicity-related impact categories 268–269, 285–286
– ecotoxicity
– – chemicals and environment 280–282
– – persistence and distribution inclusion to quantification 283–285
– – protected objects 279, 280
– – quantification without relation to exposure 282–283
– human toxicity
– – harmonised LCIA toxicity model 277, 278–279
– – problem definition 269–270
– – simple weighting using occupational exposure limit and indicative values 270–273
– – supplementary exposure estimation characterisation 273–277
transportation processes for distribution 157
Transport Emission Model (TREMOD) 90
trippage rate (TR) 105–106

*u*
Umberto (Ifu, DE) 132
uncertainty analysis 336–337
Union for the Coordination of Transmission of Electricity (UCTE) 85
unit-world-box model 276
USES Dutch model 276–277
USEtox model 278, 279, 284–285

*v*
valuation (weighting), assumptions and value 45–46

*w*
waste disposal
– options, comparison 116–117
– of product, modelling 114–116
wind power 80
World Business Council for Sustainable Development (WBCSD) 238
World Resources Institute (WRI) 238